NETWORK-BASED
MANAGEMENT SYSTEMS
(PERT/CPM)

Information Sciences series

EDITORS

ROBERT M. HAYES

Director of the Institute of Library Research
University of California at Los Angeles

JOSEPH BECKER

Director of Information Services
Interuniversity Communications Council (EDUCOM)

Consultants

CHARLES P. BOURNE

Director, Advanced Information Systems Division
Programming Services, Inc.

HAROLD BORKO

System Development Corporation

Joseph Becker and Robert M. Hayes:
INFORMATION STORAGE AND RETRIEVAL

Charles P. Bourne:
METHODS OF INFORMATION HANDLING

Harold Borko:
AUTOMATED LANGUAGE PROCESSING

Russell D. Archibald and Richard L. Villoria:
NETWORK-BASED MANAGEMENT SYSTEMS (PERT/CPM)

Charles T. Meadow:
THE ANALYSIS OF INFORMATION SYSTEMS, AN INTRODUCTION TO INFORMATION RETRIEVAL

Launor F. Carter:
NATIONAL DOCUMENT-HANDLING SYSTEMS FOR SCIENCE AND TECHNOLOGY

Robert M. Hayes:
MATHEMATICS OF INFORMATION SYSTEMS   In Preparation

NETWORK-BASED MANAGEMENT SYSTEMS (PERT/CPM)

Russell D. Archibald

M. J. Richardson, Inc.
Long Beach, California

Richard L. Villoria

Richard L. Villoria & Associates
Los Angeles, California

JOHN WILEY & SONS, INC.   New York · London · Sydney

10 9 8 7

Library of Congress Catalog Card Number: 66–25216

Printed in the United States of America

ISBN   0   471   03250   6

Information sciences series

Information is the essential ingredient in decision making. The need for improved information systems in recent years has been made critical by the steady growth in size and complexity of organizations and data.

This series is designed to include books that are concerned with various aspects of communicating, utilizing, and storing digital and graphic information. It will embrace a broad spectrum of topics, such as information system theory and design, man-machine relationships, language data processing, artificial intelligence, mechanization of library processes, non-numerical applications of digital computers, storage and retrieval, automatic publishing, command and control, information display, and so on.

Information science may someday be a profession in its own right. The aim of this series is to bring together the interdisciplinary core of knowledge that is apt to form its foundation. Through this consolidation, it is expected that the series will grow to become the focal point for professional education in this field.

Preface

Planning, scheduling and control of complex, interrelated business activities and of the resources (time, money, manpower, machines) required to execute those activities are the main concern of this book. More specifically, the book is about a class of management systems—systems based on the concepts of network planning and critical path analysis, plus certain of their applications to business. Such systems are known by a wide variety of abbreviated names, for example, there are PERT (Program Evaluation and Review Technique) and CPM (Critical Path Method). We treat these variations generically to avoid hairsplitting discussion in favor of or against any particular "system."

The book is intended for any person who is, or hopes to be, responsible for getting things done through the efforts of other people, whether he holds the title president, manager, project engineer, administrator, foreman, or any other of the myriad titles in existence. Network-based systems are of immediate practical use to these people as tools to get things done in less time at less cost than would otherwise be possible.

The network approach to action planning is a major advance in improving management planning and control effectiveness and is designed specifically to deal with the accelerated pace of today's development programs and the uncertainties associated with them. The decision-making process has come to require increasing amounts of qualitative and quantitative data, with the result that the need for new aids to sound decision-making has been recognized. No management tool can *make* decisions, but tools such as network planning can provide the basis on which to build a realistic, economical management information system which will permit more informed decisions to be made.

The need for such information systems has grown out of the operating environment of the contemporary large development project where diverse organizations and disciplines are combined for the purpose of pursuing a common mass effort. (It should be noted that these systems can be, and have been, effectively applied to projects of many types and sizes.) This environment has often included extreme time pressures—whether the project is a military weapon system, an urban renewal project, a massive power development or processing plant, or a hardware development program for the exploration of space, etc. A continually advancing and ever more complex technology is a major characteristic of this environment. Managers skilled in the technologies and management methods of just a few years ago are often hard-pressed to remain effective in this dynamic contemporary management situation. Systematic planning methods which permit the implementation of advanced management information and control systems offer at least a partial answer to the demands of organization complexity, the pressures of optimizing time, cost, and manpower constraints, and the uncertainties of rapidly advancing technology.

If the contemporary management environment has stimulated the development of such systems by presenting a clear and ever-present technological need, the products, processes, and disciplines of the technology itself have lent substantial impetus to the development of such systems by contributing to the means. Mathematical analysis, systems engineering and analysis, and above all, the computer have all left significant imprints on the developing science of management. The development of the high-speed digital computer with its vast information storage and retrieval capabilities has literally made possible the development of many of the new management systems, including the network-based system. As discussed in Chapter 10, the use of computers is not necessarily essential to the effective implementation of the network-based system in some situations; however, in the extrapolation of the technique into an integrated total management system, capable of summarizing vast amounts of information at various hierarchical levels without loss of integrity, the computer is essential.

Given the need and the means, the final ingredient is understanding the elements of the system and the capability to translate this understanding into sound, mature system design and implementation. That is the object of this book.

Our approach is based on our extensive experience in the use of network planning for industrial and governmental work and in teaching the subject to a wide range of persons, from executives to field superintendents to undergraduate students, in settings as widely disparate as a college classroom at UCLA, a hotel seminar, and an executive suite.

Part I presents the fundamental ideas involved and describes in some depth the model network-based management system. Part II discusses the various factors which must be dealt with in tailoring a model system to a particular environment found in a given organization. We have not tried to set forth pat answers to the many difficult questions which arise in this area; rather we have attempted to assist the reader to adapt the principles to his environment by discussing those topics we have found to be most crucial to effectively implementing a network-based management system. We have found very little in the available literature which deals with these subjects, although we are convinced that they are as important as the fundamental concepts to get actual results from network planning.

In Part III, our objective is to impart a better understanding of the application of network-based systems by describing a variety of situations. Our hope there is that the reader will not only see how results have been obtained in an area close to his particular interests, but also how he may gain new insight into his own responsibilities by exposure to a few different conditions.

Part IV includes a detailed examination of some serious pitfalls we have encountered in a number of organizations, together with two chapters of a theoretical nature which we hope will stimulate continued development of this type of management system. In our last chapter, we give some of our ideas with respect to the direction and dimension of future developments. The six appendices are included to make the book a more complete, self-sufficient text for both on-the-job indoctrination and classroom instruction.

A final word to instructors in the technique. The twelve chapters of Parts I and II, along with their associated exercises, are designed to provide sufficient material in appropriate increments for a one-semester course. A suggested outline for such a course is included in Appendix D, and each chapter has been planned to constitute a unit of understanding.

Many individuals contributed to this book. We particularly want to thank Professor Robert E. Hayes of UCLA's School of Library Science. As series editor, Dr. Hayes has made critical reviews of the manuscript at various stages which have been most helpful. Thanks too, go to John Magee and the *Harvard Business Review* for their permission to use Mr. Magee's articles on Decision Trees, to *Factory Magazine* for case material and networks which originally appeared in that magazine, to Elmo Pace of Pierose Maintenance Corp.; to Richard Wrestler, CPM System Division, Informatics Inc.: to James A. Campise, Computer Sciences Corp.; to H. S. Coumbe, Brown and Root, Inc.; and to others, for case material. Significant contributions were made by Joel M. Prostick, Informatics Inc., with his material on the role of the computer and network integration; by W. J. Erickson, System Development Corp., with his chapter on simulation and gaming; and by J. David Craig, IBM Corporation, with his material on precedence diagramming. Our thanks too, are extended to the several ladies, particularly, Dorothy Villoria, Cindy DuPuis, and Carol Kaiser, who contributed to the preparation of several generations of manuscript as well as performing myriad other tasks necessary to produce a readable text.

RUSSELL D. ARCHIBALD
RICHARD L. VILLORIA

Los Angeles, California
August, 1966

Contents

Introduction: an overview

This book concerns a special class of management information systems developed primarily for project management, but which is being increasingly used in areas beyond project management in American industry.

Project management is a difficult job, and traditional organizational structures have often proved to be awkward or ineffectual in carrying it out. The project manager is responsible to top management for getting the job done on schedule within allowable cost. However, he must often accomplish his goal by employing the efforts of many separate organizations and persons, most of which are not under his direct control. In a construction project, for example, these would include architects, engineers, inspectors, subcontractors, purchasing agents, vendors, and suppliers, in addition to the people directly on his payroll. The same conditions exist in research and development, and on other types of projects. The network-based system has been developed primarily to assist the project manager in planning, scheduling, and controlling the work under these conditions.

The manager's primary task is to direct and coordinate toward one goal the work of various groups involved in the project, yet the complexity of today's operations forces him to divorce himself from matters of detail, to deal only with the broader aspects of the problem. He is inclined to think and act only in generalities, because he lacks the techniques or management aids which would enable him to comprehend the whole operation in detail. He does not always know which activities are critical and so require special attention, or what effect a delay or failure in one activity will have on others following it or on the success of the project as a whole. What is needed is a master plan which will provide the project manager with an up-to-date picture of the operation at all times, and which would follow a uniform system understood by all. The network-based system is designed to fill these needs.

1

MANAGEMENT AND DECISION-MAKING

Herbert A. Simon has stated that decision-making is synonymous with managing. Further,

Decision making comprises three principal phases: Finding occasions for making decisions; finding possible courses of action; and choosing among courses of action.[1]

Each of the three principal phases of the decision-making or management process described by Simon requires information. The subject of this book is a system which will supply certain information to the manager in a timely, useful, and understandable manner. A management information system, as we define it, includes the collection, transmittal, storage, retrieval, synthesis, partial analysis, and display of information. The system does not include any of the three phases of actual decision-making; it does aid the manager in this process by providing him with some of the information he requires. To the extent that certain rules can be established to "locate any activity which is more than two weeks behind schedule," for instance, the network-based system can perform certain "programmed" decision-making functions, however.

DIFFERENCES IN MANAGEMENT INFORMATION SYSTEMS

These systems have one major objective in common: to provide the manager at each level with the information he needs to make decisions related to his responsibilities. Beyond this common objective, there are many differences among systems which may be classed as "management information systems." The first broad distinction divides them into three categories:

Product-related Systems

Usually technical information data are involved in these systems, related to the physical performance and quality characteristics of the product, whether it be equipment, facilities, services, or data. Examples include specifications, analytical reports, test data, reports, and drawings.

Operations-related Systems

Technical and nontechnical information is handled in these systems, concerning the actual operations involved in producing the products: when, where, and how the products will be created, and what resources will be needed in their creation. Examples include job plans and sched-

---

[1]Herbert A. Simon, *The New Science of Management Decision*, Harper and Row, New York, 1960, p. 1.

ules; accomplishment reports; manpower, financial, marketing, and facilities needs; contract status; funds and labor expenditures.

Administration-related Systems

These systems usually involve nontechnical (in the product-engineering sense) information related to providing at the appropriate time and place, by the appropriate methods the organizational, financial, manpower, material, and facilities resources needed to carry out the operations. Examples include personnel, financial, capital facilities, public relations, and legal.

Within each of these categories, various kinds of management information systems having different characteristics will be recognized. Some of the characteristics which vary from system to system are timeliness, accuracy, processing, summarization, mechanization, method of display, and relevancy.

THE NETWORK-BASED MANAGEMENT INFORMATION SYSTEM

The network-based management information system is classified as operations-related. The basic time-oriented system tells the manager when and how (in terms of sequence of operations) a project-type effort will be performed, and the time/cost/manpower system informs him of the resources (at least funds and manpower) which will be required to accomplish the project. These systems can be more timely, though perhaps less accurate, than other methods; a high degree of raw data processing is done by the system to provide useful information for the manager. Summarization and selection, such as slack path location or milestone selection, is possible, and the system can be highly mechanized even to providing printed graphic displays and very brief exception reports. The relevancy of the information produced depends on the proper construction of the network plan, itself a product of the system which graphically displays significant information.

Network Planning and Critical Path Analysis

The concepts of network planning and critical path analysis are developed in detail in the following sections and chapters. However, a brief overview is presented here, to convey an understanding of the relationship of the network-based system to other management information systems.

Figure 1 shows a bar chart (a), which has been redrawn as a network plan (b), in which the elements (activities) are arranged to show the logical sequence which must occur. The representation in Figure 2 means that activity A must be complete before any one of the three dependent activi-

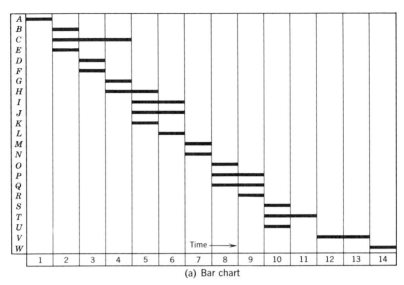

(a) Bar chart

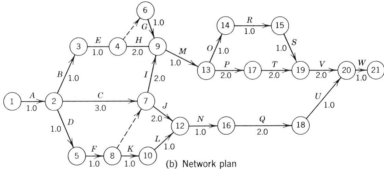

(b) Network plan

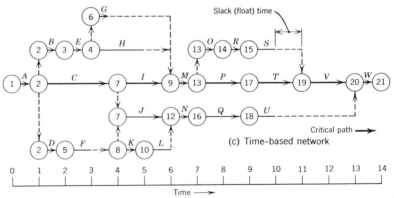

(c) Time-based network

FIGURE 1.   Comparison of bar chart with network plan. The heavy black line indicates the critical path, and the horizontal dashed lines indicate slack time.

ties, *B*, *C*, or *D*, can begin. Similarly, the representation in Figure 3 shows that *all three* activities, *F*, *G*, and *H*, on which *I* is dependent, must be complete before activity *I* can begin. Breaking down a project into discrete activities and events, and arranging these in logical sequence in the network is the first of three major phases involved in using the network system: the planning phase. (Note: Although Figure 1 shows the network

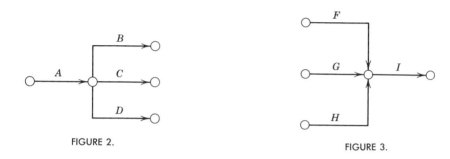

FIGURE 2.

FIGURE 3.

being derived from a bar chart for illustration, this is not the usual way networks are obtained.) Chapters 2 and 3 deal with this planning phase in some depth.

The second major phase in using the network system is *scheduling*. Chapters 4 and 5 deal with this use of the network for time scheduling, and Chapter 6 discusses use of it for resource scheduling. Briefly, the time-scheduling phase consists of estimating the duration of each activity and then totaling these times, proceeding logically through the network, to determine project duration. By reversing this procedure and starting with the end event in the network and subtracting the activity durations, one can derive useful schedule information including the identification of the *critical path*, the longest chain (or path) of activities and events through the network. Its length determines the project duration, and all activities not on this path may have permissible delay time called *slack* or *float*. Figure 1(c) illustrates the results of our calculations redrawn on a time scale, showing the critical path and the slack time.

The third major phase in the use of the network-based systems is the *monitoring* and *control* phase. As the project progresses, the actual time of completion of each activity is reported, changes in the plan and resource expenditures are recorded, and a new analysis determines the impact of progress and change on the future plan and schedule. Periodic updating and analysis of the plan and schedule provides effective management control of the operation. Chapter 7 discusses this aspect of the system.

SYSTEMS DIRECTLY RELATED

The network-based system is only one of many needed to manage a complex enterprise. Other operations-related systems which are directly involved include cost estimating and accounting, work authorization and control, contract information, and production control. Others can be identified in some enterprises. All such systems are interrelated to one degree or another, and in later chapters we will discuss the dependence of the network-based system on such related systems.

SUMMARY

This introduction has presented a brief overview of network planning and critical path analysis, and has described how a network-based management system fits into the complete set of management systems which are found in most enterprises. The subject of this book is thus placed in perspective against the background of the operation of the total enterprise.

PART 1

Network planning: what it is—
how it works

1

Fundamentals of the technique

Several characteristics of the network planning technique distinguish it radically from earlier management information systems: it employs a different planning technique, can handle uncertainty, and has capacities for prediction and simulation. Of these, the unique characteristic is the application of network theory to planning, which is the backbone of the system. As we shall see, the organizational discipline inherent in a network plan makes possible all the other dynamic aspects of the fully articulated, network-based management system.

In this chapter we will explore the origins of the network as a management tool. We will discuss the elements of the network, and how they combine to form a coherent picture of the numerous complex relationships among the activities of a large program. Finally, we will discuss the basic ways to interpret a completed network plan, and to use it to manage and control a program thus planned. The reader will realize the simplicity and obvious logic of network planning, and might immediately feel that its application in his area of interest would be an elementary step. This enthusiasm is partially justified. Network planning is simple in principle; the methodology is logical to the point of being almost self-evident. But the reader should not be fooled. Although the concepts involved are appealingly fundamental, the actual use of network planning in modern industry can be a problem of startling magnitude and complexity. Because of this, in the early attempts at network planning, management systems experts and systems engineers often handled implementation problems. Often, due to their lack of familiarity with organizational realities, they produced plans which were difficult to use and which sometimes failed. The difficulties involved in using the system prompt this cautionary word to the reader, as well as the group of chapters in Part II concerned with the problems of implementation.

EVOLUTION OF THE NETWORK CONCEPT

The application of network theory to industrial problems is not new. Scientists and engineers have been visualizing networks and using network concepts since the earliest days of the industrial revolution. The present applications of network theory are a modern manifestation of an evolutionary process begun years ago. Managers of industrial processes will recognize in network planning many of the same terms they have always employed in planning their work. The industrial engineer will see similarities between the network theory used in PERT, his own process flow charts and industrial programming techniques such as line of balance charts, which have been in use since before World War II.[1] Similarly, the mathematician will recognize in the network algorithm a familiar topological approach often employed in modern industrial programming.[2] Contemporary use of networks as planning tools is actually an offshoot of the use of production scheduling sheets and control charts. These production control and planning tools, essentially bar charts, are the basis for many of the management information and control systems, such as milestone charts, which preceded the network-based system. Thus, the evolution of the network-based system can be traced back to the early work of management scientists such as Fredrick Taylor, whose time and motion studies are familiar to every student of industrial engineering, and to Henry Gantt, whose Gantt charts (bar charts) form the basis for so many modern production scheduling systems.

Origins of the Technique—Gantt Charts

As most modern managers know, Gantt charts have been widely used in industry since Henry Gantt invented them around 1900. The Gantt chart is a series of bars plotted against a calendar scale; each bar represents the beginning, duration, and end in time of some segment of the total job to be done, and together the bars make up a schedule for the whole program. Contemporary managers have used them extensively to plan, and, with certain modifications and additions, to monitor progress

[1] For a concise description of this technique, containing excellent illustrations which reveal the similarities to PERT networks, see NAVEXOS P1851, "Line of Balance Technology," Office of Naval Material, Department of the Navy, 1962.

[2] The mathematics of PERT have been the object of much controversy. There are innumerable discussions, articles, papers, and reports on this aspect of the technique. The authors believe that most of this specialized discussion is irrelevant to the practical application and tends to obscure the true nature and value of the technique as a management tool. Hence, although Chapter 5 contains a thorough treatment of the mathematics of PERT, discussion of the esoterics of the mathematical basis for the technique has been avoided elsewhere in this text.

against the original plan. The advent of "concurrency"[3] and large-scale systems engineering projects, however, soon revealed some fundamental weaknesses in the bar chart as a management tool in an increasingly dynamic and changing management environment. These weaknesses include:

(a) The inability of the chart to show interdependencies which exist between the efforts represented by the bars. This is a serious deficiency when planning a program in which various tasks are scheduled with a large degree of concurrency. Often, using only a bar chart as reference, a manager will overlook critical interdependencies between two or more tasks because the conventional bar chart cannot display such relationships.

(b) The inflexibility of a bar chart plotted against a calendar scale, which prevents it from easily reflecting slippages or changes in plans.

(c) The inability to reflect uncertainty, or tolerances, in the duration times estimated for the various activities. In contemporary management, this deficiency can be critical. Developments in technology have created projects of unprecedented size and complexity, in which primary time and costs are uncertain. This is illustrated most dramatically, of course, by modern weapon system and space system programs, which are characterized by extensive research, development, and engineering efforts, and by relatively insignificant production in the traditional sense.

An Important Step Forward: Milestone Charts

The modern management environment has tended to eliminate, or to seriously impair the effectiveness of the Gantt chart as a management tool. Attempts were made to modify the Gantt chart by adding new elements to it, thus extending its capability to meet contemporary needs.

One relatively successful attempt of that kind forms an important link in the evolution of the Gantt chart into the PERT or CPM network: it is the milestone system, used extensively in the military and industry for the management of major weapon system programs prior to the advent of PERT. Milestones are key events or points in time which can be identified when reached as the program progresses. The milestone system provides a sequential list of the various tasks to be accomplished in the program. This innovation was important because it recognized the *functional* elements of the program, reflecting more accurately what is now known as the program work breakdown or product indenture structure. (This subject will be dealt with in some detail later in the book.) The milestone

---

[3]"Concurrency" is the term introduced by the U. S. Air Force to describe the concept of concurrently designing, building, and testing weapon systems.

system approach increased *awareness* (if not effective display) of the inter-dependencies between tasks. The list of tasks and milestones was displayed on charts *adjacent* to a time scale. Symbols on the time scale indicated the dates each milestone was scheduled, completed, slipped, etc. The milestone system also made it possible to accumulate these data on data processing equipment, and to code the data for various "sorts" or arrangements. Thus data could be presented to management in various ways; by organization, by project, by status, etc. The milestone system proved quite effective and was (and, in fact, is still) widely used, in spite of certain limitations. These limitations include:

(a) The relationship *between* milestones was still not established. Milestones were merely listed in chronological sequence, not related in a logical sequence. Hence, the all-important interrelationships were still not displayed.

(b) True *computer* use was not achieved although the use of data processing equipment provided a greatly enhanced sorting and listing capability. The system did not allow for measuring the *effect* of changes and slippages, but merely improved the reporting of them.

Even with these limitations, the significance of the milestone system in the evolution of the network-based management system was considerable. The milestone system was an important early recognition of the need for awareness and discipline at the engineering staff level; it forced a detailed, logical, sequential, and *event-oriented* planning of all the various segments of complex programs. It pioneered the use of data processing facilities to handle the enormous bulk of management data produced in large research and development programs. In the subsequent development of PERT and CPM, the use of digital computers from the start coupled with the use of network theory made it possible to eliminate virtually all the deficiencies of previous systems discussed here, and to add a capability dimension uniquely suited to solving the problems of modern management.

From Gantt Charts to Networks

At the same time that the milestone system was being perfected and widely applied, the network-based management system was emerging. In 1956, E. I. DuPont de Nemours undertook a thorough investigation of the extent to which a computer might be used to improve the planning and scheduling, rescheduling and progress reporting of the company's engineering programs. A DuPont engineer, Morgan R. Walker, and a Remington-Rand computer expert, James E. Kelley, Jr., worked on the problem, and in late 1957 ran a pilot test of a system using a unique arrow-diagram or network method which came to be known as the Critical Path Method.[4]

Then, in 1957, the U. S. Navy Special Projects (SP) Office, Bureau of Ordnance, established a research team composed of the members of SP and the management consulting firm of Booz, Allen, and Hamilton. The assignment was project PERT (Program Evaluation Research Task), aimed at finding a solution to what was at that time a typical situation.[5] The Special Projects Office was faced with a development program on POLARIS that was a product of the times: a huge, complicated, weapon system development program, being conducted at or beyond the state of the art in many areas, with activities proceeding concurrently in hundreds of industrial and scientific organizations in different areas. The Special Projects Office needed and wanted a new method to provide its management with:

(a) Information on the progress to date and the outlook for accomplishing the objectives of the Fleet Ballistic Missile (FBM) program.

(b) A measure of the validity of established plans and schedules for the optimum accomplishment of total program objectives.

(c) A means of predicting the impact of actual *or proposed* changes in plans on total program objectives.

What emerged was an original approach to the solution of some of the fast-multiplying problems of large-scale systems projects in which techni-

---

[4]Kelley and Walker reported their work in a paper entitled, "Critical Path Planning and Scheduling," which they jointly presented to the Eastern Joint Computer Conference in 1959. The technique was also described by Walker and J. S. Sayer in "Project Planning and Scheduling," Report No. 6959, E. I. DuPont de Nemours and Co., Wilmington, Delaware, March 1959.

[5]Progress of the research task was documented in two reports, "Program Evaluation Research Task Summary Report, Phase I, and Phase II," Special Projects Office, Bureau of Ordnance, Department of the Navy, Washington, D. C., July 1958. The publication of these reports, which received an unprecedented circulation in industry and the military, and the subsequent apparent success of PERT on the Polaris program stimulated interest in the technique. Reports of experiments with the technique and variations of the technique throughout industry and the military began to appear in numbers. By early 1961 literally hundreds of articles, reports, and papers had been published on PERT and PERT-like systems, making it perhaps the most widely publicized, highly praised, sharply criticized, and widely discussed management system ever invented. Enthusiastic proponents of the technique, eager to identify with progress, spawned a multitude of acronyms in addition to PERT, including MAPS, SCANS, TOPS, PEP, TRACE, LESS, and PAR. Although many of these systems had minor differences, they were all network-based. About this time, responsible industrial and military leaders became increasingly concerned about standardization, and various high-level efforts were made to minimize the differences and to develop a more-or-less standard system and nomenclature. This has now largely been accomplished, with PERT and CPM emerging as the standard.

cal innovation, complex logistics, and concurrent activity must be integrated: PERT, an integrated management planning and control technique.[6] PERT was employed on the POLARIS program and is credited (along with a dedicated, project-oriented, management) with making it possible for the Navy to produce an operational ballistic missile-firing nuclear submarine years ahead of schedule. (The actual benefits of PERT in POLARIS are subject to definite question; a great deal of press agentry surrounded the PERT effort at that time.) Subsequently, the original network concept has been adapted and extended in hundreds of management situations, where it has set the pattern for management information and control systems of the future.

Figure 1 illustrates the evolution of the network plan in both CPM and PERT. The differences between the two techniques can be explained by differences in the environments in which each was evolved and applied.

The CPM arrow-diagram network developed from more detailed bar charts which were *job-* or *activity*-oriented. Linking the jobs or activities together in a sequence of dependence, often without special identification of the connecting points, produced the arrow diagram. The environmental factors which had an important role in determining the elements of the CPM technique were:

(a) Well-defined projects.
(b) One dominant organization.
(c) Relatively small uncertainties.
(d) One geographical location for a project.

The CPM (activity-type network) has been widely used in the process industries, in construction, and in single-project industrial activities.

The PERT (event) network evolved from a combination of bar charts and milestone charts, on which milestones were identified as special events, or particular points in time, which were of interest to management. Milestones are useful for progress evaluation; with them, it is possible to determine if a job is ahead, behind, or on schedule *while* it is in progress. The influence of milestones caused PERT networks to evolve as heavily *event*-oriented. Some of the environmental factors which affected the development of the PERT technique were:

(a) Massive programs with hard-to-define objectives.

(b) Multiple and overlapping responsibility divided among organizations.

---

[6]As project PERT developed, the letters P-E-R-T, which had been used as an acronym for the original *research task*, became the means of identifying the *review technique* which was a result of the research task, and PERT became Program Evaluation and *Review Technique*.

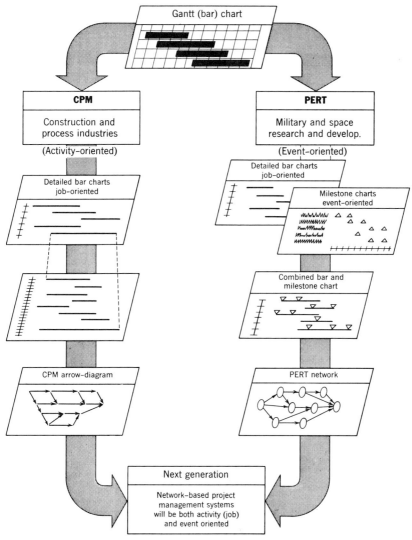

FIGURE 1. Evolution of the network plan.

(c) A large degree of time and cost uncertainty.

(d) Wide geographic dispersal and complex logistics.

The PERT technique has proved most applicable to large-scale research and development, and systems engineering programs, and in other industrial activities involving a large degree of uncertainty, such as new product development and marketing.

BASIC NETWORK ELEMENTS

There are two basic elements in the network plan: the line or arrow which represents a time-consuming *activity*, and the circle or rectangle which represents the *event* or node marking the beginning or end of an activity. When all the activities and events in a program are linked together sequentially in proper relationship, they form the *network. The network is the basic planning document in the network-based management system.*

Events

An "event" is defined as a discrete point in time. An event denotes the specific starting or ending point for an activity or group of activities.

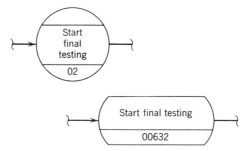

FIGURE 2. An EVENT marks a definite, discrete beginning or end point for an activity or group of activities. Events are usually represented by circles or rectangles containing information about that event.

Events do not consume time or resources, and are normally represented in the network by circles or rectangles containing descriptive information about the event (see Figure 2).

Activities

An "activity" is defined as the work necessary to progress from one event (point in time) to another. Activities are operations which consume time, money, or manpower and are characterized by a specific initiating (predecessor) event, and a terminal (successor) event. In a network, activities are represented as solid lines joining these events, with an arrowhead indicating the direction of flow, or time-dependency (see Figure 3). There are pitfalls in the selection of activities just as in the selection of events, particularly if the planner tries to define an activity without reference to its predecessor and successor events. It is generally best to start with a list of clearly defined events and to introduce the proper activities between these events, to avoid improper definition of the activities.

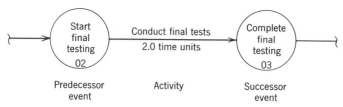

FIGURE 3. An ACTIVITY is an operation which consumes time, money, or manpower resources. An activity is usually represented on a network as a solid line proceeding from one event to another, left to right across time.

The importance of a clear and unambiguous understanding of these two basic elements cannot be overstressed. Precise event definition is important, because occurrence of events indicates what actual progress has been made on the program. Likewise, rigorous attention to activity definition is crucial to successful use of network planning for allocation of resources. Activities must be reportable items of work for valid progress evaluation.

Dependency Relationships

All of the activities in a program are related to each other in various ways. These relationships are called *dependencies*. Activities can be related to one another because they employ common resources:

(a) *Facilities and Equipment*—Activities which must employ common facilities or equipment and cannot do so concurrently are dependent upon one another—one must be complete before the next can start.

(b) *Funding*—Activities which cannot start until certain funding activities have been accomplished are dependent upon the completion of those activities.

(c) *Manpower*—Activities which must use the same manpower are dependent, one upon the completion of the other.

Most dependency relationships result simply from the fact that an activity cannot begin until the product of the preceding activity is available. We design, fabricate, and test, in that order. Or we trench, form, and pour concrete. In some cases, it is necessary to resort to the use of a "dummy" or zero-time activity (usually shown as a broken line) to represent the logical dependencies in the network. This is fully explained in Chapter 3.

HOW NETWORKS ARE USED

It should already be clear that the network plan overcomes many of the difficulties inherent in bar charts and milestone charts. The network plan can depict all interrelationships between activities and can reveal in the planning stages the factors which constrain the accomplishment of any given activity or group of activities in the program. In addition, the network shows all the events, or milestones, of interest to management and thus facilitates progress reporting and evaluation. As we shall see, the network facilitates assessing the impact of slippages or changes in any activity in the network *in terms of total program objectives,* by providing an uninterrupted flow toward the end objective. The ability to handle uncertainty in the program plan is increased by the network, because it provides a framework for estimating a *range* of duration times (optimistic, most-likely, and pessimistic) for each activity in the network plan, if such uncertainty is a problem (see Chapter 4). The network also provides a common reference document for all levels of people and organizations involved in a program. Often in the past, the quantity, type, and format of vital program information varied widely from one program level to another, reflecting the disposition and needs of responsible individuals at each level. As a result, the upward flow of data through the management hierarchy was frequently time-consuming and costly. Many times the validity of the data would become distorted or simply too dated before it reached prime decision-makers. These problems of a common language and timely communication are very significantly eased by the network system method.

The networks are used for planning, plan integration, time analysis and scheduling, and resources analysis.

Planning. To employ networks for planning, the overall project is broken down into its natural planning elements (usually including major end items), and at some level of breakdown elements are identified as activities and related events. These are arranged in the sequential order dictated by dependency, resource constraints, or external factors.

Integration of Plans. In large projects, separate organizations can plan their portions of the networks independently. Then, through proper identification of *interface events* (common events which link two or more networks), the total project network can be integrated. Chapter 12 discusses this important use of the network technique further.

Time Analysis and Scheduling. As described earlier, simple procedures for time analysis and scheduling give effective results to the network plan user. Typically, time savings of 20 percent or more are obtained, as described in Chapter 10 and Part III.

Resources Analysis. Using the network as a framework with all activities identified, the planner can make a "picture" or model of the program available for management's use in resource allocation. One of the unique capabilities of the network-based management system is its facility for realistically allocating dollars, manpower, and facilities among the activities of the network. Because all constraints are displayed on the network, resources can be allocated or budgeted with full consideration given to all program needs; this reduces underbudgeting or overbudgeting on individual activities. The network also facilitates accurate estimating of resource requirements by providing detailed visibility. In addition, by using the time analysis described above based on known constraints, management can determine *when* resources will be required, as well as *where* and *how much;* this makes budgeting and funding much easier. In some situations, resources (time, manpower, and dollars) are allotted to each activity. This is possible in situations such as construction projects, in which there is little uncertainty and in which activities can be defined accurately. In other circumstances, activities may be grouped for the budgeting of manpower and dollar resources. In a typical use on a large research and development program, it is often impossible to segregate manpower and costs accurately by activity. Therefore, when the activity duration times have been estimated, the activities are usually grouped into "work packages." Manpower and dollar estimates are then made for these groupings. (This "work package" concept will be discussed more fully in Chapter 2, and the factors of cost and manpower will be considered in detail in Chapter 6.)

The Critical Path Concept

Adding together the estimates of activity duration times along all the paths of the network reveals the path that will consume the most time in reaching the end event. This path is known as the *critical path*, and is one of the most important concepts in network-based systems. Isolation of this sequence of activities gives the manager some of the most vital program information he will need to plan and manage the program properly. If the network is constructed properly, and activity duration times are estimated without reference to scheduled dates, this path will contain no slack time. This tells the manager several things:

(a) Slippage of an activity along this path will cause a corresponding slippage of the end event, or program objective unless the slippage is recovered on some part of the remaining critical path; this happens because this sequence of activities contains the pacing activities for the total program. Inability to isolate these critical pacing activities on large programs has been one of the most important causes of program slippages in the past.

(b) The activities on this path, being the most critical from a schedule standpoint, are the activities most in need of management attention. This means that the manager can manage by exception, a goal in all times and in all places. Management by exception is particularly important on today's vast programs, in which an inability to rank program information in this manner leads to much misplaced management attention.

(c) The activities on the critical path are the activities that will be most responsive to management's efforts to improve the total program. Acceleration along this path will have a direct effect on the expected occurrence of the end event, or completion of the program. The critical path and its significance will be discussed in considerable detail in later chapters. It is one of the manager's most important tools.

Slack Paths

When the critical path has been isolated, it becomes apparent that all of the other paths of the network contain, by comparison, slack time in varying amounts. These paths, called slack paths, are also very useful in evaluating the program plan. Obviously, ranking of these paths by the amounts of slack time they contain, starting with those with the least and ending with those paths that have the most, extends the management by exception principle to all areas of the program. Not only does the manager know which activities are most in need of his attention by examination of the critical path, but he also determines which activities are next most critical in descending order, by examining the slack paths of the network. Slack paths also give the manager additional vital data:

(a) Knowing the amount of slack time associated with the activities on a given path, the manager can determine how much the activities on that path *can* slip or be deliberately delayed without causing delay of the end event.

(b) By discovering the amount of slack time associated with the various paths, the manager can decide how to best allocate resources along any particular path within the range of times allowed by the slack condition. Hence, resource allocation can be made as flexible as possible within the range available, to permit more nearly optimum allocation of resources among all of the activities of the network.

Use of Computers

Analysis of large amounts of program data, and subsequent presentation to management in an easily understood form are key elements in the effective use of any management information system. This is no less true with the network-based system. Computing and analyzing network data on a basic level requires only the most elementary mathematical knowledge, but manipulating a large quantity of even simple calculations by hand can require a considerable amount of time. Small networks of up to 100 events, for example, can be rapidly and economically handled by manual computation as explained in Chapter 5, particularly when only schedule or time data are needed. Obviously, however, manual computation becomes increasingly burdensome when the networks are larger and more complex, and when more data are desired. For this type of network the aid of a high-speed digital computer is most valuable. A minimum computer program would receive as input the activity data appearing on the network, event data, time estimates, other resources data associated with the activities, any scheduled dates, and fixed-resource budgets, as well as instructions for sorting or arranging the output. By following a routine similar to hand computation, the computer compiles and prints out all the information we have discussed above in minutes, even for very large networks with thousands of activities. The computer can also perform a variety of additional functions, depending on the sophistication of the input data and the outputs desired. Chapter 9 discusses the role of the computer in some detail.

Monitoring and Control through Progress Evaluation

Once the program plan or network has been prepared, and the initial schedules established, the next steps are to update the network, evaluate progress, and exercise management control to correct problems which are revealed.

Network updating consists of reporting the dates of actual completion of activities plus any revisions in the plan. Management then makes schedule forecasts based on the last completion dates. This updated program information is fed back on a regular schedule for processing, evaluation, and translation into management action. Such periodic evaluation of existing plans by comparison with actual, current, operating conditions gives the program manager a continual check on the critical areas of his program, and an up-to-date estimate of the probability of meeting program objectives. In addition, it permits timely consideration of alternate courses of action, and evaluation of their impact on end objectives. To provide effective program control, this regular reporting, analysis, decision-making, and direction cycle should start as early in the program as possible, and should be continued throughout the life of the program. The frequency of this evaluation varies from daily to monthly or quarterly. In most engineering and construction situations, it is either biweekly or monthly.

Advanced Uses of Network Plans

One of the fascinating aspects of network plans is the potential they offer. All their past and present uses build a solid foundation for far-reaching extensions of the system. A number of users have already vastly improved planning by applying trade-off or optimization techniques to time, dollars, and labor (Chapter 6). However, the ability to allocate resources over multiple projects (Chapter 12) means that it would be possible to develop a total corporate planning system.

Simulation and gaming with network systems (Chapter 18) are still theoretical topics, but they offer significant long-range benefits to business management. The relationship of the network system to the management function, and the outlook for further development of the system, perhaps even leading to a valid general system theory for the management of human enterprises, are discussed in the concluding chapters of the book.

SUMMARY

Chapter 1 has described the origins of the network-based system, including the management information needs which specifically dictated the design characteristics of such systems. It provides an introductory description of what networks and their related elements are, how they work, and how they are used. In subsequent chapters of Part I, the topics introduced here will be developed in detail to provide a thorough understanding of the system elements in a logical sequence.

2

Systematic planning for network systems

Chapter 1 provided an overall view of network planning by identifying the basic elements of the technique, and describing generally how events, activities, the network, data processing, and the critical and slack path concepts are integrated to produce a useful management tool. In this and subsequent chapters in Part I, we will discuss each of these basic concepts in detail, so that the reader can develop a full understanding of, and competence in the principles and mechanics of the network-based management system.

As we have seen, the basic method is simple. Using the dependency network as the primary planning tool, a manager is able to describe fully and accurately the most complicated relationships involved in accomplishing a project. It is unlikely, however, that a manager could set up a network for a complex research and development project, such as a contemporary weapon system or space system project, without first defining clearly and unambiguously the goals to be reached. In other words, before a manager can describe in detail the way in which his objectives can be reached, he must answer the following questions with a statement of premises:

(a) What are the project objectives?

(b) What are the major elements of the work to be performed, and how are these elements related to one another?

(c) Who will be charged with the various responsibilities for accomplishing project objectives?

(d) What organization of resources is available or required?

(e) What are the likely information requirements of the various types and levels of management to be involved in the project?

Furthermore, it is unlikely that a completely new physical organization will be specifically designed to complement the network logic and provide the optimum environment for the execution of the project. In the more common situation, the project will be superimposed on an existing

23

company organization, where a number of traditional planning and control structures will already exist. Common examples of such structures would include the familiar division of types of labor (e.g. , production, management planning, accounting, and engineering) by departments or operating divisions; fiscal control systems and accounting procedures; and other technical and administrative planning and control structures. Where such a functionally oriented organization already exists, it will exert an important influence on the resolution of the questions just mentioned. However sound the reasons for the existence and perpetuation of their traditional planning and management structures, when viewed in terms of implementing the integrated network-based management system for the accomplishment of a project or projects within an existing organization they may present potential functional conflicts which could hamper the effective use of the network system. This situation, far from being the exception, may be considered one of the most common limiting factors in the successful application of the network-based system. There are many reasons for the persistent existence of such obstacles to project planning, ranging from the obsolete and irrational to the most indispensable and compelling. (Further comment on such problems is given in Chapters 8 and 17.)

For the moment, however, we will not attempt to evaluate the relative merits of alternatives of industrial organization, but will treat whatever situation we may encounter as a fact of life to be accommodated in our planning. Later, it will be appropriate to consider the impact of contemporary management technology on industrial organization theory, and evaluate alternatives in that light. This chapter will discuss the inherent problems in large projects which must be accommodated in the pre-planning stage, and the methods which can be used to isolate and deal with these problems. The reader should realize, of course, that when a planner is preparing a network plan for a relatively small, completely familiar project, he can often put down the original plan in network form without extensive pre-planning. If the project is large and complex, however, and involves a large and complex organization, it is essential that systematic and thorough pre-planning precede the actual preparation of networks. Detailed pre-planning or project definition of the kind described in this chapter is typically used when applying network planning to large projects; the effectiveness of the networks derived depends upon the attention given to this vital planning area.

NEED FOR SYSTEMATIC PRE-PLANNING

However effective a management system may be, and however dependent it may be for its effectiveness on a pure or undiluted approach in implementation, it cannot be superimposed on an existing organization. Instead, the network system must be tailored to the organization that will use it. It must be specifically designed to reflect the organization's objectives, the nature of its system operations, and its managerial and supervisory requirements. The management system should be compatible with, and, if possible, complement the organization's existing accounting and other systems and policies. It must respond to the real information needs of managers at all levels and must achieve its basic objectives without damaging the existing planning and command structures. The integrated approach provided by network-based planning and control often emphasizes, or even reveals for the first time communication problems resulting from functional specialization which tends to separate vital planning and control functions, or even divisions of the planning itself. Recognizing that these planning and control structures sometimes conflict, and that they must be made to work harmoniously together towards the project objectives becomes a very necessary part of the implementation of network planning on large projects.

Objectives of Project Planning

In planning for use of a network-based system on a particular project, the manager must first attempt to bridge the gap which has traditionally existed between the organizational functions of planning and setting objectives, and those of contract funding, cost accounting, and accounting based on other resources. Such barriers have often hampered timely correlation of original plans and estimates with current actual expenditures, current plans, decision-making, and direction. Project planning attempts to:

(a) *Define the project objectives.* Identifying project objectives is the first step in setting up a network-based management system. Establishing objectives provides the basic framework from which valid networks can be prepared, and against which the network system can be correlated.

(b) *Define the form of the organization required.* Once the planner has defined project objectives, he must identify the required organizational structure within the existing larger organization, or create and correlate it with the project objectives.

(c) *Determine what other planning and/or control functions are involved.* The planner must identify the existing or required planning and control functions, such as cost accumulation and accounting systems, and identify the elements of work by contract which will be used; he must also define the ways they will work into project objectives.

(d) *Define the information needs of the various types and levels of management involved.* The planner must know the formats and types of management information to be produced (costs, manpower, progress against schedule, etc.). In addition, he must find out how much detail is to be provided to each type and level of management.

(e) *Specify the way in which the project will be accomplished.* Finally, the planner must prepare detailed resource application plans and schedules based upon all of the foregoing, to provide for step-by-step acquisition and use of the manpower, facilities, money, etc., needed to accomplish project objectives. The resource schedules must show how the work will be accomplished, and when the various end products will be forthcoming.

Accomplishing these planning objectives in an orderly, systematic way is the best way to achieve true project control, whether using the network system or not. The planning and control system developed for management of a project will be only as valid as the premises upon which it is based. With effective pre-planning, however, the system can be an enormously efficient tool in planning and control.

Elements of Project Planning

Before defining the elements of project planning specifically related to use of the network plan on a project, it is useful to clarify the distinction between project planning, on the one hand, and product engineering (product planning) which is not normally included in the scope of a network-based project management system. Figure 1 shows the differences between these types of planning. As it indicates, product planning has a definite impact on project planning, but product engineering planning is concerned only with the technical characteristics of the products to be produced, and usually produces various types of technical specifications which the products must meet, such as performance or reliability characteristics. Although it might be desirable for such technical planning to be an integral part of the network-based system it is not usually incorporated in present systems. Integrating such elements into overall program planning is an ultimate goal of a complete management system, but there are many difficulties inherent in such planning which have not yet been resolved. In a later chapter we will examine some of the problems and potentials of integrating this important added dimension into management planning, but to understand and use the network-based system at present, we will be concerned primarily with the project-planning aspects of management illustrated by the lower half of Figure 1.

The basic elements of project planning are:

(a) *Project definition*—developing a detailed breakdown of the work

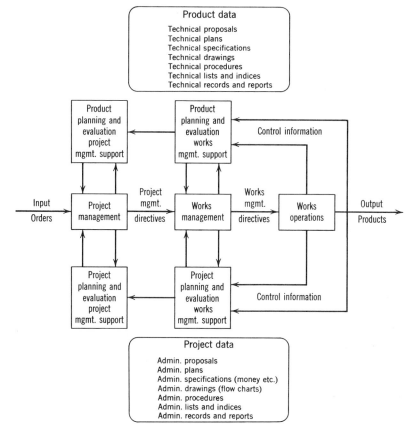

FIGURE 1.   Large business program model (Source: Frank Little, The Boeing Company).

to be performed to produce the products. Deliverable products are equipment, services, facilities or data.

(b) *Organization definition*—defining in detail which organizations will be responsible for the project, and which will actually carry out the activities.

(c) *Work plans*—developing detailed network plans to describe the production process.

(d) *Schedules*—allocating work plans over a calendar scale.

(e) *Resource estimation and pricing*—estimating manpower and dollar requirements for elements of the project breakdown and work plans.

(f) *Resource accounting*—collecting and accounting for resource expenditures.

(g) *Contract administration*—authorizing and controlling resource expenditures.

(h) *Management reporting*—accumulating, coding, organizing, selecting, and summarizing management information in formats which reflect management's needs and serve program objectives.

In the network-based management system, these various and sometimes conflicting elements of planning are integrated into the basic *planning structures* listed below. *The development and correlation of these planning structures is the vital first step in the development of a network system.*

Project Definition: The Work Breakdown Structure
Organization Definition: The Project Organization Structure
Work Plans: The Network Plan Structure
Schedules: The Calendar Time Structure
Resource Estimation and Pricing: The Estimating Structure
Resource Accounting: The Chart of Accounts Structure
Contract Administration: The Funding and Authorization Control Structure
Management Reporting: The Report Structure.

All these structures must mesh together carefully, and so the correlation of these planning structures is a central concept of the system approach. In the past, however, it has been difficult to achieve this goal. In the following paragraphs we will discuss each planning element in turn, and develop the concept of the Work Breakdown Structure as the basic foundation upon which all of the other planning structures rest.

DEVELOPING THE PLANNING STRUCTURES

The reader can now understand the need for establishing objectives and developing the major planning elements, before beginning to prepare networks and other detailed plans. We can therefore proceed to discuss the principles and mechanics involved in developing these objectives and planning elements. (Discussion of the actual preparation of networks will come in Chapter 3.)

The Indenture Level Concept

The first planning structure we will examine is the list of the various tasks which must be completed to achieve project objectives. This indentured *Work Breakdown Structure* is the most important single planning structure, because it establishes the project objectives and provides the framework upon which all of the other planning structures can be correlated. It is also the basis for the project network plans.

The concept of indenture levels should not be alien to the average reader of this text, because it is ingrained in planning and organizing throughout our civilization. We need only glance at the most common means of describing the organization of our government, a corporation, or any other grouping in which some elements are subordinate to others, and in which subelements combine to make up to the whole. The ordinary organization chart depicts indentured functions, departments, branches, etc., which combine into divisions arranged in a hierarchy, together constituting a whole. Other examples come to mind:

(a) In a book such as this one, the whole is indentured into successively smaller sections: parts into chapters, chapters into paragraphs, and paragraphs into subparagraphs, corresponding to the ways the expressed ideas form subordinate parts of the whole.

(b) In biology the hierarchy of phylum, genus, and species uses the level-of-indenture concept to bring order and coherence to the seemingly chaotic process of classification.

(c) The manufacturing process uses the level-of-indenture concept to indicate that raw materials are combined to form parts, which are then assembled into successively higher levels of hardware, finally coming together into a single end-item or system at the end of the process. Figure 2 gives two examples of this.

In modern major weapon system and space system projects, the indenture level concept or Work Breakdown Structure makes it easier to identify the elements of the programs, and has been used extensively for this purpose. The acquisition process for such programs (design, development, test, manufacture, integration, and checkout) has presented modern management in government and industry with the most formidable planning, communication, and control problems ever faced, including such factors as:

(a) Advanced technical requirements which introduce large degrees of uncertainty into the programs.

(b) Large size and complexity.

(c) Long lead times, and the exigencies of the cold war, which force long-range, detailed planning for the future.

(d) The need to communicate and integrate plans among hundreds of more-or-less independent industrial and governmental organizations dispersed over a wide area.

Therefore, experience acquired in applying PERT to the major weapon system projects such as POLARIS and MINUTEMAN has resulted in widespread use of the Work Breakdown Structure in the defense industry. For this reason, in this chapter we will use the defense industry approach and example to explain the Work Breakdown Struc-

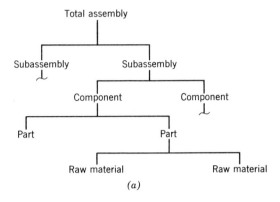

(a)

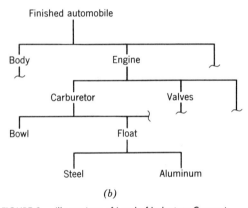

(b)

FIGURE 2.   Illustrations of Level-of-Indenture Concept.

ture. There are many industrial activities which are not concerned with weapon system programs, of course. The concepts developed for weapon system projects are directly applicable to any project-type effort, however, and will probably be applied to an increasing number of industrial projects, with the continuing expansion and sophistication of industrial technology.

Project Definition: The Work Breakdown Structure

As stated earlier, the Work Breakdown Structure is the basic planning structure which provides the framework for developing network plans and indicates the structure of management reports. In addition, the planner can relate all the other planning to the elements of the Work Breakdown Structure, to satisfy organizational requirements and eliminate barriers to communication.

The planner begins the breakdown by identifying the total effort, and then the finer and finer subdivisions of that effort. In Figure 3, the highest element is the "Missile Weapon System" at level 1. There are no standard terms used for subdivisions of work below the project level, and so they may be termed "system," "subsystem," "subtask," "level 1," "level 2" (as in the example shown), "assembly," "subassembly," "component,"

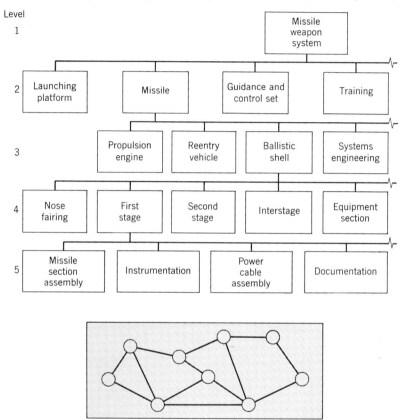

FIGURE 3. Generalized Work Breakdown Structure (Source: DOD/NASA PERT Cost Systems Design Guide).

or other words commonly used by management. The underlying principle of the breakdown is always the same, however. The top element in the hierarchy is divided into its major elements (in this instance, "launching platform," "missile," etc.) at the level below. Each of these elements is then subdivided into its component elements. The planner continues this work breakdown to successively lower levels, reducing the scope, complexity, and dollar value of the elements at each level, until he reaches the level at which the elements represent manageable units for visibility, planning, and control.

Additional examples of the Work Breakdown Structure concept are shown in Figure 4, and in Figure 10, Chapter 13. The reader should note that the Work Breakdown Structure is clearly *product-oriented,* but not necessarily *end-item-oriented.* For example, the breakdown will include

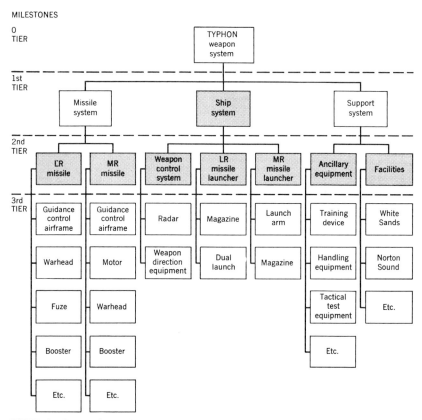

FIGURE 4. Proposed TYPHON PERT milestone program structure: TYPHON Work Breakdown Structure (Source: BUWEPS PERT Milestone System—A Tool For Program Management).

products which are undoubtedly end items, such as equipment and facilities; it will also include "products" such as systems integration, services, training, and data, which are just as essential to overall project objectives, but which are not necessarily deliverable items. There are important criteria for the selection of elements to be included in the Work Breakdown Structure. All those included must be definable segments of the work to be accomplished, and every element of that kind must be included; the planner must also be able to draw networks representing each element identified, indicating the way each product will be produced. Once the planner has set up the Work Breakdown Structure for the project he can:

(a) Establish the scope and number of networks to be prepared.

(b) Assign organizational responsibilities to the elements of the Work Breakdown Structure.

(c) Specify and correlate contract administration, fiscal planning and control structures.

(d) Specify the management reporting structure and design the reports.

At the lowest level of the Work Breakdown Structure, individual networks will be prepared which represent all the related events and activities required to accomplish each of the goals for that level. The reader should note that the number of levels identified in the Work Breakdown Structure will vary from case to case. The only criterion is that the lowest level be one at which management feels it has sufficient visibility for planning and control. Factors which will affect this choice include:

(a) The complexity and time span of the program.

(b) The magnitude of resources such as cost and manpower used in the program.

(c) The customer's internal management structure and his reporting requirements.

Further, the number of levels required to describe elements of the Work Breakdown Structure may vary even within the same program. For example, at level 2 on the Work Breakdown Structure illustrated in Figure 3, the element identified as "Missile" may be divided into six or more levels to reduce it to manageable elements for planning and control. By contrast, the element identified as "Training" at the same level may require only one or two further subdivisions to achieve the same level of visibility. Once the manager has reached this basic planning level all across the Work Breakdown Structure, he will break out one additional level of detail, either before or after the networks are drawn. (In the latter case this is often done on the networks themselves.) In this last subdivision "work packages" will be identified, representing the basic units of work which will be used to estimate cost and resources, and against which actual costs will be collected and compared with estimates for cost control. (Identification of work packages will be discussed later in this chapter.)

Organization Definition: The Project Organization Structure

The reader should note that the breakdowns of levels of indenture in the Work Breakdown Structures in Figures 3 and 4 resemble organization charts. It is essential that he realize, however, that there is a very fundamental difference between the two types of charts. Rarely will an actual organization correspond exactly to even a small portion of the

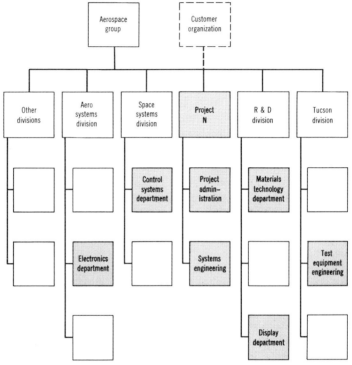

FIGURE 5. Developing the project organization structure. Key: functional organization in white boxes; project organization in shaded boxes. (Source: Hughes-PERT Guide Hughes Aircraft Company).

breakdown described by the Work Breakdown Structure, because the breakdown represents a completely product-oriented organization of the effort required to accomplish project objectives. (See Figure 5.) Many industrial organizations have evolved into a mixture of both functional and project-type structures; on a major advanced-technology project, for example, it would probably be necessary to integrate many such organizations. Whatever the specific organizational characteristics in a given case, the responsible and performing units must be correlated with the elements of the Work Breakdown Structure.

Having once identified the responsible and performing units, the planner must next indicate that identification on the Work Breakdown Structure, usually by using the organization cost-accounting code number. In this way, regardless of the combination of functional and project-type units in the organization, the responsible units can be identified without disrupting the existing planning and command structures. This straightforward association of the organizational units with the elements of the Work Breakdown Structure allows the planner to pinpoint work assignments essential to realistic resource allocation and pricing, work authorization, and cost accounting procedures. In addition, he learns what types of management reporting are required for the project.

The reader should realize that the Work Breakdown Structure is best prepared and evaluated *before* organizational responsibilities are assigned, because the existing organizational structure should not unduly influence its development. We have seen that the normal organization will not usually fall into a neat pattern at any level of the established Work Breakdown Structure; if the existing structure is permitted to influence the indentured project definition, conflicts may result which would hinder subsequent planning and control. Conversely, using the Work Breakdown Structure as the basis for all subsequent planning will have the effect of imposing a desirable planning order which will anticipate potential conflicts in objectives and facilitate control.

Work Plans: The Network Plan Structure

One of the most important functions of the Work Breakdown Structure is to provide a disciplined framework for the preparation of the actual network plans. The definition provided by the Work Breakdown Structure builds the framework in three ways: it provides a basic list of the networks that will be required; it defines the basic events (milestones and interfaces) which must be included in each network; and, it gives the visibility needed to select work packages.

Milestone and Interface Event Identification. Let us ignore for the present the function of the Work Breakdown Structure in establishing the network framework, to see how the planner arrives at the definition of basic events. When every major element of work at each level has been defined by the Work Breakdown Structure, it is possible to examine each element more closely, and to isolate certain important events involved in its accomplishment. The planner is specifically interested in identifying significant *milestone events* and *interface events* which occur in each element.

Milestone events are those specific events (points in time) which management has identified as important reference points in accomplishing the project; progress of the work according to the plan is monitored primarily against these events. The plan must therefore reflect occurrence of these milestones. Milestone events appearing near the top of the Work Breakdown Structure might be scheduled dates which the customer has imposed, perhaps without reference to plan, or target dates set by management for completion of certain segments of the work to achieve profit incentives written into the project contracts.

Interface events are commonly defined as events which mark a change in responsibility for an event from one network segment to another. This means that events in one segment of the work defined by the Work Breakdown Structure and reflected in a network plan may constrain or be constrained by an event in another segment. For example, occurrence of an event marking the start of an assembly operation in one segment may depend upon having the components; however, another organization may be responsible for providing the components, and the activity may occur in another segment of the work plan. In such instances an identical event must be noted in both segments of the plan to show that they join together or are interdependent at that point. If such interdependencies are not noted by the use of interface events unrealistic planning can result, particularly in cases in which accomplishment of one task might be adversely affected by serious slippage in another. At the upper levels of the Work Breakdown Structure such interdependencies or interface events might include those which involve the customer, such as approval of designs, release of funds, authorizations of work, or availability of facilities. At lower levels, an interface event may mark delivery of equipment from one organization to another, transfer of paper, or availability of resources.

To provide these milestone and interface events at the lowest level of the Work Breakdown Structure as a framework for preparing networks, the planner must identify them in each segment of the Work Breakdown Structure, and must carry them down from level to level. Further, this identification and carrying-down process must be done in a systematic fashion, reflecting the discipline of the Work Breakdown Structure.

Use of Master Phasing Charts. One excellent and convenient way of identifying and carrying down milestone and interface events to the lowest level of the Work Breakdown Structure is to prepare *master phasing charts* representing each segment of the structure. The master planning charts are simply bar charts similar to those discussed in Chapter 1, prepared for each segment at each level of the work. The bar chart representing the top level of the structure contains a bar for each segment identified at the level below; at the next level the bar chart for each segment contains one bar for each subdivision of that segment at the next lower level, and so on, until the lowest level has been reached. On each bar it is possible to note particular points in time, and so to identify the milestone and interface events described above. (The point in time selected will be a "best guess" at this stage, which will be evaluated later by time analysis of the detailed network plans.) At the highest level the significant milestones are identified on the bars. At each level below, the milestones for that level are similarly identified. In addition, the milestones identified at higher levels are carried down into the lower levels, and indicated in the job segment and point in time in which they occur. In similar fashion, the planner identifies and carries down the interfaces between segments at each level, and between levels. Thus, when the Work Breakdown Structure has been analyzed to the lowest level, all the significant milestones and all the vital interfaces have been assembled and grouped into the lowest-level subdivisions.[1]

At this point, the first two functions of the Work Breakdown Structure in network preparation have been completed. If they wish to do so, the planners can use the lowest-level subdivisions of the work breakdown to define the scope in developing network plans for individuals. The milestone and interface events within each subdivision on the lowest level comprise the skeleton for the network to be developed. The lowest-level subdivisions make up a list of networks to be prepared; the events identified in each subdivision represent a basic list of events for that network. In some instances, however, it may be more desirable to postpone actually preparing the networks until completing the third function of the Work Breakdown Structure: the identification of work packages.

The Work Package Concept. Earlier in this chapter we referred to a final subdivision of the Work Breakdown Structure into the basic units of work. This is done to facilitate cost and resources estimating, and to give a standard for comparing actual costs with estimates, for cost control. Identifying these added work subdivisions is an innovation involved in extending the basic PERT technique into the PERT Cost System; the latter includes dollar and manpower factors and is the basis for correlating these resources with the network. The basic units of work are known as *work packages:*

A work package is an activity or a group of activities that have a common charge account number in the chart of accounts.

This definition is important to remember, because it implies several con-

---

[1]Time-phased milestone networks have been proposed, and, in some cases, used in place of master phasing charts. However, this often results in a confusing welter of levels of networks which are difficult to use.

cepts vital to an understanding of work packages, and to their successful use in including cost in the network-based system.

A work package may include many activities and most often does. In a typical advanced-technology project such as a major weapon system development there may be upwards of 50,000 events in the total network plan. It would obviously be sheer folly to try to estimate and control costs for each of the activities involved in such a project; in any event, no cost control system would consider it either necessary or desirable to do so. In selecting units of work to be identified as work packages the planner should consider the cost-accounting needs of the responsible and performing organizations, and the cost or personnel size of the work package which will give the desired visibility. On a small project involving a few thousand dollars the size of a work package may be measured in hundreds of dollars. By contrast, on a large multimillion-dollar project, work packages may be "worth" $100,000 or more. How costly the work packages will be depends on the size of the project, and, more important, on the visibility requirements of management in a given case.

Work packages are often functional breakdowns of the lowest-level items of the Work Breakdown Structure, such as fabrication, assembly, testing, or excavation, but they may also be designed to enable management to segregate costs by organization, skills, end items, or other criteria. In the example in Figure 6 the cost accounting for the project is essentially functionally oriented, separated into electrical engineering, mechanical engineering, manufacturing, and testing segments. The chart-of-accounts structure will include discrete charge account numbers for each of these functions. The charge account numbers may be designed to collect costs for "testing" wherever they occur in the program in a single account, or they may be coded to segregate testing costs for each item on the lowest level of the Work Breakdown Structure. In another instance, the customer may require cost accounting by hardware items, perhaps subassemblies or components of end items identified at the lowest level of the Work Breakdown Structure. In such a case, the work packages identified would include all activities which contribute to that hardware item, regardless of function. Again, the composition, size, and number of work packages are determined by management's needs, and by other influences such as organizational accounting procedures, customer requirements, or flexibility of the chart-of-accounts numbering system available.

Work packages must be designed to be closed out at a specific point in time, so that total expenditures can be compared to the estimate. The time span of each work package must be short enough to provide enough

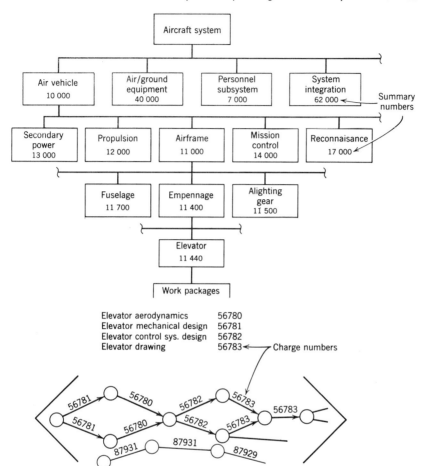

FIGURE 6.   Work Breakdown Structure with account code structure and work package/network correlation (Source: USAF PERT Cost System Description Manual, Vol. III, December 1963).

check points during the life of the project so that management can evaluate both time and cost progress as it occurs.

Once the work packages have been identified, a discrete charge account number is established for each. All activities in a work package will carry that particular charge number and no other activities will carry it, thus establishing a discipline for estimating and monitoring costs on the project. As indicated above, it may be desirable to postpone the actual preparation of networks until these work packages have been identified, and then to construct networks for each work package or related group of packages. Factors affecting this decision will be the number of activities in the work packages and the complexity of the chart of accounts. In any event, the identification of activities on the network will include the applicable chart-of-accounts code number.

Schedules: The Calendar Time Structure

After the planner has developed networks showing the sequential actions necessary to accomplish the work, he must develop the calendar time structure and relate it to the Work Breakdown Structure. The planner does this through analyzing times on the network plans, a process described in detail in Chapters 4 and 5. After completing several analysis cycles, during which the networks and even the Work Breakdown Structure have probably been considerably revised, the manager will be able to establish an acceptable set of dates. These dates, when approved, form the schedule for the calendar time structure of the program, and each element of the Work Breakdown Structure can then be assigned dates for starting and completion. Milestone and interface events are also assigned dates.

Resource Estimating and Pricing: The Estimating Structure

Within one organization there may be several widely differing procedures for estimating and pricing labor, material, and other resources. Typically, such practices have not evolved as independent systems designed to allow the best estimating but have emerged because of the demands of proposal preparation practices and the organizations responsible for pricing. Procedures may range from literally pricing hardware units "by the pound" and allocating resources within the dollar constraints thus derived, to sophisticated, completely computerized estimating routines which accept detailed raw data and extend it by means of prescribed routines.

Often, existing estimating routines are fundamentally incompatible with the integrated-project management approach, and must be carefully modified and correlated with it to achieve valid resource estimates. As we have indicated, in integrated project management all project planning, including resource estimating and pricing is accomplished with reference to the Work Breakdown Structure. Specifically, at the lowest level of the Work Breakdown Structure resource requirements and prices are estimated for previously identified work packages.

Estimates derived in this way can be easily combined and compared with the Work Breakdown Structure and then correlated with other planning structures (organization, detail network plans, time schedules, work authorizations, and expenditure accounting) to arrive at the integrated time/cost/manpower plan required by management. Existing estimating and pricing methods do not often lend themselves easily to that approach. Often, too, government demands pricing by contract line items and other categories which are incompatible with the Work Breakdown Structure; this will further complicate costing.

Therefore the existing estimating structure, whether composed of historical records or prescribed pricing practices applied on a case-by-case basis, must be reconciled with the work-package resource estimating required by the integrated system. Often, the planner can change the format of historical pricing data so that it can be collapsed into work-package increments. Conversely, in some situations it may be more desirable to develop work-package estimating techniques which can be "shredded out" by manual or machine accounting techniques to maintain traditional systems. In either situation, the relevant interfaces between differing practices must be identified and incorporated into the overall estimating process.

Resource Accounting: The Chart-of-Accounts Structure

The chart-of-accounts structure establishes the classifications of accounts for which costs will be collected. In addition to the cost-accounting requirements of the project itself, there will usually be a number of other accounting functions not directly related to project requirements which will be served by the chart of accounts; these other functions often involve the larger accounting requirements of the company. Project cost-accounting requirements can usually be satisfied by assigning a discrete cost account number, or "charge number" to each identified work-package element in the project. These can be cross-referenced to the company accounting system by building into the charge number certain significant codings which will meet the requirements of the company chart of accounts. In addition, the chart-of-accounts structure will be considered in selecting the work packages themselves. The important thing for the planner to remember concerning the chart-of-accounts structure is that correlating the project work breakdown elements effectively with the company accounting system will result in an integrated network management system, which is a practical tool for cost and resources control.

Contract Administration: The Funding and Authorization Control Structure

In any project, funds must be available before work can be authorized; therefore the contract administration procedures, including machinery for releasing funds and authorizing work, must be identified and correlated with the elements of the Work Breakdown Structure. Like the accounting structure, this fiscal or contract administration structure is apt to already exist in the organization, or to be dictated by the customer; it must be correlated with the Work Breakdown Structure so that it can function normally with maximum relevance to project objectives.

The contract administration structure authorizes funding of work packages, makes organizational assignments, work plans, and time schedules, and estimates and prices resources related to the project contract structure. These functions can become complicated when more than one contract is related to one project.

Work authorized must fall within the contract work statement, and total authorizations and budgets must be within contract limits. To assure following these contract elements in a manner consonant with project objectives, and including a specific capability for reporting contract-status information in the network system, contract code numbers must be established. These contract code numbers can be keyed to the elements of the Work Breakdown Structure to provide a convenient cross-reference to the contract administration function.

Management Reporting: The Reports Structure

The management reports structure is the culminating objective of the project planning elements discussed thus far. The Work Breakdown Structure provides the basis for summarizing management information to report to the various levels of project management. The project organization structure specifies the recipients of such reports and their information requirements. The fiscal and chart-of-accounts structures determine what types of information management reports will contain. When these basic limits have been defined, the primary objectives in developing the report structure and designing the form for reports are:

(a) Use of familiar formats and terms to denote the types and levels of managerial responsibilities, so managers can translate information derived from reports into timely and effective decision-making.

(b) Making it possible to relate the various levels of summary to one another, so that management gets the information it needs at each level and can trace information revealed at each level to its source at levels below, when it is necessary to do so.

(c) Facilitating ranking of information at each level by its immediate significance to managers at that level, to exploit as fully as possible the "management by exception" capability inherent in integrated management systems.

CORRELATING THE PLANNING STRUCTURES

To this point we have described the development of a number of planning structures, which are preliminaries to actual execution of the project. In the process of carrying out the planning steps described so far, the planner will have developed a considerable body of management information related to defining *what* tasks are to be performed, *who* will be held responsible for their accomplishment, and *how* the work will be performed. In establishing the Work Breakdown Structure, and correlating subsidiary planning structures with it, the planner will have also established the basic record keeping methodology: the chart of accounts, the reporting structure and requirements, the networks to be maintained, and the various elements of data which comprise the *work-package unit record*. All this information must be combined in a manner facilitating summary and translation into succinct, usable management reports.

The Work-Package Unit Record

We have discussed how identifying work-packages and assigning charge account numbers to them relates the work to be done to the chart of accounts, to allow financial control. In addition, by using other code numbers the planner can indicate the organization responsible for a particular work element, the contract applicable, or whatever other kinds of information management may desire.

As indicated in Table 1, correlating management information at the detail level creates what may be called a "unit record" for each work package. While the work-package unit record may never actually exist as one consolidated document, it exists in effect in the correlation scheme or in situations employing data processing as a unit record stored on magnetic tape, providing a source for the various sorts of management information concerning work-packages.

Assigning one additional identifying *summary number* to each element in the Work Breakdown Structure completes the job of tying the planning structures and associated information together. A composite picture of the total effort at any level of the Work Breakdown Structure can be identified by that summary number. When handled by computer, such selective summarization of management information at any level becomes a routine operation which can be performed at almost any time. Figure 7 illustrates information correlation and summarization. It pictures a dynamic management planning scheme in which all elements are developed in a logical, consistent manner and interrelated in a manner assuring the validity of the whole and the parts. Altering any element in the plan must affect the entire structure, and so the structure is dynamic.

TABLE 1
Work Package Unit Record

Item	Identification code	Source
Work Breakdown		
Structure	Summary number	Project plans
Dollars		
Estimated	Estimate number	Estimating procedures
Budgeted	Work order number	Work authorization system
Actually expended	Charge number	Cost accounting system
Labor		
Estimated	Estimate number	Estimating procedures
Budgeted	Work order number	Work authorization system
Actually Expended	Charge number	Cost estimating system
Organizations		
Responsible	Organization code number	Work authorization system
Performing	Organization code number	Work authorization system
Time		
Estimated	Activity numbers or	Network plan
Scheduled	preceding-succeeding	
Actually expended	event numbers	
	for each activity in	
	the work package	
Functional identification		
of work items (activities)	Functional code	Network plan
Other	Special codes	As appropriate

Further, this structure is not discarded after it is used as a basis for detailed planning; it must be maintained in an up-to-date condition to assure management continued visibility of the overall project control system.

SUMMARY

Chapter 2 has discussed the systematic planning discipline which must precede preparation of detailed network plans; it included description of the background and characteristics of the planning structures which underlie the network-based system. The chapter emphasized develop-

Project Work Breakdown Structure

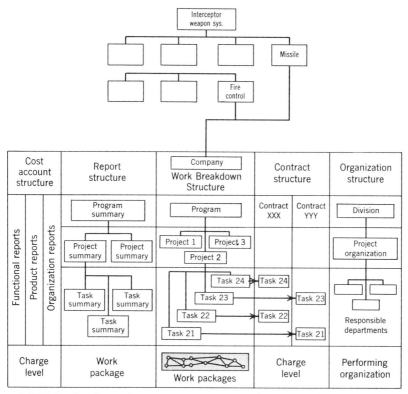

FIGURE 7. Correlating planning structures (Source: Hughes-PERT Guide, 16 July 1962).

ment of the Work Breakdown Structure as the background for correlation of all other planning structures and for preparation of networks. In Chapter 3, the next planning step, actual construction of the network plans, will be discussed.

3

Preparing the network plan

Detailed work plans must be represented as dependency networks in the network-based system. The need for preparing networks carefully by using valid criteria and a disciplined approach cannot be overemphasized. The network displays graphically the detailed work plan for accomplishing the various tasks in a project. The network shows the planned sequence in which events and activities must occur, as well as their interrelationships and dependencies. The completed network serves as a basis for estimating the duration times for individual activities as well as for the total project, for scheduling, for establishing priorities, and for integrating time, cost, and manpower information. It also performs the essential function of communicating the work plan to all management levels.

WHO SHOULD PREPARE THE NETWORK PLANS?

As we will show in Chapter 8, the organizational form and the function of the individuals involved in network planning operations will vary from case to case. Therefore, it would be difficult to specify who within an organization should prepare networks. One can state some general qualifications essential to the people responsible for network development, however.

A network plan presents the work process by which project objectives and requirements will be realized. Therefore, the persons developing the networks should be those who will be responsible for doing the work, and those who are most familiar with the work objectives and requirements at the level where it will be directly supervised. This is particularly important in the early planning stages of a project. Failure to observe this principle is the most common cause of ineffective use of the system. If a central group is to coordinate the overall effort among various organizations (as often occurs on large projects) this group should also be represented in the initial planning sessions within the responsible or performing organizations. When this fundamental is overlooked, the resulting networks may be unrealistic, and may never really be accepted by the managers who are to use them. In addition, the subsequent integration of plans may be difficult or impossible. When management abdicates the responsibility of actively participating in planning and delegates this function entirely to less knowledgeable or less responsible personnel, it risks losing control of a project at the outset, or using the network system ineffectively. Independently prepared plans, while perhaps reflecting local intelligence, may ignore overall project objectives and incentives. Therefore, detailed plans should be prepared at the performance level within overall guidelines laid down by higher management, because it is at the performance level that experience and knowledge are most appropriate. However, planning must not become so detailed that it jeopardizes overall management visibility and control.

ESTABLISHING THE NETWORK FRAMEWORK

To maintain the disciplined, systematic approach to planning discussed in Chapter 2 and assure the validity of the detailed work plans, the planner should not undertake networking haphazardly. Prior to constructing the actual networks, the managers should tentatively decide how many networks are to be prepared, the scope of each, and the identity of the essential milestones and/or interface events. On larger projects the number of networks and the scope of each will probably be defined by the tasks identified at the lowest level of the Work Breakdown Structure. Hence, when these tasks are grouped into convenient areas of effort to be portrayed on separate networks the result is the *network list;* this list of separate networks should identify the organization responsible for constructing each network. Assuming that the planners have used the methodology for identifying interface events and milestones described in Chapter 2 and elaborated in Chapter 12, they will also have a preliminary *event list* for each proposed network. At this point the event lists will consist of minimum event identifications which must be included in each network. Finally, the planners can list all known activities or jobs within each network to facilitate rapid development of the network. Armed with information about the number of networks to be prepared, the scope of each, the essential interface events or common linkage points betwen them, and the important milestones in each, the planners can proceed to actually prepare network plans, confident that the results will reflect the overall objectives established by the preplanning operations. On smaller projects where the scope of work and organizational logistics permit, all the tasks in the project may be represented in a single network. In such cases, interface events would not be required, except to reflect external constraints. The scope of the network would, however, be defined by the number of tasks identified; defining the scope will aid in proper network preparation and in location of essential milestones in the plan.

In the case of large projects involving many organizations, complex and numerous networks with many interfaces, and the massive total planning effort that would characterize such a project, it may be desirable to formally establish the network framework by preparing several lists: a network list, an event list for each network, and an activity list, which is an additional aid and first step in networking. Each of these lists would contain certain basic information to be recorded on the networks.

Network List

The network list would enumerate the tasks identified at the lowest level of the Work Breakdown Structure. The planner may prefer to prepare a separate network for each task, or for several tasks grouped into one network. Each network thus identified would specify:

(a) The name of the network, including any identifying codes such as the work breakdown summary number.

(b) The responsible organization.

(c) The individual responsible for approving the network and accountable for both the plan and the milestones and interfaces included.

(d) The individual responsible for preparing and maintaining the network if different from (c).

Event List

The event list would contain those milestones and interface events which have been identified for the task or tasks represented on the network. These might be supplemented by other events to be included in the network. The entry for each event listed would consist of the following:

(a) An event description.

(b) Any special identification of its type, such as interface, milestones, level.

(c) The responsible organization, and the responsible individual if known at this stage.

(d) The scheduled occurrence date, if any.

(e) The temporary event number, if needed for reference. (Formal numbering will be done after the network has been completed.)

Activity List

If the planners believe it will facilitate networking, and if they have such information reliably available at this stage, they may identify obvious individual activities on a "working" activity list. Like the network list and the event list, the activity list helps avoid overlooking some activities; it can be checked off as the network is drawn. Each listing would contain:

(a) The activity description.

(b) The performing organization.

(c) The tentative resource estimates (time and perhaps dollars and manpower) might be included at this time, but this is usually not the case.

Armed with these lists, the planners can build the network plan rapidly. Although lists are not essential, they can substantially aid networking by preventing time-consuming revisions to the network when additional events and activities come to light after dependencies are established. The planner may find it easier to follow the sequence and interrelationships of events by scanning the lists. The reader should note, however, that such lists do not have to show sequence or dependency; these factors will be established as the network is created.

GUIDELINES FOR NETWORK CONSTRUCTION

The mechanics of network construction are quite simple. The methodology and logic are less rules of procedure than they are observations of common sense. Some writers on the subject have stressed a multitude of rules which can become confusing; we will try to avoid such over-complication here.

Forward or Backward Planning

The planner can build up the network physically in one of two ways. One approach would be to select a specific event marking the start of the project (for example: "Contract Go-Ahead"), and proceed forward, establishing events and activities in chronological sequence until the end or terminal event marking completion of the project is reached. Although this approach can produce perfectly valid networks in situations such as construction projects, experience in large research and development projects indicates that it may produce certain difficulties. These difficulties usually come when trying to establish valid dependency relationships because this type of planning is not oriented towards the end event. In other words, the planner often only knows what the end event or objective is, and he may find it difficult to visualize how to get "from here to there," so to speak. We must keep considering: "what comes next?" and "what activities can proceed concurrently?" as we move through the network. Figure 1 illustrates this process.

The alternative to forward planning is backward planning, or starting with the end event (often a milestone or interface event), and proceeding backwards through the network, adding the activities and events *which*

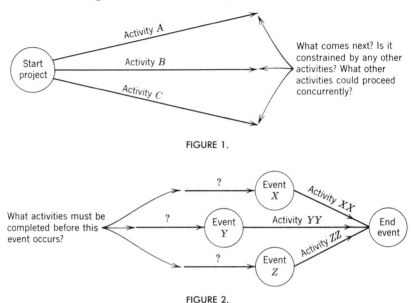

FIGURE 1.

FIGURE 2.

must be completed before the object event can occur. The planner continues this process until he reaches the initial event in the project. Many planners consider this method more advantageous because it makes it easier in several ways to develop a more valid and objective network. In backward planning thinking and selection are reversed, so that the question associated with each step in the process becomes: "what activities must be completed *before* this event occurs?" Figure 2 illustrates this process. This eliminates the more subjective decision about *what occurs next* encountered in forward planning, and substitutes the more objective question of what *must have already occurred* for the objective event to occur. This approach will usually result in an initial network with fewer subjective biases both about dependency relationships and opportunities for concurrency.

Often, we find planners employing a combination of these techniques, working back and forth across the network until the entire plan is fleshed out.

Networking Terms and Concepts

As simple as networking is, there are a number of common terms and concepts in it which often confuse the uninitiated. None of these is particularly complex itself; if the planner is aware of possible difficulties and has a terminology and method for dealing with them, he can handle them easily and save hours of work redrawing networks to eliminate errors in logic. These various terms and concepts are presented in the paragraphs which follow.

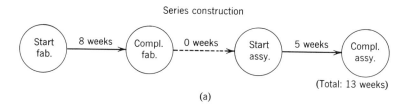

(a)

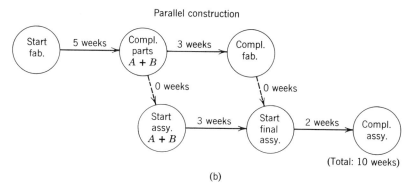

(b)

FIGURE 3.

Series Versus Parallel Paths. A network may be seen as a combination of series and parallel paths of activities and events. Parallel paths may be a means of shortening the time required to do a series of tasks. The planner might transform what appears to be a *series* of activities as shown in Figure 3(a) into a *parallel* or concurrent effort with a shorter overall duration, by rigorously examining the actual dependency relationships involved (see Figure 3(b)).

Some of the ways to shorten the time required to complete an activity sequence are:

(a) To apply additional resources to the activity to reduce calendar duration time (for example, overtime scheduling).

(b) To take maximum advantage of available opportunities for concurrency or parallel planning.

(c) To reduce the scope of work.

For optimum planning the manager must use skillful and knowledgeable judgment in determining the details of work to be done, the resources available, and the points at which series or parallel planning will produce maximum benefit in scheduled accomplishments.

Constraints: Dummy or Zero-Time Activities. The constraint line is a necessity in developing logical networks which reproduce the true dependency relationships. The line is often called the "dummy" or "zero-time" activity. Such an activity would be introduced between two events in a network to show that a dependency relationship exists, i.e., that the latter event cannot occur until the former has; no real work is represented by the activity, however. Such activities consume no time (hence, zero-time) or resources. To distinguish them from "real activities," which represent real work consuming time and resources, these dummy activities are often represented on the network by broken, rather than solid

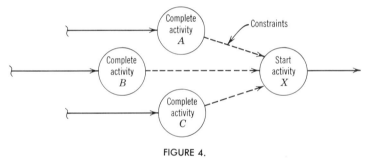

FIGURE 4.

lines. An example of the use of dummy constraints is shown in Figure 3 (b). Another example of the use of dummy activities would be a situation in which the start of several activities is constrained by the completion of a single activity (or vice-versa): Figure 4 illustrates this method.

Partial Dependency. One of the most common errors encountered in networking logic is the error known as "partial dependency," or the erroneous depiction of a dependency relationship between two activities that are only partially related. For example, in Figure 5 the start of activity *D* is shown as dependent upon the completion of activities *A*, *B*, and *C*. If the planner examines the activities involved more closely, however, he might discover that *all* of activity *A* does not actually have to be complete before *D* can begin: once *A* has started and certain portions of it have completed *D* could start assuming *B* and *C* are complete. Hence, activity *D* is not fully dependent upon activity *A* as shown, but it is only *partially dependent* upon that activity. The network does not therefore show a true picture of the situation. This situation can be remedied by redefining activity A.

In the revised version of the network shown in Figure 6, activity A_1, represents that portion of the original activity A which must be complete before activity D can start; activity A_2, represents the remainder. Redefining activity A, and introducing the constraints between events 10 and

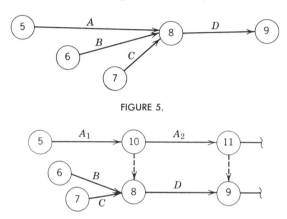

FIGURE 5.

FIGURE 6.

8, and events 11 and 9 eliminate the partial dependency error and preserve the true network logic. If we attempt to do without the constraint here, by showing A_1 going to event 8 and A_2 starting at event 8, we produce another false dependency; that representation says that A_2 cannot begin until A_1, B, and C are all complete, but that is not true.

Discrete "Starts" and "Completes." It may sometimes be desirable to identify each activity by a discrete predecessor and successor event, that is, to provide a discrete "start" event and "complete" event for each. Figure 7 shows this type of network. In some PERT uses discrete starts and stops have been advocated to assure proper activity identification.[1] Other system planners have attempted to eliminate this approach entirely in order to simplify networks and keep the number of events to a minimum. In practice, a compromise between these two extreme positions is generally the most useful. There are certain networking situations where introducing additional start or complete events can be extremely useful, and other situations where including them would create more

---

[1]See *Minuteman PERT System Indoctrination Course I and Course II*, Ballistics Systems Division, (AFSC), U. S. Air Force, 1961.

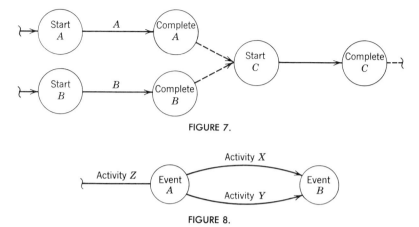

FIGURE 7.

FIGURE 8.

problems than it would solve. For example, in the network shown in Figure 8 the occurrence of event A permits activities X and Y to proceed concurrently. Figure 8 might be an acceptable representation of concurrent activities for some very small networks which are to be manually analyzed and scheduled. In larger networks, particularly those employing data processing equipment, the identical start and complete event for more than one activity should not be used. If they were used it would not be possible to identify each activity by its predecessor and successor events, because both activities would have the same predecessor and successor event numbers. The computer is always programmed to identify each activity by the pair of event numbers, and so the computer would show the two as the same activity. Another difficulty arises in terms of reporting the occurrence of events. Suppose, for example, that event A on Figure 8 was an important milestone: would the occurrence of event A signify the completion of activity Z, the start of activity X, the start of activity Y, a combination of these, or all of these? It is apparent that this would be decidedly unclear. At the cost of adding a few events, however,

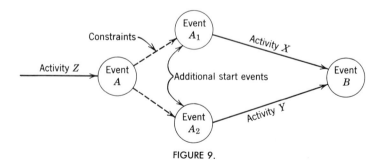

FIGURE 9.

the planner could clear up this ambiguity. Figure 9 shows how the network could be revised.

There is another kind of situation in which dummy activities and events can aid in coherent expression of the plan; this is the kind in which several activities constrain a single event. Figure 10 illustrates this type of situation. The network appears straightforward, but there may be some difficulties hidden in it. Suppose that events A, B, and C mark the

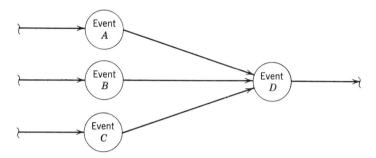

FIGURE 10.

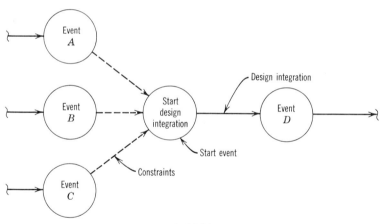

FIGURE 11.

completion of activities constituting the design of components of a system, and event D marks the completion of the activity of integrating these component designs into a single system design. In that case, the network erroneously indicates the single activity, "design integration," as *three* activities. If time, cost, and manpower data are to be assigned to the "design integration" activity, this arbitrary fractioning of the activity unduly complicates the estimating and subsequent monitoring process. If the planner made two of the activities "dummies" it would still not solve the problem, because the design integration process is dependent upon *all three* predecessor events. The obvious solution is to add a start event for "design integration" with the dependency relationship shown by constraints. Figure 11 illustrates this arrangement.

One final kind of situation calls for the use of constraints. This type concerns the intended reporting structure. When this structure is *event-oriented*, primarily relying on the reporting of *event occurences* to monitor progress, having no predecessor and successor events for all, or at

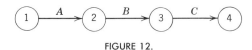

FIGURE 12.

least for particularly significant activities, can be a problem. Consider the case shown in Figure 12, where progress is to be monitored by the reporting of event occurrences. In that case, the occurrence of event 2 marks both the completion of activity *A* and the start of activity *B*. If we could assume that activity *B* starts automatically when event 2 is reported as having occurred (signifying the completion of activity *A*), there would be no problem. However, the planner usually cannot make such an assumption; this leaves us not knowing whether the occurrence of event 2 did or did not mark the start of activity *B*. If activity *B* were of significantly long duration (18 weeks or so) we would not be aware of the slippage (had there been one) until the occurrence of event 3 was not reported according to plan. By that time it might be too late to remedy the problem contributing to the slippage or to take compensating action.

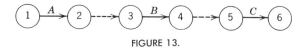

FIGURE 13.

An alternate, more informative version of this network under such conditions would be the one illustrated in Figure 13. In this second version we have added some events and constraints to achieve the visibility needed for event reporting. The completion of activity *A* would be signified by the occurrence of event 2, but we would immediately be aware of any slippage of the start of activity *B* if event 3 was not also reported as having occurred.

It is apparent that the more heavily event-oriented our thinking and planning is, the more important it is that we include some of these discretionary events and constraints. Conversely, if our planning is heavily activity-oriented (as in the typical CPM application) and our reporting scheme contains more activity information,[2] such considerations may be trivial. Other valid objections to extra events and activities are that they could tend to increase the complexity and readability of large networks, and that they increase processing time and cost. In practice, it is best to consider both the event- and activity-oriented approaches, using judgment to obtain the benefits of both, and to include dummy activities and start and complete events selectively.

Use of Connector Points. Large networks will often contain many *merge events* and *burst events*. A merge event is an event which is immediately constrained by two or more activities. Figure 14 represents a merge event. A burst event is an event constraining two or more activities. Figure 15 shows a burst event.

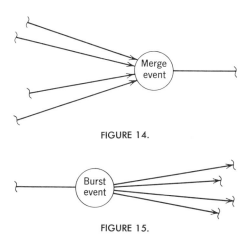

FIGURE 14.

FIGURE 15.

Networks with many such events on which all constraints are shown can become overcrowded and difficult to follow. The reader can see this difficulty in the network in Figure 16. An alternative to such a confusing representation is the use of *connector points* to eliminate excessive crossing lines, and to clarify the presentation of the dependency relationships. Figure 17 shows a network incorporating connector points. Care must be exercised in using this device, however, because it is easy to overlook a dependency when analyzing the network.

Avoiding Network Loops. A problem known as "looping" often occurs on complicated networks. A "loop" occurs when the network doubles back on itself because of the insertion of an activity going back in time. The activity going from event C to event A in Figure 18 is a loop. A loop

---

[2]See discussion in Chapter 5.

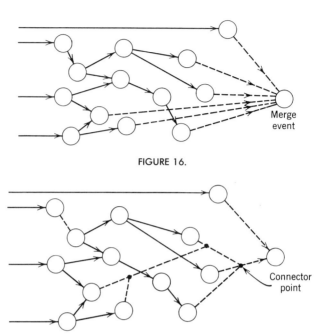

FIGURE 16.

FIGURE 17.

may show up when the planner tries to show the repetition of a particular activity before the next activity can start.[3] When a loop occurs in a network being processed on a computer, the processing may lock into an endless circle as indicated in Figure 18. Some computer programs will detect and identify loops, but in any case, looping in the network results in lost processing time and costly network changes.

Avoiding Redundancy in the Network. A frequent problem in large networks, which increases complexity, are *redundant dummy activities.* These erroneous network representations result from a confusion about dependency relationships. Figure 19 shows an example of redundant

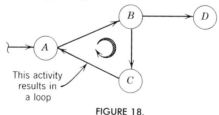

FIGURE 18.

[3]A loop may also result from connecting arrows inadvertently, transcribing data inaccurately, or from duplicating event numbers. Event numbering is discussed later in this chapter.

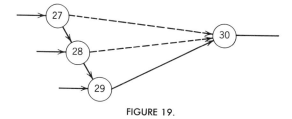

FIGURE 19.

dummy activities in a network. In that example event 30 is constrained by events 27, 28, and 29; however this same constraint relationship is effectively depicted by the sequence 27–28–29–30, because event 30 cannot occur before event 29, or 29 before 28, or 28 before 27. Activities 27–30 and 28–30 are therefore redundant. It is not necessary to depict them to show dependency relationship, and so they should be removed to avoid cluttering the network with redundant information.

Event Symbols. We have shown that events are specific definable accomplishments in the project plan, recognizable as occurring at particular points in time. They may be shown on a network as circles, ovals, squares, rectangles, or any other closed shape, or by combinations and variations of these, in order to distinguish between different kinds of events. Some examples of differing ways in which events may be represented are shown in Figure 20. The reader will think of other possibilities.

Use of Interface Events. We have touched on the nature and use of interface events at various points in this chapter and Chapter 2. However, some further elaboration and example is needed to really clarify this concept.

Because they represent common linkage points between networks, interface events must appear on all networks affected. The particular

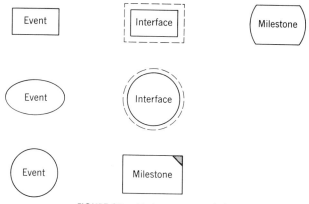

FIGURE 20.   Various event symbols.

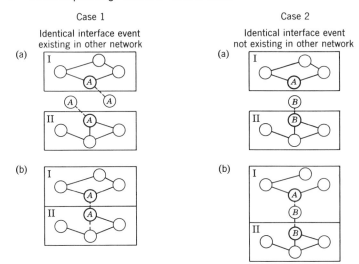

FIGURE 21. Handling interface events. In case 1, events must carry identical description, including event identifier number. Case 2 shows the identical descriptor B added to network I; events A and B occur at the same time but have different descriptions.

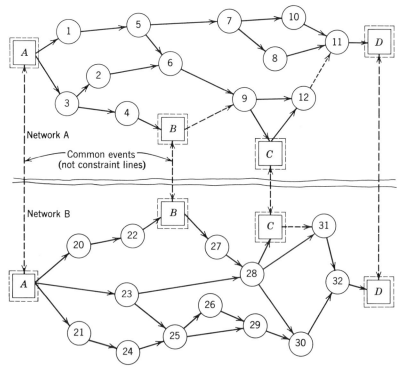

FIGURE 22. Example of integration of two separate networks.

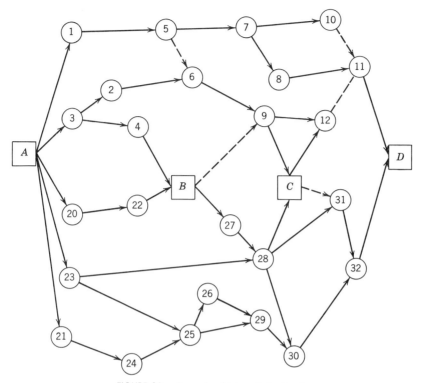

FIGURE 23.　Example of integrated network.

interface events may be integral to the plan, and already included in the sequence of events, or the planner may need to add them to connect the related networks. Interface events must carry the identical event description and other identification wherever they occur. Figure 21 shows how an interface event might be handled for two separate cases.

Large projects, consisting of large, complicated tasks involving many organizational units, will often require a number of detailed working-level networks, each showing only a portion of the total effort. Individual networks by task or organization may often be desirable where physical size is a factor in reproduction, network maintenance, handling, or display. In such cases, using interface events judiciously will assure that the individual networks are all tied together logically at the appropriate interface points; network analysis will then include the overall project interdependencies and constraints. Figure 22 illustrates the use of interface events to integrate two networks. Figure 23 indicates the net effect of

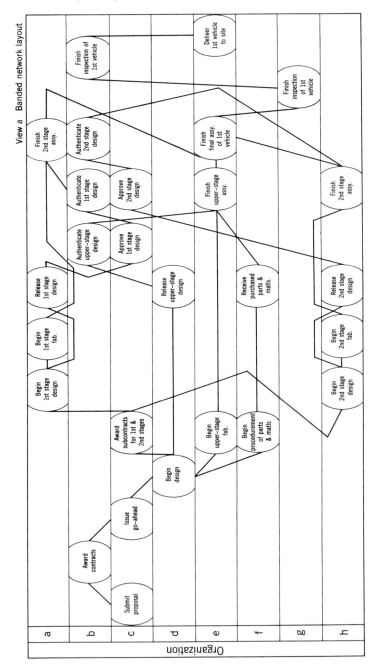

View a Banded network layout

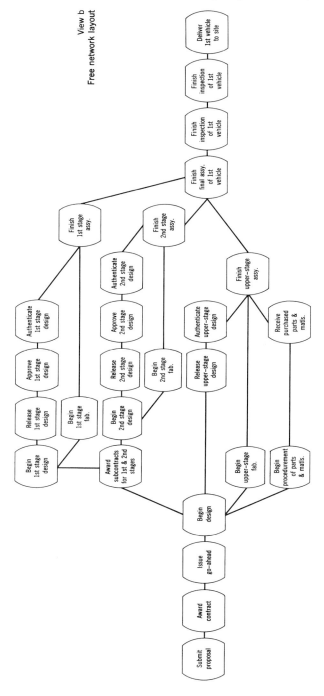

View b
Free network layout

FIGURE 24. Free versus banded networks.

this integration for network analysis purposes. (We will say more on the subject of network integration in Chapter 12.)

Physical Layout of Network Charts. There are several ways to present a network for display purposes, each of which has both advantages and disadvantages. The main kinds are "free" networks, "banded" networks, and "time-scaled" networks. All the networks we have considered this far are free networks. This means that events have been located on the network where space is most convenient, and so the physical length of activity lines has had no significance. Most preliminary networks will use this layout method because it concentrates on dependency relationships and conserves space. By contrast, a banded network is one in which events are grouped by organization, type of work, function, or other criteria and laid out so that related events are located in horizontal bands. Figure 24(a) illustrates a network banded by organization. Although banded network charts make it easier to recognize the specific responsibilities of particular organizations, they usually consist of mazes of crisscrossing, zigzag lines that may obscure the flow of work towards the end objective. Figure 24(b) shows the relative clarity of the free layout compared with the banded layout shown in 24(a). Another way of achieving the same effect (i.e., differentiation between groups of events) is by coding the events, or using characteristic event symbols for each group. Finally, networks are often time-scaled to indicate graphically the position of each event and activity against a calendar scale. Although this can be an effective display because of its similarity to the bar chart, it is completely impractical for larger networks because it is too expensive to keep up-to-date as changes occur. In addition, it is often difficult to place all events at their appropriate point in time. While it may be of some value to be able to see important milestones on a calendar scale, the important consideration must always be that all events occur in correct chronological sequence. For smaller networks or portions of networks, it is useful to lay out the network a second time on a time scale for display purposes after a time analysis (described in Chapter 5) has been completed. Illustrations of the time-scaled network layout are shown in the Introduction and in Part III. Such networks are frequently used for construction projects.

Summary of Network Guidelines

A summary of the more mechanical aspects of network guidelines includes the following points:

(1) *Events* Circles, squares, or rectangles can all be used to repre-

sent events. Special shapes can be used if meaningful. Descriptions should be placed on the network.

(2) *Activities* These are shown by lines (broken for constraints) with arrowheads touching the succeeding event. Descriptions and organizations should be placed on the network.

(3) *Time Flow* The general time flow should always be from left to right. The planner should not loop activities back in time; this can cause great difficulties.

(4) *Legibility* The network does not have to meet drafting room manual standards: legibility is all that is required. Because many changes and editions will probably be made, overly elaborate drafting of the network is a waste of time.

(5) *Title Block* There should be some standard format for the title block on all networks.

(6) *Approval* The person responsible for approving the network should sign it as required by policies and procedures.

(7) *Changes* A method of quickly identifying the change number or issue number is necessary.

(8) *Materials* Networks can be drawn by:
  (a) pencil on vellum.
  (b) typed stick-on decals for events, and tape for activities on clear plastic film.
  (c) grease pencil on clear plastic although this is difficult to reproduce.
  (d) decals and tape on wall-board or special display boards.
  (e) blackboard and chalk.

(9) *Reproduction* Charts can be reproduced by using:
  (a) continuous blueprint or Xerox; size can also be reduced.
  (b) photographs of the wall-board and blackboard type; use a Polaroid camera, then transcribe to vellum.

(10) *Display* Charts can be hung:
  (a) on the wall.
  (b) on a "clothes-line."

Precedence Diagramming

An alternate method of drawing a network plan is described in Appendix B. This method, called precedence diagramming, is completely activity-oriented and has certain advantages as well as disadvantages, compared to the more widely used event-activity method.

AN EXAMPLE OF NETWORK DEVELOPMENT[4]

We are now going to see how the guidelines are employed to actually develop a network plan. As an illustration we will imagine a hypothetical development project for a missile weapon system. For simplicity, we will restrict the scope of our initial networking task to achieving delivery of

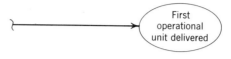

FIGURE 25.

the first operational system at a launch site. The end event, or ultimate objective of the work plan will be selected and placed at the far right of our network worksheet, and we will employ the backward planning approach described above to build up the network plan. Figure 25 illustrates this beginning formulation.

The Work Breakdown Structure indicates that the operational system consists of five major subsystems: the missile, a launch site, equipment to fire the missile, equipment to maintain both the missile and associated ground support equipment, and the personnel subsystem capable of oper-

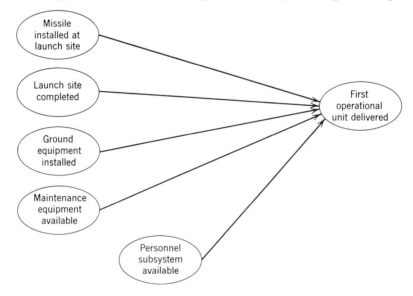

FIGURE 26.

[4]This example is based on a network example presented in *Volume 1, PERT/Time System Description Manual*, U. S. Air Force (AFSC), September 1963.

ating the site, maintaining the missile and equipment, and launching the missile. With these five major subsystems in mind, we can refer to our event list. It prompts us to show the end objective constrained in the manner indicated in Figure 26.

To assess this first step in our network logic we may ask, do all five of these subsystems have a direct relationship to the end event, or are they interrelated? If we examine the work to be accomplished more closely we might revise the network to depict the interrelationships more correctly. Figue 27 shows a possible revision.

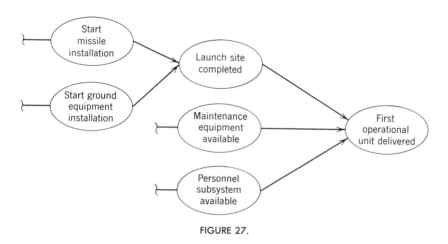

FIGURE 27.

The reader should note that the Missile and Ground Equipment events have been redefined. In the revised network they represent the start of installation activities as well as the completion of other activities yet to be defined. We will also want to examine the three activities constraining the end event more closely, to determine whether or not they represent actual work. If not, we will indicate that they are dummy activities by making the lines dotted or otherwise identifying them as zero-time activities.

Figure 28 shows how the planner continues the process of redefining and adding detail backwards to the beginning of the project. The reader should be aware that at this stage, the "construction complete" event is connected to the "missile" and "ground equipment installation start" events, and the "launch site completed" events, indicating that construction must be completed before any of these events can occur.

At the point of network development represented by Figure 28, it may be easier to set up the series of activities leading to each of the events now

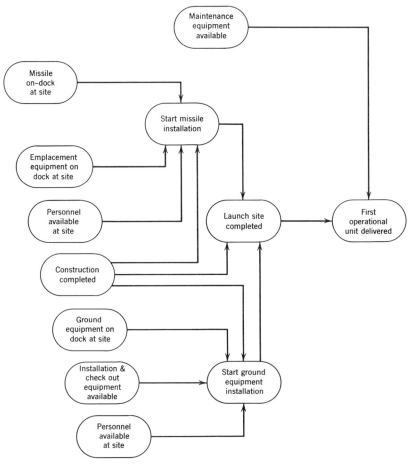

FIGURE 28. Example of network development.

shown on the network. The planners may follow a sequence of thoughts similar to the following, in planning towards the Missile On-Dock event:

What is necessary to start loading the missile? Probably a transportation vehicle, the missile itself, and equipment to place the missile on the vehicle, as a minimum. All of these must be complete and ready, and should consequently be shown as complete events.

What is necessary to have the missile complete and ready to load? It will have to undergo testing, for one thing. This should probably be shown by a start event.

What will be necessary to be able to start the testing operations? The operation will call for test equipment of various types, a test site, and an assembled, and otherwise physically complete, test missile.

As such activity and event definition progresses, the planners will have to give more attention to interdependencies between major elements of the project. Interdependencies may exist between elements such as the missile, missile handling equipment, ground equipment, and the installation and checkout equipment. The design of the handling equipment, for example, must necessarily take into account the dimensions, weight, and other characteristics of the missile itself. The people designing handling equipment must have such preliminary missile design data before they can begin their detailed design work. Prior to that, however, there probably could be some general design work, with provision for a final missile design review before starting the final handling-equipment design.

In Figure 29 the sample network has been partially replanned and expanded to show the principal interdependencies. The reader should note that internal events in a given function have been added to make it easier to show interrelationships with other functions. When these interdependencies have been entered on it, the initial layout of the example network is complete. This sample network is a very simple preliminary network plan; such a gross representation of a complex program will not allow detailed work planning and resource allocation. At this point the planners must therefore solve one of the most difficult problems in network planning: they must decide what *degree of detail* should be shown on the network.

DEGREE OF DETAIL

Let us assume that the planner has shown all the events and activities he had originally listed on his preliminary network and has added the other events and activities necessary to depict the flow of work correctly. Does he now have too little or too much detail on the network? In other words, is the *degree of detail* adequate to assure management visibility and control of the project? Trial and error and experience provide the only reliable answers to these questions. An adequate degree of detail in one situation may not be so in another. No rules can be given—we can only offer guidelines for adapting to a particular situation. Too little detail will limit control capability; too much will inundate management with detail that could obscure the information needed for control, and increase the cost of the planning effort. It is important to minimize detail in preliminary networks; too much detail on a preliminary network can delay completion and initial analysis. The planner should try to make an initial analysis of the preliminary network as soon as possible, because this first analysis will best guide him in deciding where more or less detail is needed. The gross network with its associated analysis may also be an

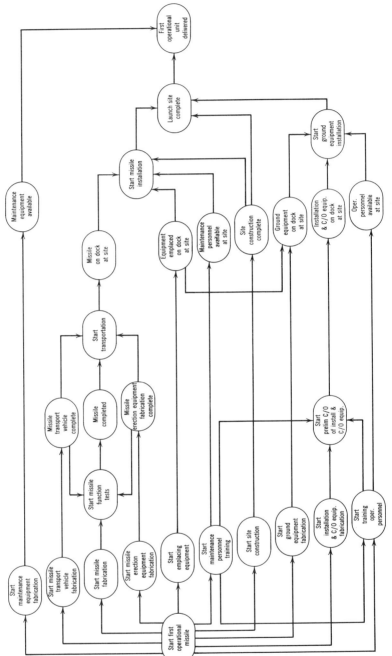

FIGURE 29.   Sample network expanded to preliminary work plan.

ideal vehicle for top management evaluation of overall project constraints and incentives.

Adding Detail to the Network

There are several ways to add detail to the network. The first, and most obvious method, is to break up all the network activities into smaller activities with shorter duration times. This approach is severely limited particularly when the network extends over a long period of time, because even on small projects many changes will be made in the basic plan as the project progresses. Numerous network changes are much more difficult to incorporate into a network of, say, 2,000 activities, than into one with as few as 200.

A second method of adding detail consists of breaking up activities in selected areas of the network, such as the critical path, or in particularly uncertain areas of work only. This selectivity provides maximum visibility and control for critical and uncertain activities without the drawback of a corresponding degree of detail in areas requiring less management attention.

Finally, a third approach is to add detail gradually, concentrating on activities in the immediate future (for example, 3 to 6 months) while incorporating less detail further in the future. As the program progresses the planner adds more and more detail to work further in the future. This is a natural and powerful approach because it concentrates visibility in the area in which it is most needed: the work immediately ahead. In addition, this approach corresponds more realistically to the project management situation, in which the initial plan is continually subject to refinement and change as experience and new knowledge are gained. Using this method, the planner can take maximum advantage of current experience in choosing between alternate future courses of action.

RECORDING DATA ON THE NETWORK

The completed network is a very effective place to record data connected with the plan. The overall plan is immediately evident and available, and writing information in the event symbols or on the activity lines minimizes transcription errors and helps prevent the omission of data. When the network is to be processed by computer, the data transmittal work sheets from which input data cards are punched can be filled out directly from the network (see Chapter 9 on input sheets).

Event Numbering

One of the most important items of data to be recorded on the network will be the discrete identification numbers assigned to each event on the network. These numbers and the dependency relationships which have been established provide a topological map for the computer to follow in processing and analyzing the network.

When plans call for hand computation of the network, it will help to rank event numbers in what we will term $i < j$ sequence (or "i less than j" sequence). This simply means that for every pair of events on the network in predecessor-successor sequence, the i, or predecessor event number, must be less than the j, or successor event number. Thus, in the

FIGURE 30.

example in Figure 30, the event number i must be less than the event number j. This rule holds for all activities in the network. In addition, a given event number must be used only once in a given network. It is common practice to use every tenth number (10, 20, 30, 40, etc.) as event numbers when numbering in $i < j$ sequence, to allow space for insertions and changes. To number the network in this manner, the planner should make all predecessor event numbers less than the number of a given successor event number; numbering a given successor event only after all predecessor event numbers have been assigned helps in this process. A simple event number checkoff table should be used when numbering all but the smallest networks; the planner crosses off each number as it is used. This prevents accidental use of the same number twice and shows which numbers are available for later insertions in the network. When events are numbered in the $i < j$ sequence it facilitates rapid manual processing of the network, because it becomes possible to move in an orderly way towards the end event, going from event to event "by the numbers." This type of sequence also minimizes the possibility of dropping an activity in the process of numbering.[5]

When computers are used, the speed and search capacities of the machine make it possible to disregard the discipline of $i < j$ numbering and to assign numbers randomly. The computer program will use either a "ranking" technique or a topological analysis to establish the order in which the data will be processed. It is most important to avoid duplicating event numbers when using nonsequential numbering, however, and the use of a checkoff table such as the one discussed above is highly recommended. If numbers are duplicated, one of the serious problems that can result is looping. A loop caused by duplicate event numbers

---

[5]The reader should refer to the description of network analysis in Chapter 5.

would appear as shown in the network in Figure 31. In this example, the ranking routine moves from predecessor event (a) to successor event 430, then moves on and encounters what appears to be the same event. Since the sequence 460–430 has already been established as the predecessor-successor sequence for the first activity, before the sequence 430–460 is

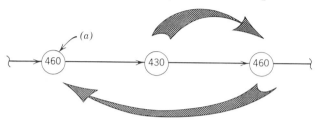

FIGURE 31.

encountered, the path-tracing will be locked in the endless loop as depicted in the figure. A method of event numbering which has been found very effective for some situations is the use of a *coordinate numbering system*. In this system the sheet of paper on which the network is to be drawn is given horizontal and vertical coordinate numbers similar to those used on an ordinary road map, and the location of the event on the sheet automatically numbers the event. This prevents duplicate numbers, and has the advantage of allowing the planner to locate events quickly when troubleshooting a network. If more than one sheet or network is to be integrated, the sheets are numbered and all events have this number as the first 1 or 2 digits. Many hours of frustrating searching to locate events, which can result from completely random numbering of events, can be avoided by the use of the coordinate numbering system. Figure 32 illustrates this approach, which is somewhat cumbersome if extensive revisions are made to the network.

Other Network Records

Other types of data which may be recorded on the network as a record and a source for preparing input transmittal sheets include:

(a) Event descriptions (in event symbols).
(b) Milestone or interface identification (such as those described in Fig. 20).
(c) Activity descriptions (on the activity lines).
(d) Activity duration time estimates (on the activity lines).
(e) Activity charge numbers (on the activity lines).
(f) Responsible organization (on the activity lines).
(g) Other resource estimates (manpower) (on the activity lines).

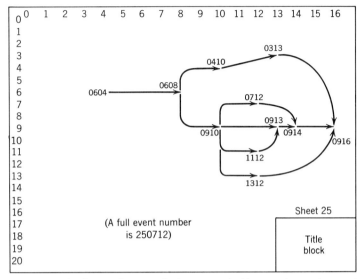

FIGURE 32.   Example of a coordinate numbering system. In this example, events are not represented by a special shape such as a circle, but merely occur at the junction of the activity arrows. Event numbers are indicated near these junctions.

In addition, the planner can use the network as a permanent reference document for display of progress against plan (by checking off events or activities as they occur), scheduling (by adding dates or timescaling), and for recording changes in the plan. These uses are discussed more fully in later chapters.

REVISIONS AND CHANGES

An important advantage of the network system is the relative ease with which revisions and changes can be reflected in the plan and subsequently evaluated as part of it. By deleting, adding, or rearranging activities and events, the planner can make small or large changes, evaluate alternative courses of action, and even simulate them, before committing resources. This flexibility is greater in a free network, and much reduced in either the banded or time-based network arrangement.

For these reasons, the authors recommend avoiding complicated event identification schemes or complex layout techniques. Using the simplest, most legible techniques fosters maximum network flexibility for reflecting revisions and changes. Success in maintaining currently valid networks throughout the life of the project is directly tied to ease of maintenance.

SUMMARY

Chapter 3 has described the concepts and mechanical considerations involved in actually preparing network plans. The importance of choosing the persons who prepare the networks and the degree of detail, and the mechanics of successful network preparation techniques have been emphasized. Some of the more common pitfalls in a successful networking effort have been highlighted. In the next several chapters we will see how the network is used for scheduling, resource planning, and program monitoring.

4

Time estimates

To use the network as the basic action-planning tool the manager identifies in sequence the milestones and related activities that must occur to accomplish the project objectives. When we identify these milestones and the interrelationships indicated by their associated activities, we are specifying the procedure for accomplishing objectives, and, by implication, the resources that will be needed.

Graphic representation of the plan on the network as a result of disciplined planning is only part of the story, however. When the planner has decided *how* the job is to be done, he must next develop the schedule. The fully developed network plan must indicate *how long* it will take to complete the program planned, and when resources such as time, manpower, dollars, and facilities will be needed to best achieve the project objectives. The network-based system can be an effective tool if planned against the background of a schedule; however, the potential of the technique is realized more fully when planning is independent of predetermined dates, so that the network can be used as the basis for the schedules. Independent planning avoids the possibility of unrealistic planning to accommodate arbitrary schedules; it also makes subsequent reporting and control of the project against the plan easier.

The first step in scheduling is to obtain time estimates for each activity in the network from the responsible persons, and to record these estimates on the network. The requirement that the work be planned at the appropriate levels is an important feature of the network-based system. As we have seen, top management should define the project and its objectives, assign responsibilities, and identify work packages; the managers do so by a "top-down" approach, using the Work Breakdown Structure. This top-down approach provides the broad policy guidelines and the framework for detailed work planning. However, once these working-level elements have been defined, the individuals who know most about the technical details should be responsible for defining the work in detail by preparation of networks, and for estimating the time and other resources required; in most cases these individuals will be the responsible line supervisors and operating managers who will have the day-to-day job of producing the work. This requirement assures realistic planning of individual tasks without the distraction of overall planning. The top-down approach and the structure of the management system itself provide for integrating and summarizing these individual work plans.

TRADITIONAL ESTIMATING TECHNIQUES[1]

In the past the usual manufacturing operation was predictable. Using techniques that had been established for five or ten years, planners were not required to *estimate* production times as often as they were required to carefully *assemble* reliable historical data. When setting up a production line they could count on a set of standard production times established statistically by experience to define the production schedule. These times were reliable enough to use as commitments for every operation involved in a project; even the uncertainty in related engineering effort was insignificant enough to have only a negligible effect on the reliability of the commitments. Under such circumstances planners did not make an effort to handle uncertainty in scheduling and pricing because none was required.

Today's environment is a drastically different situation: uncertainty is a pressure pervading all aspects of management. The contemporary advanced-technology project demands advances that cannot depend on accident or chance; breakthroughs are increasingly the result of steady, planned, technological pressure under economical and political conditions which demand advances—on ever-shortening time cycles. We have arrived at a point where we must *schedule* creativity and invention, as well as the production which follows right on its heels, and must attempt to predict with some accuracy when operational hardware will be delivered and what it will cost.

The manager's recognition of technological uncertainty in his plans for a contemporary advanced-technology project does not rid him of the uncertainty problem. Under recent accelerated development schedules vast cost and schedule overruns have continued to be characteristic and persistent problems, indicating an added dimension to the uncertainty problem: that of human judgment in estimating. When we examine this

---

[1]The authors are indebted to Ernest O. Codier of General Electric Company's Light Military Electronics Department, whose monograph, *PERT Application at LMED*, provided the idea and much of the substance for the examination of the deficiencies of traditional estimating techniques which follows.

aspect of the problem, we find that a multitude of complex, interrelated forces affects these judgments, not always in ways that foster reliability or objectivity. The greater the technological advance, the greater the uncertainty. Management tries to find the accurate time estimate for completion of a piece of work or an entire project, but must work within restrictions such as the customer's estimate of the date he needs the product, the price he is willing to pay, and the competitors' estimates. The estimator today must balance factors such as time and cost probabilities, competition, existing work-loads, available resources, and the need for the work. The individual estimator may find his objective judgments impaired by human tendencies to interpret schedule demands, to be overly optimistic in the face of uncertainty, to build in the "fudge" factor, and to assure compliance with the schedule on this basis. The problem is most encountered at the individual level, because total project estimates are based on the sum of all the detailed individual estimates. This all-too-common problem of judgment is beautifully illustrated by Ernest O. Codier's amusing hypothetical situation, drawn from his experience at General Electric:

There is a traditional man/manager gaming model for time estimating. This game runs approximately as follows: you pick up the phone, and with casual and disarming friendliness inquire, "Say, George, just off the top of your head, how long will it take to get out 14 frabastats?" Now George is a very competent fellow and he knows instinctively that it would take eight weeks to fabricate 14 frabastats provided everything happened the way it *should* happen. But he does not give you this answer immediately because he has to stop and figure this thing out. In the first place, he knows that there will be some normal amount of difficulty involved, and he will add a factor to take into account this average uncertainty, say, two weeks. Now he has ten weeks. But George is not misled by the informality of your request; he knows that ultimately, in one form or another, this is going to show up as a commitment to you to provide 14 frabastats in a specified time. Furthermore, George knows the nature of the business, and he knows you. He knows that time estimates are traditionally too long, and that somewhere along the line he can expect to get cut back. So at this point, he adds the fat, which is his considered opinion as to how severely this time estimate will get cut later on—say, three weeks. He has now arrived at 13 weeks, and this is the figure you get. When this proves to be too long, as it invariably does, since it is in fact too long, the time will get cut as George expects. By one or another mechanism, you will work the problem backwards and ask for performance two and one-half weeks sooner. You have done pretty well as a manager, and have cut out only approximately the amount that George anticipated; he makes routine grumbling noises, because this is proper form, and $10\frac{1}{2}$ weeks becomes his commitment. Now, precisely what do you know at this point? Not very much. You do not know how much time this job should take under ideal circumstances, and you have no measure of potential trouble sources which you, as George's manager, might be able to help him smooth out. Furthermore, if George actually delivers 14 frabastats in $10\frac{1}{2}$ weeks, you are not really sure whether this was an outstanding effort, or whether it was just an average performance.

Now let's take a somewhat more aggravated case, the frabastats are an absolutely vital portion of a crash program related to a missile shoot that must occur on such-and-such a date. You're pretty sure George can do better than he's doing. In all managerial honesty you have got to see that this thing is done faster, and you need to impress George with this necessity. So under these circumstances you cut George's time more heavily. This is known as setting a challenging goal. "Look, George, we have a tough one this time; you've got to cut the program in half—seven weeks is all you can have." George sees that he is below the irreducible minimum of the basic requirements for the job, and he screams, "But there is not a chance in a hundred, I can't do it that fast." So George is afraid for his life—he works really hard, and ends up delivering your frabastats in $8\frac{1}{2}$ weeks. This, on its merits, is really an outstanding performance, but is still a week and a half later than you had required. At this point, you, as the manager, have Hobson's choice. You suspect that this was a good job, but can't commend George because he didn't deliver as ordered. If you chastise him too harshly, he will get disgusted at being reprimanded for good performance and leave the company, and you don't want to risk this. On the other hand, if you let him get away with it, he will recognize that your orders are subject to interpretation and that he merely needs to increase his fat-factor. At this point, everybody in the game has to readjust his coefficients. This is a hell of a way to run a business.[2]

If this seems to be simply an amusing anecdote, the reader needs to be assured that it is also a fairly accurate representation of the estimating problem at the detail level; the planner must deal with this problem if he wants to achieve the total program objectives. How can management help improve George's original estimates and eliminate the lack of objectivity shown in the example? Perhaps it is most important to understand the elements of George's thinking when he is making the estimate. First, George is a competent professional who knows his job and is reasonably well motivated. He is as aware as anyone else that there is a degree of uncertainty in the job at hand, and is going to be very reluctant to commit himself to deliver at a particular time if he thinks he will be embarrassed by that commitment—so he adds the "fat factor." At the same time, if he is really pressured, he is just as likely to underemphasize possible setbacks to come up with an overly optimistic estimate of the time needed to complete the job. Neither end of this scale of uncertainty is a really satisfactory estimate. The network-based system, specifically the PERT approach, attempts to overcome this problem by letting George use his professional judgment in an objective way to estimate a *range* of times for completion of the job; this range realistically reflects his technical understanding of the uncertainties involved.

[2]*Ibid.*, pp. 15–17.

THREE TIME ESTIMATE TECHNIQUE

To gain a basis for estimating a range of times and to enlarge the pre-dictive capabilities of the network, the planner may use an optional form of time estimation which specifically provides for handling the uncertainties involved in today's accelerated projects. He records three estimates—an optimistic, a most-likely, and a pessimistic—for each activity; these three estimates form the basis for determining the uncertainties involved and the probability that expected events will occur as planned. These duration time estimates can be revised periodically to reflect actual expenditures or changes in the rate of time use.

The Optimistic Time Estimate. This is the estimate of the shortest possible time in which an activity can be completed under optimum conditions. The optimistic estimate assumes that the activity is accomplished in the ideal environment, free of even the normal amount of delays or setbacks.

The Pessimistic Time Estimate. This is the estimate of the longest time it might take to complete an activity. The pessimistic estimate assumes that everything goes wrong, all of the possible delays or setbacks occur, and everything in general goes badly (short of strikes, acts of God, etc.).

The Most-Likely Time Estimate. This is the estimate which lies between the optimistic and pessimistic; it assumes normal conditions will be encountered in the activity. The most-likely estimate assumes that the *normal* amount of things will go wrong, the *normal* amount of things will go right. It anticipates a satisfactory rate of progress but no dramatic breakthroughs: in short, "business as usual."

The reader should realize that the differences between the optimistic, pessimistic, and most-likely estimates are measured in terms of factors such as people making errors, misunderstanding instructions, and encountering unforseen technical problems. Estimators exclude certain factors from the estimation process when using the three time estimates:

(a) Estimates should be based on the actual or potential manpower expected to be available for each activity. They should not include any major manpower changes above this expected level. For example, possible idle time due to labor strife would not normally be included, although alternate plans should be made if it seems that they may be needed. Estimates should be based upon a uniform workweek, consistent with schedules and budgets if available at this stage. Unless specifically authorized to do so, estimators should not include vacation or other absenteeism, overtime, or extended workweeks.

(b) Estimates should be based upon the best present information about the technology, techniques, and tools available, as well as about the current scope of the work. Planners should make no allowances for technological breakthroughs, dramatic labor- or cost-saving developments, drastic revision of the scope of the job, or improved methods except for standard learning-curve improvements.

(c) Managers should not allow for failure on the part of interfacing activities, such as delivery of parts, release of facilities, funding, or work authorizations.

(d) The possibility of acts of God, such as fire, flood, or local war should not be considered in estimating.

(e) It is as important not to expand time estimates to fit anticipated dollar budgets as it is to *schedule* with the network and to avoid expanding or limiting time estimates to fit arbitrary schedules. As we shall discuss in Chapter 5, the network system can be an effective cost- and budget-planning tool when the planner uses his judgment and pays attention to the validity of estimates.

When a planner uses three time estimates, he derives from them a single value to represent the *expected* activity duration time. This is found by using a simple formula:

$$t_e = \frac{a + 4m + b}{6}$$

which approximates a mean (not average) value for the three estimates. In the formula, the mean value derived is represented by the symbol t_e (*expected* activity duration time); and the optimistic, most-likely, and pessimistic time estimates by a, m, and b, respectively. The derivation of this formula was presented in the original *PERT Summary Report, Phase I.*

Our problem is to estimate the expected value and variance of an activity time from the likely, optimistic, and pessimistic times discussed above. Furthermore, we feel free to use a non-normal model of the distribution of activity times as a tool in this estimation.

We recall that for unimodal frequency distributions, the standard deviation can be estimated roughly as one-sixth of the range. Hence, it seems reasonable to estimate the standard deviation of an activity time as one-sixth of the difference between the pessimistic and optimistic time estimates.

The estimate of the expected value of an activity time is more difficult. We do not accept the likely time as the expected value. We feel that an activity time will more often exceed than be less than an estimated likely time. Hence, if likely times were accepted as expected values, an undesirable bias would be introduced. Our apprehension of this bias is supported by estimates already obtained. In many cases the likely time is nearer the optimistic than the pessimistic time. In such a situation one feels that the expected time should exceed the likely time.

We shall introduce an estimate of the expected time which would seem to adjust at least crudely for the bias that would be present if likely times were accepted as

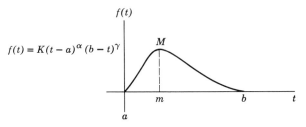

FIGURE 1. β distribution. Values of α and γ obtained by (1) setting M—m of estimates, (2) setting σ (beta distribution) $= \dfrac{b-a}{6}$.

expected times. As a model of the distribution of an activity time, we introduce the beta distribution whose mode is at the likely time, whose range is the interval between the optimistic and pessimistic times, and whose standard deviation is one-sixth of the range. The probability density of this distribution is

$$f(t) = (\text{constant}) (t - a)^\alpha (b - t)^\gamma \quad (\text{see Figure 1})$$

in which the optimistic and pessimistic time estimates are respectively a and b (the left- and right-hand ends of the range); the "constant," α and γ are functions of a, b, and the likely time M (i.e., the modal time). To reduce this probability density function to the standard form of the beta distribution, we introduce the random variable x as related to t, as follows:

$$x = \frac{t - a}{b - a}$$

The probability density of x is

$$f^\circ(x) = \{B(\alpha + 1, \gamma + 1)\}^{-1}x^\alpha(1 - x)^\gamma$$

Since the modal value of t is M, if r denotes the mode of x,

$$r = \frac{M - a}{b - a}$$

If $E(x)$ and $V(x)$ are respectively the expected value and variance of x, straightforward computation leads to

$$r = \frac{\alpha}{\alpha + \gamma}$$

$$E(x) = \frac{\alpha + 1}{\alpha + \gamma + 2}$$

$$V(x) = \frac{(\alpha + 1)(\gamma + 1)}{(\alpha + \gamma + 2)^2(\alpha + \gamma + 3)}$$

Since the variance of t is $(b - a)^2/36$, the variance of x is $\frac{1}{36}$. We eliminate γ from the equations for r and $V(x)$ after substituting $\frac{1}{36}$ for $V(x)$, and we obtain

$$\alpha^3 + (36r^3 - 36r^2 + 7r)\alpha^2 - 20r^2\alpha - 24r^3 = 0$$

TABLE 1

r	E(x)	(4r + 1)/6
0	0.2053	0.1667
$\frac{1}{8}$	0.2539	0.2500
$\frac{1}{4}$	0.3228	0.3333
$\frac{3}{8}$	0.4075	0.4167
$\frac{1}{2}$	0.5000	0.5000

We can now compute the parameters in $f(t)$. Given M, a, and b, the above formulas enable us to calculate in succession r, α, γ, (using the relation between r, α, and γ) $E(x)$, and finally $E(t)$ (because the equation of transformation between t and x implies that $E(t) = a + (b - a)E(x)$).

Let us study the relation between r and $E(x)$. Numerical calculation with use of the above formulas gives the first two columns of Table 1.

If $E(x)$ is plotted as a function of r, it is seen that the relation between these variables is approximately linear. A simple linear approximation, namely

$$E(x) = (4r + 1)/6$$

is given in the third column. We shall accept this approximation because it is accurate enough for our purposes, and because the prior computation requires among other operations the solution of a cubic equation. From the relations between $E(x)$ and $E(t)$ and between r and M, the linear approximation reduces to

$$E(t) = (a + 4M + b)/6.$$

This formula will be used in the FBM° system analysis.

This result was derived under the assumption that the beta distribution is an adequate model of the distribution of an activity time. The choice of the beta distribution was dictated by intuition because empirical evidence is lacking. Hence, it will be appropriate to consider the reasonableness of the result.

The result can be reduced to

$$E(t) = \frac{1}{3}\left(2M + \frac{a + b}{2}\right).$$

This means that $E(t)$ is the weighted mean of M and the mid-range $(a + b)/2$, with weights 2 and 1 respectively. In other words, $E(t)$ is located one-third of the way from the likely time M to the mid-range.

We can consider whether the weights 2 and 1 in the above formula seem appropriate. Perhaps it would be intuitively better to move the likely time only one-fourth of the way towards the mid-range. Perhaps one-half or one-tenth. Our own opinion is that one-third seems reasonable. However, we intend to code into the computer routine this ratio as a parameter. Hence the weights can be changed if authoritative judgment dictates. Furthermore, by making computer runs with various parameter values one can test the sensitivity of the computed result to this choice of weights.

°Fleet Ballistic Missile (POLARIS).

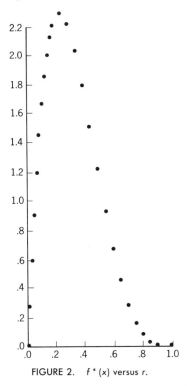

FIGURE 2. $f^*(x)$ versus r.

As experience with the development of the FBM accumulates, we can compare actual activity times with estimates of M, a, and b. This will enable us to reconsider the weights used.

As further evidence of the appropriateness of the beta distribution model, we have presented the graph of $f^\circ(x)$ for the case in which $r = \frac{1}{4}$ (Figure 2). This represents the case in which the likely time is one-fourth of the way from the optimistic to the pessimistic time. This graph reveals that most of the activity times will be symmetrically distributed about $r = \frac{1}{4}$ between $r = 0$ and $r = \frac{1}{2}$, but that there is enough probability above $r = \frac{1}{2}$ to bring the expected time up to .32. In our opinion this graph is as reasonable as that obtained from any other commonly occurring probability density function.[3]

A DISCUSSION OF TIME ESTIMATING IN THE
NETWORK-BASED MANAGEMENT SYSTEM

One important advantage of the network approach is that it produces a more realistic estimate of the total length of a project. Characteristics of the network-based system which make this possible are:

---

[3]Appendix B, *PERT Summary Report, Phase 1*, Special Projects Office, Bureau of Naval Weapons, Dept. of the Navy, Washington, D. C., USGPO, 1962–640679, 25 cents.

(a) Its ability to show realistically and comprehensively the interdependencies between the subelements.

(b) Its subdivision of the major elements into many smaller activities, which can be estimated more accurately.

As an illustration of (b), consider a flight plan for the trip from Los Angeles to New York City. We could make a gross estimate of flying time by measuring the total distance and dividing this by the expected average ground speed. But if we were to look more closely at the actual airway route to be traveled, and break the trip up into 100-mile legs, we could obtain a more accurate estimate. Using the variables below to make individual time estimates for each leg and then summing these estimates would yield a considerably more reliable total. In estimating each leg (equivalent to an activity in a network plan) we might consider variables such as local wind conditions, flight altitude, aircraft weight and ground speed, and weather detours or delays. Similarly, a planner can estimate more realistically a project plan which is broken into meaningful activities.

Single versus Multiple Time Estimates

The desirability of the single time estimate compared with three time estimates for each activity has been discussed a great deal. Consideration of the background of different uses indicates that the desirability of one time estimate compared with three depends upon the environment and the nature of the application.

The single time estimate is appropriate in networks used in the process and construction industries. On a construction project, for example, we can expect that the activities or jobs are well known and that past experience provides a basis for reliable and accurate time estimates. Therefore, using a single time estimate for each activity is a valid and realistic practice.

In the defense and space industries, network planning is used primarily in research and development, on programs with a large degree of uncertainty. Since many of the activities in such projects have never been carried out before or have been carried out only a few times under very different circumstances, using three time estimates to reflect this uncertainty has definite advantages. The greatest advantage comes in the initial uses of the technique, however; when the planner is sufficiently familiar with the networks, and has gained confidence in using them, he can often use a single time estimate.[4]

---

[4]Governmental regulations originally specified three time estimates, although the National Aeronautics and Space Agency specified single estimates. Department of Defense agencies now generally make three estimates optional.

Using Tolerances in Real Situations

C. D. Flagle has recognized that one of the basic limitations of the Gantt chart is that it restricts management to a fixed single time estimate. He says:

> It takes little experience or imagination to perceive that under the progress bar system of review, the time estimates become goals, and the ego involvement of personnel in the various operations is extended to the attainment of these goals. This is characteristic of Gantt's techniques, for he correctly recognized the potentially large increments of performance to be realized by creating ego involvement in the task. It must be admitted however, that such involvement is not an unmixed blessing, a fact which becomes apparent at the stage of planning in which performance time requirements must be estimated.[5]

Obtaining objective and accurate estimates for scheduling and review is always a difficult matter. It is understandable that estimators tend to be unrealistically optimistic or overly cautious, often depending upon whether they want to impress their superiors when the estimates are made or when the jobs have been completed.

Accepting time uncertainties as unavoidable in many operations to be scheduled is one of the fundamental advances of the network-based system. When the estimator is allowed to provide three time estimates—optimistic, pessimistic, and most-likely—some of the justifiable barriers to making accurate commitments on inherently variable activities or events are removed. In addition, substituting a range of times between them eliminates the use of extreme estimates. As Flagle goes on to say, "the challenge is to achieve this end without destroying the basic simplicity of the Gantt Chart and its subtle effectiveness in establishing performance goals."[6] Proper use of the network plan, with three time estimates for an activity when appropriate, achieves this goal. The most important benefit of using three time estimates is a more realistic estimate. By reducing the psychological barrier to objective estimating, planners can encourage the engineer, scientist and job foremen to cooperate with one another. The validity of using three estimates should not be judged solely on the basis of the estimates being averaged in a certain way, or of a certain technique being used to derive a measure of probability versus time; these factors are irrelevant to a discussion of the value of the network approach as a whole.

[5]C. D. Flagle, "Probability Based Tolerances in Forecasting and Planning," *Journal of Industrial Engineering*, May 1961 (first presented in 1956 in the Johns Hopkins Informal Seminars in Operations Research).

[6]*Ibid.*

Obtaining Valid Time Estimates[7]

Although the specific procedures which will be employed in developing time estimates will vary from situation to situation, some guidelines and rules can be suggested.

A time estimate is simply the individual contributor's experience reflected in a numerical guess, and therefore cannot be viewed as a commitment. Often, the estimator can and will be wrong; it is highly improbable that estimators might consistently have such extensive experience that they will never make errors. A real commitment exists only when a specific penalty will be imposed for failing to meet the submitted estimate. Most companies officially recognize only one commitment: the date required by the contract with the customer. The interfaces between the organizations within the company are similar to commitments; they are used only for convenience within the company, however, and are not so binding as commitments to customers. The internal dates can and do slip around. If the individual estimator knows that management is abiding by the premises suggested here, he will give three time estimates much more readily and accurately than one time estimate, to which he becomes committed.

Another important development to watch for in making time estimates is the so-called balanced estimate, in which there is an equal spread (for example 2–4–6 or 5–10–15) on each side of the most-likely estimate. Sometimes the planner will find balanced estimates given for most estimates on a project. Experience shows that actual times, when reported against a balanced time estimate, are frequently closer to the pessimistic time than to the most-likely time; this indicates that the estimator did not allow for possible trouble. Therefore, although estimates may not necessarily be skewed toward the pessimistic, the planner should review the estimating rationale if he receives too many balanced estimates, to be sure that the pessimistic time has accounted for everything.

Another problem occurring in time estimation is that some estimators have a strong tendency to rattle off three estimates such as "4, 5, and 6 weeks," or "3, 5, and 8 weeks," without hesitating. This kind of impulsive estimating is dangerous and unsuitable, and probably indicates that the estimator is not considering the problem sufficiently. The planner can guard against this impulsiveness by indoctrinating estimators with the

---

[7]The following discussion of guidelines for obtaining time estimates is based on experiences at Remington Rand, described by K. L. Dean in *Fundamentals of Network Planning and Analysis*, PX 1842 A, Management Services Department, Remington Rand UNIVAC Military Department, July 1961.

technique,[8] and by insisting that the estimators get the optimistic time first, the pessimistic time second, and the most-likely value last.

When obtaining time estimates the planner should skip around the network rather than follow a specific path. There are several good reasons for this:

(1) Each activity should be considered an independent effort, isolated from any preceding or succeeding jobs. There is a tendency to forget this independence when the planner obtains estimates for activities in sequence; in such cases the estimator may hedge his own estimates because he assumes that the one for the next activity can absorb the difference, and in this way may spread his estimates over two or more activities.

(2) As indicated earlier, computing the expected activity duration times is not difficult. Technical people are often qualified to make these calculations as they see the critical path developing. In such instances, experience shows that activities on the critical path are generally estimated optimistically.

One of the extremely difficult questions to answer about activities is, what is a normal duration for an activity—a day, a week, or a month? The answer is not clear-cut, because the duration of the activity depends upon the judgment of the person who selected the work segments to be described as activities. For the average research and development program four to eight weeks activity duration time (most-likely time) is common, although a two-week period is not unusual. Less detail is generally required on networks prepared as part of a proposal so a somewhat larger time span, say eight to twelve weeks, may be appropriate. In the usual research and development project, time is best measured in weeks, although in some other situations measuring in days may be more practical. Networks used for construction projects generally use working days, but sometimes measure hours or shifts.

It is important for management to have confidence in the estimates supplied by the individual estimators. Therefore, it has often been useful to have the estimator review the tasks and his time estimates with his supervisor, and to reevaluate some estimates on the basis of mutual understanding and the supervisor's experience, if necessary. Such a review strengthens the network estimation technique because it gives different levels of management a common language to use in arriving at a common understanding. Both the estimator and his supervisor emerge from such reviews with a clearer understanding of the estimator's job.

After such a review has been made, however, the estimator should not

---

[8]See the discussion of training in Chapter 11.

change a time estimate unless there is a change either in the scope of the work or in the resources allocated for the job (including man-hours available within a given time span, as in overtime authorization): time estimates should not be cut arbitrarily, although the estimator may be continually pressured to shorten his schedule. However, if a particular plan shows that it will take 25 weeks to complete a given project, this does not *necessarily* mean that it takes 25 weeks to complete that project; it does mean that *under that particular plan* it will take 25 weeks to complete the project. Alternate plans may be devised, however, involving more or less money, risk, or intelligence, or a different combination of resources and responsibility. If management wishes to shorten plans these alternatives must be available to support the revised plans. However, it is easier to consider these alternate plans involving varying resources when using network planning systems than when using any other management system.

A final word will summarize what we have been saying. If the scope of an activity is redefined so that it involves less work, it may be best to revise the estimate. Similarly, if the managers can show that additional resources will reduce the required time, and they know these resources will be available and will be used, they can justifiably reduce the time estimate. However, the original time estimates are usually more reliable in the long run than those based on later second-guessing; to arbitrarily reduce time estimates provided in good faith is to regress to the questionable practice of "setting a challenging goal." Some companies do adjust time estimates, but we assume that in such cases there is a high degree of latitude and flexibility in the labor force so that management can make increases or decreases to meet a specified time requirement. These adjustments are not generally made in businesses with a large ratio of highly skilled or professional people.[9]

The preceding paragraphs have given rules and suggestions to use in obtaining time estimates. It may seem to the reader that the manager has little control over this estimating process, which resembles a cat-and-mouse game between estimator and manager. How good are such estimates? Must a manager rely on them? Does he have an alternative? The answer is yes, he should learn to rely on them, and yes, there is an alternative.

In learning to live with the estimates, the planner should remember

---

[9]An exception to this is found in certain segments of the aerospace industry in which all classes of workers are very mobile; even at the highly skilled and professional levels turnover has been as high as 50 percent every year. This condition probably has had some impact on project planning in those companies.

that one object of the network technique is predicting problems. If activities are subdivided into short, controllable elements, and one gets out of hand because of faulty estimating or other problems, timely corrective action is indicated; the manager can act before the condition spreads and endangers the entire project. At the risk of taking a Pollyanna-ish view, the authors assert it is better to be 100 percent wrong on a two-week job than 10 percent wrong on a 52-week job. Poor estimates on small jobs, particularly if they point up potential trouble spots, are not as dangerous as poor estimates on long jobs where the problem may not be apparent until the project is in serious trouble. Also, on large projects overly optimistic estimates and overly pessimistic estimates tend to cancel each other out.

To go on in this vein would be to oversimplify the problem however, and the authors do not intend to do this. It is a matter of fact, though, that the estimating process can be substantially improved by experience and education, and by developing a common framework and language for all management levels. The U. S. Navy has conducted extensive studies of estimates in connection with its use of PERT on the Polaris program. Comparing actual and estimated times, they found at first a pronounced tendency for estimators to underestimate original schedules; but, over a period of about two years as estimators gained experience, their estimates became increasingly realistic and reliable. In the network-based system operating personnel are asked to estimate in more detail over longer periods of time than ever before. Naturally there will be a steep learning curve. As managers develop experience and a wealth of data is obtained and retained, the long-term benefits of the network system discipline become immeasurable both to the individual man and to management.

SUMMARY

Chapter 4 has described the basic time-oriented use of network planning, and has discussed the importance of accurate time estimates to the success of the system. Single time estimates were compared with the three estimate system. In Chapter 5 we will discuss the procedures for calculating schedules and identifying the critical path by use of the network plan and the activity time estimates.

5

Network analysis and scheduling

When acceptable time estimates for each activity in the system have been obtained and recorded on the network, the planner has sufficient basic information to make a simple but effective time analysis of the basic plan. Once completed, this analysis will accurately picture current planning at the detail level, and its implications for the overall project objectives. It also gives a basis for evaluating current plans in terms of imposed schedules, and for establishing and using a new schedule.

The steps in this analysis can be grouped into three computation sequences: the *forward pass* through the network, the *backward pass* through the network, and the *slack analysis,* which includes the calculation of slack or float and the determination of the critical path.[1] A time analysis cast in these terms produces a variety of data which can be used to schedule the project.

Before proceeding further, it will be necessary to define more formally some of the terms we will encounter, however. Some confusion results from the fact that the same or similar terms or data can be applied to events, activities, or both. Table 1 summarizes the data derived from network time analysis and shows the most commonly used symbols for both activity and event data. Table 1 lists the basic types of data in categories, which we have identified as I through V. Categories I and II describe data which are produced in the form of calendar dates or lengths of time from particular base points expressed in days, weeks, etc., from time zero. Category III describes data which are produced in the form of time intervals, expressed as a number of days, weeks, etc. Category IV describes a particular organization of the data in Categories I through III to indicate the location and nature of the critical or pacing activities and events in the network. Category V concerns probability values.

The terms *forward pass* and *backward pass* are simply convenient ways

---

[1]Some applications also compute probability in the ways discussed later in this chapter.

TABLE 1
Data Derived from Network Time Analysis

Data Type	Data Description			Data Object	Symbol
I	Earliest Expected Start Time			Activity	EES
	Earliest Expected Completion Time			Activity	EEC & T_E
	Earliest Expected Occurrence Time			Event	T_E
II	Latest Allowable Start Time			Activity	LAS
	Latest Allowable Completion Time			Activity	LAC & T_L
	Latest Allowable Occurrence Time			Event	T_L
III	Slack			Event	S_E
	or			Activity	S_A
	Float:	Total		Activity	TF
		Independent		Activity	IF
		Free		Activity	FF
IV	Critical Path			Event and/or Activity	(Various; identified by least slack value in network)
V	Probability of an Event Occurring at a Particular Time			Event	Pr

to describe the manual or machine process of proceeding forward through the network along each path, and the reverse process of tracing paths backward through the network. These procedures will be illustrated in detail as we go along, but first, it will be useful to know what information is needed to perform the calculations.

INPUT DATA

As we stated in Chapters 3 and 4, the planner should record all the basic data needed to perform a time analysis of the plan on the network. Commonly referred to as input data, these include data associated with activities and those associated with events. Table 2 lists the various types of input data. In addition to the data indicated in Table 2, the input includes the *network logic,* all the dependency relationships which have been identified between activities and events in the network. As we explained in Chapter 3, this information is provided by identifying each activity by its predecessor and successor event number. When the same number designates the end of one activity and the beginning of another,

TABLE 2
Types of Input Data

DATA ELEMENT DEFINITION	IDENTIFICATION
Activity Data:	
Beginning Event Number	1) Preceding Event Number or 2) Predecessor Event Number (E_p) or 3) "i" number.
Ending Event Number	1) Succeeding Event Number or 2) Successor Event Number (E_s) or 3) "j" number.
Activity Duration Time(s)	Either single (D) or triple time estimate (a,m,b). Units may be hours, shifts, days, weeks. t_e is weighted average.
Scheduled Dates	Scheduled or planned start or completion dates may be inserted for selected activities. Identification varies, often T_D or T_s.
Activity Description	Short word description of activity.
Activity Codes	1) Organization Responsibility. 2) Cost accounting category or number. (NOTE: Other special activity references and codings may be employed for obtaining specialized sorts of the basic data.)
Event Data:	
Event Number	Up to 10-digit code used to label the event. It may be coded to correlate with the Work Breakdown Structure.
Scheduled or Planned Event Occurrence Dates	Such dates are often inserted; they are similar to scheduled activity completion times, but relate to the event and therefore affect all activities starting or ending with the event. Identification $(T_D$ or $T_s)$ varies. Often used for milestones and/or interface events.
Event Description	Short word description of event.
Event Codes	Additional codes may be used to provide special references for report purposes (for example, cognizant department, milestone or interface identification).

it is possible to reconstruct the network logic in a computer or manual analysis.

FORWARD PASS COMPUTATIONS

The *forward pass* through the network is made by computing the earliest expected occurrence dates for each event (T_E), and/or the earliest expected starting or completion dates for activities (EES and EEC). It is done by first setting a starting date for the network itself; it may be a calendar date such as that stated in a contract, or one expressed for planning purposes as a baseline date, such as time zero or week zero. This baseline date is the earliest expected occurrence date for the first event (the T_E for the event), and therefore the earliest expected starting date for each activity proceeding from that event. In the example shown in Figure 1 the baseline date has been given a time-zero value, but it could also be a calendar date, such as June 5, 1978. In any case, the earliest expected start date (EES) for each of the three activities proceeding from event 1 (activities 1–2, 1–3, and 1–4, respectively), has been established as day zero. Similarly, if the planner is more concerned with event data, he can say that the earliest expected *occurrence* time (T_E) for event 1 is day zero.

The next step in the forward pass is to establish the same information for each subsequent event and activity in the network. These data are obtained by simply proceeding forward along each path of the network, adding the estimated activity duration times $(t_e$ or $D)$ to the EES for the activity to obtain the earliest expected completion date (EEC) for the activity (in terms of events, T_E plus t_e of the succeeding activity $= T_E$ of the succeeding event). For example, we can add a few activities to our diagram, and establish the activity duration times, as shown in Figure 2.

The EEC for activities 1–2, 1–3, and 1–4 in Figure 2 can be established by adding their activity duration times to their EES values, which we have already established as day zero, as follows:

Activity	EES	+	Duration	=	EEC
1–2	0	+	3	=	day 3
1–3	0	+	6	=	day 6
1–4	0	+	2	=	day 2

The expected occurrence dates (T_E) for events 2, 3, and 4 will then be day 3, day 6, and day 2, respectively. Continuing this process along all the network paths will establish these data for all subsequent events and activities.

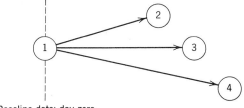

Baseline date: day zero

FIGURE 1.

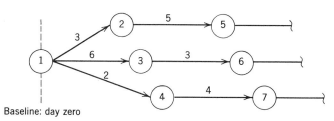

Baseline: day zero

FIGURE 2.

Activity	EES or T_E	+	Duration (t_e)	=	EEC or T_E
2–5	3	+	5	=	day 8
3–6	6	+	3	=	day 9
4–7	2	+	4	=	day 6

In the example shown in the last chart, the EEC of the preceding activity automatically becomes the EES for the next succeeding activity. In a network with continuing parallel paths, this would always be true.

Most networks will also include many *merge* points, which introduce an exception to this rule, however. Chapter 3 described a merge point as a point at which an event or activity is constrained by 2 or more preceding activities; for example, suppose that activities 2–5, 3–6, and 4–7 in Figure 3 all constrain a single event. In such a case, event 5 is a merge

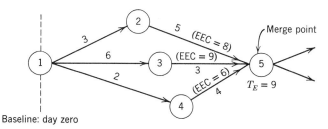

Baseline: day zero

FIGURE 3.

point, and if we use the forward pass methodology discussed we have a problem of three differing EEC's at that point (or three differing potential values for the T_E of event 5). In such a case, the EES established for the activities proceeding from event 5 (and/or the T_E for event 5) will be based upon the computation which produces the *maximum* value, or the EEC latest in time. In this example, the computation which produces the maximum EEC is through activity 3–5, yielding an EEC of day 9 as opposed to EEC's of day 8 and day 6 for the other paths. Event 5 and the activities subsequent to it are constrained by activity 3–5 as well as by the others; therefore the T_E for event 5, as well as the EES's for activities which proceed from that event, *must* be no less than day 9—the maximum of the EEC's for all constraining activities. Our network construction rules state that no activity can begin until *all* constraining activities are complete; hence, we can use the maximum length of all paths to establish the earliest start of all subsequent activities. This forward pass procedure proceeds from event to event along each path of the network until the end event in the network has been reached. The maximum EEC value for all activities leading up to the end event (the T_E for the end event) establishes the *expected overall project duration* based on the current network. This expected duration time will be equal to the longest sequence of activities through the network—the length of the *critical path*.[2]

BACKWARD PASS COMPUTATIONS

In the backward pass computations the planner establishes the *latest allowable occurrence date* for each event (T_L), and/or the latest allowable start and completion dates for activities (LAS and LAC). In the backward pass, the planner employs the path-tracing procedure characteristic of the forward pass in reverse, starting with the last event and proceeding backwards along each path to the baseline or beginning event.

---

[2]For the sake of illustration, we have said that a succeeding activity can begin on the same day the latest constraining activity is completed. This would be following a noon-to-noon schedule, which assumes, for example, that a one-day activity starts at noon on the first day (day zero if it is the first activity in the network) and is completed at noon on the second day. The activity (or activities) following could then begin at noon on the second day. This simplifies the calculations on the network for our purposes but is not realistic. The first activity would actually start at 8 A.M. on the first day and finish at 5 P.M. on the same day. The next activity would then start at 8 A.M. the second day. A number of computer programs for CPM and PERT incorrectly use the noon-to-noon schedule, a mistake most noticeable on one-day activities. Some confusion and misunderstanding in such applications results from this factor.

In the backward pass a latest allowable occurrence time (T_L or LAC) is set for the end event, corresponding to the scheduled or expected completion date for the project. In the absence of a scheduled end date the earliest expected completion date is automatically set as the latest allowable date. For example, if the latest EEC for event 10 (the end event in a network) is known to be day 14 and no scheduled end date exists, the LAC for the event will be set as day 14. Anchoring the end of the network in this way serves the same function as establishing the baseline at the beginning of the network: it provides a starting point for further computations. The next step, then, is the *reverse* of the forward pass. To establish the latest allowable start time (LAS) for all preceding activities (and/or the T_L for preceding events), we proceed *backwards* through the network from event to event, *subtracting* the activity duration times from the LAC for the activity (for events, the T_L of the preceding event is equal to T_L of the succeeding event less the t_e of the succeeding activity).

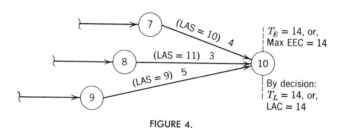

FIGURE 4.

In the example in Figure 4 the computed maximum EEC or T_E value at event 10 is known to be day 14. In the absence of a scheduled date, we may set its LAC or T_L value equal to day 14 and proceed to determine the LAS for the preceding activities.

Activity	LAC or T_L	−	Duration (t_e)	=	LAS or T_L
7–10	14	−	4	=	day 10
8–10	14	−	3	=	day 11
9–10	14	−	5	=	day 9

We can continue these calculations for the other events and activities, using the expanded network shown in Figure 5.

Activity	LAC or T_L	−	Duration	=	LAS or T_L
4–7	10	−	6	=	4
5–8	11	−	5	=	6
6–9	9	−	2	=	7

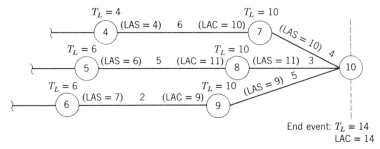

FIGURE 5.

As this sample calculation shows, there are certain similarities between the backward and the forward pass; in the backward pass a problem with a *burst* point, similar to one with a merge point in the forward pass can come up. As we proceed backward in our example, the LAS of an activity automatically becomes the LAC for the immediately preceding activity. In a network with continuing parallel paths this would always be true. However, we have noted that most networks include many merge and burst points. As in Chapter 4, a burst point is a point at which an event constrains two or more activities; for example, suppose in the example that activities 4–7, 5–8, and 6–9 were all constrained by event 6. All three activities would then have event 6 as their common predecessor and would become 6–7, 6–8 and 6–9. Figure 6 shows how such a network would appear. Event 6 is the burst point. If we have been using the backward pass methodology discussed, at this point we would have a problem similar to that encountered in the forward pass: three differing LAS's (or three differing potential values for the T_L of event 6).

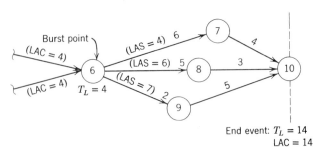

FIGURE 6.

In this case, the LAC established for all the activities coming to-gether at event 6 (and/or the T_L for event 6) will be based upon the computation which produces the *minimum* value, or the earliest LAS. In our example the computation which produces the minimum LAS is routed through activity 6–7, yielding an LAS of day 4 compared with day 6 and day 7 on the other paths. Hence, since event 6 constrains all three activities and since we have learned that activity 6–7 has an earliest allowable start date of day 4, the LAC for all constraining activities (activities leading into event 6) *must* not exceed day 4.

The planner continues this backward pass procedure along each path, proceeding from event to event until he reaches the baseline event. The minimum LAS for all activities leading up to the baseline event estab-lishes the latest allowable start time for the project if objectives in the current network plan are to be met.

SLACK COMPUTATIONS AND IDENTIFICATION OF CRITICAL PATH

Before getting involved in discussing slack (or float) we can clarify our understanding of slack concepts by reviewing and evaluating the mean-ingfulness of the data we have derived so far in our analysis.

First, we have computed a simple, logical statement for each event and activity in the program and for the end objective. Given the con-straints imposed by the known interdependencies and the most reliable duration estimates for the individual activities, we can identify the earliest possible calendar times on which we can expect each event to occur, or each activity to start or end. There is nothing particularly mysterious about this, but it *is* information which is hard to obtain with-out a network.

Second, knowing these expected times and understanding that many concurrent activities of varying lengths exist in the network, we realize that all the activities must have certain latest allowable start and end times, and the events have latest allowable occurrence times; these may be earlier than, the same as, or later than their earliest expected times.

Finally, since there is one path through the network that is longest; the other paths must be equal to it or shorter than it. And, because their latest allowable times may not necessarily coincide with their earliest ex-pected times, the ocurrence dates for these paths may be flexible. We call this flexibility *slack* or *float*. If we can precisely determine what quantities of slack or float occur in each activity we can rank the activi-ties by slack value, and isolate the longest sequence of activities: the critical path.

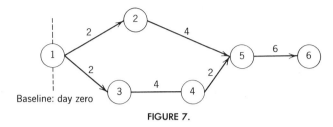

FIGURE 7.

Slack Computations

Slack, then, is defined as the difference between an earliest expected occurrence time for an event, for example, and its latest allowable occurrence time. This difference, expressed in a time unit, indicates how much the occurrence of that event can be delayed without delaying the end event in the network. Obviously, the planner can also determine the slack for an activity by comparing its earliest and latest start or completion times. Let us see how this works by looking at the simple network shown in Figure 7.

If we set the EES for baseline event 1 in Figure 7 equal to zero, we can quickly determine the EES, EEC, LAS and LAC for each activity in the network by making the forward and backward pass computations.

Activity	D	EES	EEC	LAS	LAC	Slack (LAC−EEC)
1–2	2	0	2	2	4	2
1–3	2	0	2	0	2	0
2–5	4	2	6	4	8	2
3–4	4	2	6	2	6	0
4–5	2	6	8	6	8	0
5–6	6	8	14	8	14	0

When we compare the EES and EEC of activity 1–2 with its LAS and LAC, respectively, we can see that that activity has a slack value of +2. This means that if the start of activity 1–2 is inadvertently or deliberately delayed beyond its EES of zero, it can slip up to two days before its LAC of day 4 is in danger. On the other hand, when we look at the lower path (1–3–4–5–6) in the diagram and compare the EES and EEC against the LAS and LAC for each activity, we find zero slack. Any activity slippage along this path will cause corresponding activity slippages all along the path and delay the end event. Of course, if the slack on the upper path were used up in an earlier activity, the end objective could also be endangered.

To sum up, we can compute slack as follows:

For event-oriented computations:

$$S_E = (T_L) - (T_E)$$

where S_E = event slack

T_L = latest allowable occurrence time (event)

T_E = earliest expected occurrence time (event)

For activity-oriented computations:

$$TF = (LAC) - (EEC) = (LAS) - (EES)$$

where TF = total float (or slack)

LAC = latest allowable completion time (activity)

LAS = latest allowable start time (activity)

EEC = earliest expected completion time (activity)

EES = earliest expected start time (activity)

Determining Critical and Slack Paths

Once we have determined the slack values attached to the various events and activities in the network, it is simple to identify the critical path. By definition, the critical path is the longest sequence of activities leading to the end objective. It is the path with the lowest slack value. In Figure 7, a simple network of two paths, 1–3–4–5–6 is the critical path. All the activities on that path have a slack value of 0 days, compared with the other path's consistent slack value of +2 days. If the scheduled date for the end event had been *earlier* than the expected date derived through the forward pass computations, this path would have had a negative slack value. When a network contains negative slack the path with *most negative slack* is identified as the critical path. The planner might also insert a scheduled date which is later than the earliest expected completion time, in which case the rule of least slack identifies the critical path. The critical path may contain positive slack (more than zero), zero slack, or negative slack, depending upon the scheduled dates.

On the remaining paths in a given network, there will be varying amounts of slack. Therefore, all the paths in a network can be compared with one another and ranked by order of criticality, so that management attention is focused on the most critical areas first.

The Significance of Slack

Our discussion has given a relatively uncomplicated view of both the computation and the use of slack. We assumed that all activities prior to the one for which we are computing slack will be, or have been, completed at their earliest possible times, and that all subsequent activities will start at their latest allowable times. This simplified view of an activity lets us consider the available slack as peculiar to that one activity, and seems to indicate great flexibility in scheduling that activity. But each activity is only one in a network; therefore it is sometimes useful to modify our view of slack by recognizing what impact prior and subsequent occurrences may have. Doing this, we can define three specific types of slack:

Total Slack (Float) is the excess time allowed for an activity when all preceding activities are completed as early as possible, and all succeeding activities start as late as possible. Total slack is the amount of time an activity could be delayed without affecting the overall project duration.

Free Slack (Float) is excess time allowed for an activity when all preceding activities are completed as early as possible, and all succeeding activities start as soon as possible. Free slack is the amount of time an activity could be delayed without delaying subsequent activities (but not necessarily the overall project). Free slack appears on the last activity in a path as it rejoins a more critical path; it has practical importance because an activity with free slack can be delayed without rescheduling any other activity.

Independent Slack (Float) is the excess time allowed for an activity when all preceding activities are completed as late as possible, and all succeeding activities start as soon as possible. It appears at complex junctions in the network, and is relatively unimportant.

If the reader thinks about the multitude of interrelationships possible among activities in larger networks he will see how these variations in slack can occur, and how it may be useful in some cases to be aware of them. Total slack is always greater than, or equal to free slack; free slack is always greater than or equal to independent slack.[3] While these discriminations may seem rather specialized, in some circumstances they can be valuable in scheduling resources over time.

MANUAL TIME ANALYSIS TECHNIQUES

We have now considered the concepts of the earliest expected and latest allowable times, and of slack and critical path determination in segments of networks. To illustrate these procedures and demonstrate the use of manual analysis techniques, let us consider a hypothetical project, and see how these concepts work. Our project will be represented by the network shown in Figure 8. The input data available for the network analysis include:

---

[3]In some circumstances, independent slack could have a negative value. In such cases, it should automatically be set equal to zero.

(a) All events and activities and their dependency relationships (network logic).

(b) Event numbers in ascending order through the network, to facilitate manual calculations.

(c) Activity duration time estimates.

The techniques we will use to make the analysis are simplified versions of actual computer routines; they can be performed by hand on small networks.

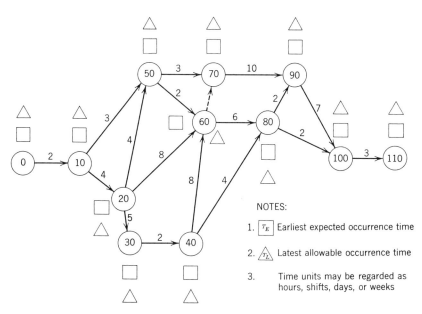

FIGURE 8.   Manual calculation of a simple network.

In Figure 8 there are two symbols, □ and △, adjacent to every event. These symbols represent the earliest expected occurrence time for the event and the latest allowable occurrence time for the event, respectively. As the planner calculates these values, he enters them in the appropriate symbol; he then isolates the critical path by a simple comparison.

These symbols can also be activity-oriented, designating either the EEC of the predecessor activity or the EES of the succeeding activity, or the LAC of the predecessor activity or the LAS of the successor activity, respectively. Because it is constantly necessary to remember which activity you are working with at a given point in the analysis, and because additional symbols are needed for an activity-oriented analysis and for recording slack values, it is not always practical to perform such an analysis on the network. To avoid covering the network with a jumble of symbols, the planner usually performs only the event-oriented analysis directly on the network; we will therefore use the event-oriented PERT terminology (i.e., T_E and T_L) in our discussion. Later we will demonstrate another manual technique using a supplementary tabular calculation sheet, which makes an activity-oriented analysis more practical.

To perform a manual analysis with the greatest ease and least error, the planner should number the events in the i less than j sequence described in Chapter 4. This allows the planner to analyze the network quickly by the numbers. In large networks processed by computer random numbering would probably be used to save time during revision.

Forward Pass—Sample Project

The forward pass through our hypothetical project will proceed from left to right through the network by the event numbers, from event 0 to 10, from 10 to 50, 10 to 20, 20 to 50, 20 to 60, 20 to 30, etc., adding the activity duration times to their predecessor T_E to obtain a T_E for the successor event. We will assume in this case that the time unit is working days, and will set the T_E for event 0 = day zero.

Event No.	T_E of Pred. Event	+	Duration (t_e) of Pred. Activity	$= T_E$
0	—		—	= 0 (assumed)
10	0	+	2	= 2
20	2	+	4	= 6
30	6	+	5	= 11
40	11	+	2	= 13

As the planner makes these computations he enters them directly on the network in the symbol provided (☐). The reader should remember that the T_E of an event which is the *merge* point for several activities is the maximum value obtained for all paths leading to that event. For example, at the merge point represented by event 60, there are three earliest expected completion times (EEC), one for each of the activities leading into the event. The three values are 12, 14, and 21 days; the T_E for event 60, therefore, is 21 days. Eventually, when all paths have been traced through to the end event (110), and T_E's have been established for each event, the T_E for the end event will indicate the total expected duration time for the project. However, except in the simplest examples, we still do not know which series of activities comprises the critical path. This is revealed by the data obtained in the backward pass.

Backward Pass—Sample Project

The backward pass is the reverse of the forward pass. The end event must be anchored, so let us assume that the scheduled completion date for the project coincides exactly with its T_E (day 41). We can then proceed backward through the network along each path, computing the latest allowable occurrence time (T_L) for each event by subtracting the duration (t_e) of each succeeding activity from the T_L of the succeeding event. The following chart shows examples of this calculation.

Event No.	T_L of Succ. Event	—	Duration (t_e) of Succ. Activity	$= T_L$
110	—	—	—	= 41 (by decision)
100	41	—	3	= 38
90	38	—	7	= 31
80	31	—	2	= 29
70	31	—	10	= 21

As the planner makes these computations, he enters them directly on the network in the symbol provided ($\triangle$). The reader should recall that the T_L of an event which is the *burst* point for several activities is the minimum value obtained. For example, at the burst point represented by event 80 in Figure 8 two potential T_L's, 36 and 29 days, are available, one for each activity proceeding from that event. Of the two values, the minimum value concept will lead the planner to establish the T_L for event 80 as 29 days. When all paths have been traced back to the baseline event 10, and T_L's have been established for all events in the network, it will be possible to locate the critical path.

Locating the Critical Path—Sample Project

When the analysis is performed by hand directly on the network, the planner can locate the critical path through the network without a slack calculation per se, by using a simple testing procedure on each activity. The procedure asserts that an activity lies on the critical path if it meets *all* the following conditions:

(1) The predecessor event $T_E = T_L$.
(2) The successor event $T_E = T_L$.
(3) The successor T_E − the predecessor T_E = activity duration time (t_e).

Let us test some of the activities in the sample network (see Figure 9). When we examine activity 30–40 we see that the conditions hold:

(1) 11 = 11: O.K.
(2) 13 = 13: O.K.
(3) 13 − 11 = 2: O.K.

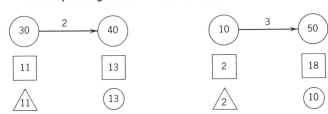

FIGURE 9.   FIGURE 10.

Activity 30–40 *is* on the critical path, because it satisfies all three conditions. Let us try another; activity 10–50, illustrated in Figure 10. In this
case, the conditions do not hold:

 (1) $2 = 2$: O.K.
 (2) Successor $T_E \neq T_L$ $18 \neq 10$.
 (3) $18 - 2 \neq 3$.

Activity 10–50 *is not* on the critical path, since it satisfies condition 1,
but not the others.

If we test each activity in turn against the three conditions, we can
determine quickly which activities are on the critical path. Figure 11
shows the network with all values inserted, and with the critical path
indicated by the heavy black line.

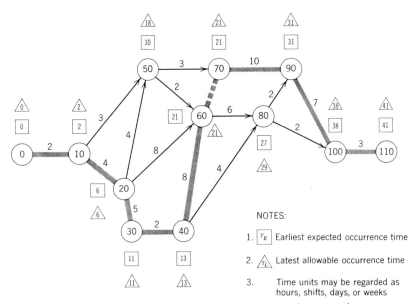

FIGURE 11.   Manual calculation of a simple network showing results.

Tabular Calculation

The preceding has illustrated a simple, fairly effective manual means of performing the basic time analysis; however, the method has certain limitations which can be eliminated. The method is heavily event-oriented, making it difficult to get detailed activity information. In addition, it does not analyze anything off the critical path. There is an alternate manual procedure which provides the additional flexibility of activity detail; it uses analysis sheets to permit a tabular presentation of the data similar to that obtained from a computer (Figure 12).

Just as it did in the hand calculation on the network, numbering events in $i < j$ sequence facilitates rapid analysis. To make a meaningful tabular presentation, the planner should list activities in a specific pattern, such as the i major, j minor sequence; this means that all the j (successor) event numbers must be listed in ascending order for the same i (predecessor) event number. Then the next higher i (predecessor) event number is listed, and the associated j (successor) event numbers are listed in ascending order with the new i number, etc. To illustrate this method, we have listed the activities in our sample network on the form shown in Figure 12. The identifying nomenclature and estimated activity duration times are entered in the appropriate columns for each activity listed. Referring to Figure 12, we can see that the two columns designated "earliest" are calculated in a forward pass through the network (or from the top to the bottom of the table) with maximums found among the EEC's entering an event; these are entered on the sheet as EES's for all activities with that predecessor event. The backward pass provides the entries for the "latest" columns, which are entered on the form from the bottom of the table to the top. The various slack columns are computed as follows:

(1) Total slack (float) = LAC − EEC.
(2) Free slack (float) = (EES of all following activities) − (EES of activity in question) − Duration of activity in question, or = T_E of successor event − EES of activity − duration.
(3) Independent slack (float) = T_E of successor event − T_L of predecessor event − duration of activity.
(Total slack ≥ free slack ≥ independent slack.)

The critical path column lists those activities with the least slack. As the reader can see in our sample project, dummy activities are handled just as other activities and may lie on the critical path. The entire process is illustrated by the data entered on Figure 12 from our sample network;

PROJECT_____SAMPLE PROJECT_____DATE_____

TIME AND FLOAT/SLACK CALCULATION SHEET											
ACTIVITY			DURATION	EARLIEST		LATEST		FLOAT/SLACK			CRIT. PATH
DESCRIPTION	PREC. EVENT (i)	SUCC. EVENT (j)		START (EES)	COMP. (EEC)	START (LAS)	COMP. (LAC)	TOTAL (TF)	FREE (FF)	INDEP. (IF)	
	0	10	2	ASSUMED 0	2	0	2	2-2 0	0	0	X
	10	20	4	2+4=6	6	2	6	6-6 0	0	0	X
	10	50	3	2+3=5		15	18	18-5 13	10-2-3 5	10-2-3 5	
	20	30	5	6	11	6	11	11-11 0	0	0	X
	20	50	4	6	10	14	18	18-10 8	0	0	
	20	60	8	6	14	13	21	21-14 7	21-6-8 7	21-6-8 7	
	30	40	2	11	13	11	13	0	0	0	X
	40	60	8	13	21	13	21	0	0	0	X
	40	80	4	13	17	25	29	29-17 12	27-13-4 10	27-13-4 10	
	50	60	2	10	12	19	21	21-12 9	21-10-2 9	21-18-2 1	
	50	70	3	10	13	18	21	21-13 8	21-10-3 8	21-18-3 0	
	60	70	0	21	21	21	21	0	0	0	X
	60	80	6	21	27	23	29	29-27 2	0	0	
	70	90	10	21	31	21	31	0	0	0	X
	80	90	2	27	29	29	31	31-29 2	31-27-2 2	31-29-2 0	
	80	100	2	27	29	36	38	38-29 9	38-27-2 9	38-29-2 7	
	90	100	7	31	38	31	38	0	0	0	X
	100	110	3	38	41	38	41	0	0	0	X
	110	–	–			41	41	BY DECISION			

FIGURE 12. Tabular calculation of a simple network. NOTE: Arrows and small figures indicate the general procedures for a few of the calculations.

a little practice with this form will familiarize the reader with the simple steps required.

Listing Results

One of the ways to use network analysis results most fully is to rearrange or sort the results in various ways, depending on the need. Results may be arranged or selected by any date, code number, or special code included in the activity description. The most common sequences in which results are listed include:

Total slack.
Expected (early) start.
Latest completion.
Event numbers:
    Predecessor (major sort) — successor (minor sort).
    Successor (major sort) — predecessor (minor sort).

The results of the sample network analysis are listed in various ways in Figure 13. Chapter 7 will present similar results prepared by computer.

FIGURE 13

Network Analysis Results Arranged in Various Sequences

Slack Sequence			Expected (Early) Start Sequence	
Activity	Slack		Activity	EES
	Total	Free		
0–10	0	0	0–10	0
10–20	0	0	10–20	2
20–30	0	0	10–50	2
30–40	0	0	20–30	6
40–60	0	0	20–50	6
60–70	0	0	20–60	6
70–90	0	0	50–60	10
90–100	0	0	50–70	10
100–110	0	0	30–40	11
60–80	2	0	40–60	13
80–90	2	2	40–80	13
			70–90	21
20–60	7	7	60–70	21
			60–80	21
20–50	8	0	80–90	27
50–70	8	8	80–100	27
			90–100	31
50–60	9	9	100–110	38
80–100	9	9		
40–80	12	10		
10–50	13	5		

(NOTE: In this sequence listing, the critical path is readily identified, followed by the next most critical, etc. Free slack always shows up on the last activity of a chain just before joining a more critical path.)

Determination of Early and Late Event Occurrence Times

When the planner wants results related to particular events, he must sort through the network analysis results and select the applicable start or completion date by comparing activity data relating to the predecessor or successor event. Thus, the easiest way to determine the latest allowable occurrence time for an event (T_L) is to arrange the results in predecessor-successor sequence; then, by comparing the latest allowable start dates of the activities and selecting the earliest from among the activities starting with a given event, the planner can identify the T_L for the predecessor event. Figure 14 illustrates this procedure.

To determine the earliest expected occurrence time for an event the planner arranges the network analysis results in the successor-predecessor sequence shown in Figure 15. The latest completion for any activity ending in that event number is taken as the expected or earliest occurrence for the successor event.

FIGURE 14
Determination of Latest Allowable Occurrence Time for Events (T_L)

Network listed by Predecessor-Successor Sequence		
Activity	Latest Allowable Start (LAS)	T_L for Predecessor Event
0–10	0	0
10–20	2	2
10–50	15	
20–30	6	6
20–50	14	
20–60	13	
30–40	11	11
40–60	13	13
40–80	25	
50–60	19	
50–70	18	18
60–70	21	21
60–80	23	
70–90	21	21
80–90	29	29
80–100	36	
90–100	31	31
100–110	38	38

FIGURE 15
Determination of Expected (Early) Occurrence Time for Events (T_E)

Activity	Network listed by Successor-Predecessor Sequence Earliest Expected Completion (EEC)	T_E for Successor Event
0–10	2	2
10–20	6	6
20–30	11	11
30–40	13	13
10–50	5	
20–50	10	10
20–60	14	
40–60	21	21
50–60	12	
50–70	13	
60–70	21	21
60–80	27	27
70–90	31	31
80–90	29	
80–100	29	
90–100	38	38
100–110	41	41

Probability of Meeting A Scheduled Date

An interesting and controversial—although not particularly practical—concept developed in the original PERT system by the U.S. Navy, is the use of calculations of probability to evaluate the likelihood of meeting certain scheduled dates.[4]

The Chance Aspects of Slack. The slack computation discussed previously produces estimates of the expected slack for future events. Actual slack will differ from the expectation; it will assume some actual value, depending on the exigencies of the research and development process and their probabilities. The theoretical analysis does allow the planner to make inferences about the actual slack that will develop, however.

The top diagram in Figure 16 shows the expected slack graphically. However, chance factors in the situation may cause the observed slack to turn out to be smaller or larger than anticipated. Actual slack smaller than

---

[4]The following paragraphs are condensed from the PERT Summary Report; see Appendix *B*.

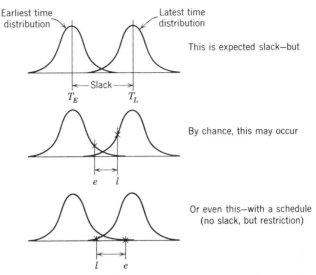

FIGURE 16. Slack—actual versus expected, based on three time estimates and original PERT use of probability distribution (see chapter 4).

the expected slack is shown in the middle diagram in Figure 16. Chance may even work perversely to make the latest time for an event fall earlier than the earliest time. Such a situation is portrayed in the bottom diagram in the figure.

In spite of the wide range of possible values that slack may actually take, many of the values are highly improbable. When a "no slack" condition exists, the manager must examine and control the situation. When the probability of a "no slack" condition is in the neighborhood of 0.5, it is important to monitor the events in question closely, because slippages in these events can jeopardize punctual accomplishment of the objective.

The Effect of an Existent Schedule. The analysis of slack discussed earlier does not explicitly take into account any scheduled dates but the objective end date. The actual situation will have scheduled dates for many of the interim events, however. If the activities have been programmed according to such a schedule, the slack analysis should take a different form. Before beginning such an analysis, the planner should appraise the feasibility of the existent schedule.

The feasibility of accomplishing an event on a scheduled date must ultimately be judged subjectively. As a guide to developing criteria for such decision-making, however, estimates of the probability of actually meeting scheduled dates can be made. The uncertainties of the future make it impossible to forecast the precise time at which an event will occur. However, the manager does know the expected times, as well as a procedure for calculating the probability of any deviation from those expectations.

Figure 17 shows the uncertainties involved in predicting the precise time for an event to occur. The figure shows the last few events in the example to which we referred earlier. Event 50 in Figure 17 might have been scheduled at time T_S, but the earliest time analysis might indicate that the event is expected to occur at time T_E with variance $\sigma T_E{}^2$. Statistical theory predicts that the probability distribution of times for occurrence of an event is closely approximated by normal probability density. It is therefore possible to calculate the probability that the event will have occurred by any future date. The shaded area under the curve in Figure 16 represents the probability that event 50 will have occurred by time T_S. The three time estimate approach discussed previously is used to establish the time spread of the probability curve.

The planner can make similar calculations for each scheduled date. The resultant series of dates might represent an existent schedule determined by any means. Management can then make statements regarding the probabilities of meeting each scheduled date if all activities are carried out as soon as possible. When the probabilities assume low values, the planner can assume that the schedule is infeasible. High values indicate the opposite—that there is a high probability that the schedule will be met. Technical managers can reappraise a given schedule in the light of the probabilities cited here. If the managers decide that a scheduled

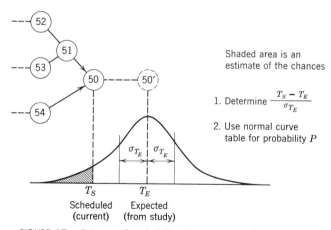

Shaded area is an estimate of the chances

1. Determine $\dfrac{T_S - T_E}{\sigma_{T_E}}$

2. Use normal curve table for probability P

T_S
Scheduled
(current)

T_E
Expected
(from study)

FIGURE 17.   Estimate of probability of meeting scheduled date-T_s.

date is infeasible, then they must alter resources and/or performance, or reschedule activities. If they choose to alter resources or performance competent technical advisors will provide a new plan, and estimating sources will give appropriate estimates. If the decision is to reschedule, a rational means of changing the schedule should be found.

The merits and statistical validity of this use of probability have stimulated a great deal of discussion. One major deficiency of the approach is that statistical probability requires a large number of identical occurrences to establish validity, but the events in a network plan may occur only once and similar events may have occurred only a few times in the past. Debate also centers around the merits of alternate methods of treating the variances in chains of activities—whether it is better to sum them as we move from event to event, or to ignore the cumulative effect.

A valid measurement of the importance of this use of probability is the degree to which managers have used it in projects in which it is provided with the PERT system. The authors' experience indicates that management is making little or no use of the probability of meeting a scheduled date. A few managers are intrigued by the possibilities of the concept, but it has not been a major factor in the decision-making process.

USING THE RESULTS OF NETWORK TIME ANALYSIS

When all the data described above are available, management has an unprecedentedly powerful array of information; by using it, managers can gain meaningful insight into their project, and a sound basis for management decisions about scheduling, allocating resources, the likelihood of achieving objectives, and the like. In addition, the information is available in a form which is uniformly communicable at all management levels and so lends itself to unambiguous interpretation.

Scheduling with the Network Plan

We have considered the network plan as a "free floating" or non-calendar related plan of the work to be accomplished. If we view the plan this way, given a calendar date for the start or baseline event we can express all subsequent activity start and complete times as calendar dates determined by computations based on the activity time estimates described earlier. When all times are expressed as calendar dates the network plan becomes a *schedule* for the project. Thus the planning methods described so far—identification of activities and events, preparation of the dependency network, time estimation, and network time analysis—are actually basic scheduling procedures.

However, *scheduling* in the sense used most commonly in an industry, is not normally such a simple matter; it will therefore be useful at this point to differentiate between *planning*, which we have essentially been discussing so far, and *scheduling*, which warrants some further consideration. Development and preliminary analysis of a network plan can be considered the basic planning phase in the use of the network-based system. When the plan has been developed and analyzed, the network model of the project must be converted into a timetable or schedule to guide management in carrying out the project; to convert it, the planner must integrate into the basic network a more rigorous time assessment, in terms of actual calendar objectives, and resources. The basic logic of the network plan may be outstanding, but its investment requirements, or its timeliness may be completely out of line with available resources or schedule constraints imposed externally (such as directed delivery dates specified by a customer or supplier). Determining the reasonableness of the network plan as originally conceived, and replanning activities in terms of time or level-of-effort comprise the scheduling phase of use of the network-based system. The planner considers resources in a general way when estimating activity times in the network planning, but these basic assumptions must be evaluated against other time and resource schedule constraints which may not fit into the original, ideal plan.

Effect of Directed Dates on Scheduling

When no one has scheduled dates for the start and end events in the network, the usual procedure (either manual or computer) is to assign a network baseline date to the start events. When this is done, the earliest expected occurrence date of the end event (T_E) (or the EEC of the last activity) is designated as the scheduled end date for the network. This permits management to compute calendar dates for each event and activity relative to the network baseline date. As we have seen, this procedure always results in a zero slack value on the critical path, with all other paths having varying degrees of *positive* slack.

The reader should realize what effect externally imposed, or directed start and completion dates have on the network plan, however. If the imposed end date, for example, was earlier than the computed earliest expected occurrence time (T_E) for the end event, the directed end date would have to be used as the basis for the latest allowable occurrence time (T_L), and the backward pass computations would result in a negative slack value on the critical path. In the sample network shown in Figure 10, imposing a directed end date of day 39 with no change in the start date (day 0) would produce a total slack value of -2 days on the activities on the critical path, and lowered positive slack values on certain other paths. Negative slack in the network indicates potential schedule slippage. An excessive amount of it (assuming directed dates are not overly unrealistic) may indicate inaccuracies or discrepencies in the network plan; opportunities for concurrency may have been overlooked, estimates may be overly pessimistic, or assumptions about the risk level in the scheduling of resources may have been too low.

Directed dates may also present scheduling constraints at other points in the program besides the beginning and the end. Earlier in the book we discussed the concept of milestones in some detail. Milestones are often intermediate scheduled or target dates for the occurrence of selected events; they must therefore be considered in scheduling the activities associated with those events. When such dates are later than the T_L's for the events to which they are assigned, some additional slack time may be available. However, the scheduled dates may be earlier than the computed T_L's; if they are, they must be considered when planning the associated activities, because they will affect the associated slack computations significantly. The planner must decide whether to let such dates affect the slack values, that is, whether to let them replace the T_L values derived in the network analysis procedure. The term *secondary slack* is used to denote slack computed against such fixed intermediate dates; some computer programs will print this out along with the total slack against the end event of the network. Unless all persons involved understand what is being done, this procedure can introduce a certain amount of confusion.

Hence, having scheduled or directed dates among the project objectives may affect the slack analysis in all or part of the network plan significantly. The scheduling process should then include careful activity scheduling relative both to other activities and to the assigned dates, so that work is accomplished in keeping with project objectives. To accomplish this, a more rigorous examination of resource constraints will be required. Integrating cost and manpower with timing in network planning and scheduling is discussed in Chapter 6; some mention of the process is needed here, however.

Effect of Resource Constraints on Scheduling

A duration time is estimated for each activity in the network plan. As we suggested earlier, the planner may assume that a range of possible durations exist for any activity; the figure given as the most-likely or normal duration time is the result of an estimation procedure influenced by a number of objective and subjective criteria. If we consider this normal duration time as the upper end of our range of duration times, we can envision a number of points indicating successively shorter durations between it and the lower end of the range. Shortening durations will increase costs, however; this cost must be in resources or quality. This duration-cost relationship is illustrated by the curve in Figure 18. The

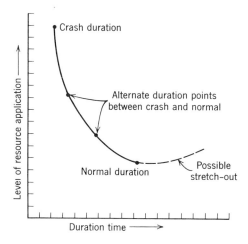

FIGURE 18.   Basic scheduling consideration—alternate durations.

normal duration at the high end of the time scale would call for the lowest level of investment in resources. At the low end of the time scale, the crash duration would require the highest investment of resources possible for the activity, within availability constraints. Between these two points one or more alternate durations may exist, each requiring different direct resource investments.

This demonstrates that besides scheduling activities according to the directed dates and available slack, planners must also consider the level of effort and resources needed on any activity to achieve the best possible mix consistent with project objectives. In addition, such time/resources planning will often be needed to show limitations imposed by allocating fixed resources; this helps the planner avoid unworkable manpower "peaking," overuse or underuse of facilities, unrealistic dependence on material from external sources, etc.

Without delving too far into areas which will be more clearly understandable in later chapters, we must add one final point to this discussion. The network planning and analysis procedures lay stress on systematically developing a workable plan for accomplishing the various activities in the project. Preparation of the network plan is a useful way to display the activities and events in the project sequentially. If he adds time estimates and calendar dates to the network, the planner can use it to evolve a schedule consistent with the time/cost objectives of the project. Because of its analysis of slack and display of critical activities, the network plan can also be used to verify the practicality of the scheduled use of resources, manpower, and funding. These topics will become clearer in the discussion of cost and manpower in Chapter 6, and will be handled more comprehensively in Chapter 7.

Using the Network to Manage

Once the project plan has been prepared it must be interpreted and used. It must be continually analyzed, evaluated, and modified as progress is made, and management gains additional insights into uncertain areas. The outputs of the system described so far may be used in two major phases. The project is first analyzed by checking the network against known schedule and resources constraints; the objective is to balance the resource allocation, scheduling, and slack distribution, to verify the validity of the Work Breakdown Structure, and to validate the networks. The results of this analysis become the project plan, representing management's best judgment concerning all project constraints. At this point the second phase of evaluation begins: it consists of applying the same evaluation techniques to the project throughout its course, considering the effects of actual accomplishments and experience, changes in project definition and objectives, etc. This second phase of evaluation makes it possible to use the system for current planning and control.

Periodically, all the network information is processed and sorted by computer, analyses are made, and the data are reported in summary form to management. Management reviews and evaluates it, makes decisions and gives new directions. As we have observed, in the planning and scheduling phase activities may be replanned or rescheduled, or resources reallocated on the basis of these results. In the control phase, management is continually able to compare the actual with the planned accomplishments, and to assess the impact of delays and changes, because the actual occurrence of events and activities is reported. This up-to-date knowledge forms the basis for corrective actions.

SUMMARY

This chapter has described the fundamental characteristics of network time analysis. The three basic steps in this process, the forward pass, the backward pass, and the slack analysis computations, were described and demonstrated. An introduction to the various ways of analyzing the results and the uses of probabilities in scheduling were presented. Some of the uses of the time analysis results were discussed. In Chapter 6 we will consider the significant effects the integration of cost and manpower data has on the basic concepts of network planning and critical path analysis.

6

Cost and manpower

Adding the cost and manpower extension to the basic time-oriented network system provides cost and manpower data directly related to the work sequence and time schedules. Such an addition is aimed at making it possible for management to predict overruns or underruns in programmed costs and manpower expenditures, and to have sufficient supplementary information concerning critical areas to act effectively.

Experts have questioned whether cost versions of the network-based system are compatible with traditional cost-accounting methods and whether such network systems should replace existing systems. At present, cost extensions of network systems are not seen as replacing existing accounting systems, but as complementing them by being useful cost-planning, control, and reporting devices for individual projects. This does not imply that reporting systems would be duplicated, but that existing systems would be adapted to provide time, cost, and manpower information in integrated formats. Hence, at best, time/cost/manpower versions of the network-based system would offer management substantial increases in cost and manpower planning and control capability with minimal changes in input methods.

OBJECTIVES OF THE TIME/COST/MANPOWER APPLICATION

The network-based system which includes cost and manpower dimensions aims at relating the activities to be accomplished to the schedule constraints, manpower requirements, and financial information. The network plan, which is a realistic and understandable model of the project, provides the framework within which to correlate these factors. The network system lets the planner link these elements of information together, and compare the actual expenditures of time, labor, and funds, with the projected ones. Using the operational time/cost/manpower system those responsible for the project could prepare and present integrated information to managers for review, to facilitate decision-making.

The general objectives of the integrated network-based system are:

(a) To provide a sound basis for the development of valid time, cost, and manpower estimates that reflect the actual resources available.

(b) To aid in determining where resources should be applied to best achieve time, cost, and technical performance objectives.

(c) To provide a system for early identification of those project areas with the greatest potential schedule slippages and/or cost overruns to facilitate action before trouble develops.

It also enhances management's ability to determine, at various levels of the organization,

(a) Whether current time and cost estimates for the entire project are within allowable (contract) limits.

(b) Whether the project plan is consistent with current schedule and cost *commitments,* and, if not, the extent of the differences.

(c) Whether planned allocations of manpower and other resources over time are realistic.

(d) Where resources can be reallocated via trade-offs among the activities of the project to expedite accomplishment of pacing activities.

(e) What impact slippages (delays) or changes in the ongoing project will have on the availability and allocation of resources.

THE BASIC TIME/COST/MANPOWER APPROACH

Network-based system design including time, cost, and manpower plans centers about a system model; the model commonly provides for the enumeration and prediction of costs related to schedules, and (as an optional extension) for resource allocation by optimization techniques.

The Enumerative and Predictive System Model

The enumerative and predictive system model is designed to provide a common framework for cost estimation, actual cost accumulation, cost forecasting, and comparison of actual with estimated or planned costs for individual or groups of activities. The basic objectives and capability of this system model may be characterized as:

(a) Comparison of expected and actual costs.
(b) Predictions of future costs.
(c) Isolation of trouble areas (i.e., projected overruns or underruns).
(d) Dynamic correlation of time, cost, and manpower information.

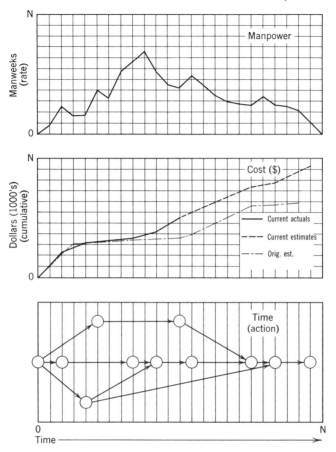

FIGURE 1.   General time/cost/manpower relationship as seen by enumerative and predictive system model.

Figure 1 shows the basic nature and functioning of the model. Time, cost, and manpower requirements are estimated for each activity or work package in the network. As the project progresses, actual expenditures are accumulated and recorded by activity or work package. This procedure enumerates costs already incurred and predicts costs *to* completion and *at* completion. In addition, distributing these costs across the activities lets management isolate the components of this cost data by activity and time period. These data form the basis for corrective actions or replanning, and can be useful as an aid to budgeting. Typical initial data requirements for the extended system (in addition to the previously discussed time data) might be:

(a) Direct man-hours and associated costs for each work package.

(b) Other direct expense items (e.g., materials, travel) to be applied to the work package.

(c) Various codes as needed to further differentiate the data and/or assist in its collection by department, function, labor classifications, etc.

(d) Estimated or committed subcontract costs.

These categories are only representative: Specific input requirements in any given case would be tailored to the specific information needs. As activities are completed, the estimated expenditures (time, cost, and labor) are replaced by actual expenditures when computing remaining expenditures. Thus, with a regular reporting cycle management can continually assess the validity of plans in terms of time, cost and manpower. In addition, quantitative forecasts of manning requirements may indicate critical periods in which manpower use may be over or under the practical limits of staffing. Management can then consider whether to hire, lay off, or reassign personnel or to provide for overtime work.

Depending upon the level of detailed inputs furnished, management can obtain a variety of system outputs for use in planning and control. Included are such outputs as:

(a) Expected manpower requirements by time, period, organization, etc.

(b) Evaluation of proposed or actual changes in resources levels.

(c) Expected direct costs in a variety of categories.

(d) Regular time/schedule outputs, including critical and slack paths, expected completion times, impact analysis, etc. (see Chapter 5).

Variations in the input and processing may produce outputs in cost categories by indirect costs, actual costs, subcontract costs, labor costs, facility costs, etc. Regardless of the format of the data, however, the basic computations and functioning of the technique remain the same and are presented relative to the network framework. Figure 2 illustrates a typical manpower loading report, showing labor hours for a particular skill by time period and performance unit.

The Time/Cost Trade-Off and Resource Allocation Models

The time/cost trade-off and resource allocation models are multiple-factor systems which can be used to determine optimum overall schedules and optimum allocations of resources among the network activities; this is done by using mathematical modeling techniques. The models

PREPARED BY CPM SYSTEMS, INC.
C.P.M.SYSTEMS,INC.   MANPOWER LOADING REPORT
GENERALIZED ENGINEERING

NETWORK
SUBNET
LEVEL   2

REPORT DATE-   SKILL   02
RUN DATE-   29NOV63
CONTRACT NO.600255GENENG   01MAY64

MONTH	PERFORMING UNIT	CHARGE NUMBER	ESTIMATED MAN-HOURS	ACTIVITY FLOAT	1 WORKING DAYS IN EACH TIME UNIT
01-64	ELEC	.7..............	80	+ .	
01-64	MECH	.7..............	16	+ 1.0	
01-64	PROC	.7..............	1928	+ .	
		TOTAL FOR MONTH	2024		
02-64	ELEC	.7..............	8	+ .	
02-64	PROC	.7..............	820	+ .	
02-64	PURCH	.7..............	40	+ .	
		TOTAL FOR MONTH	868		
03-64	MECH	.7..............	428	+ .	
03-64	PROC	.7..............	2088	+ .	
		TOTAL FOR MONTH	2516		
04-64	MECH	.7..............	220	+ .	
04-64	PROC	.7..............	100	+ .	
		TOTAL FOR MONTH	320		
05-64	MECH	.7..............	104	+ .	
		TOTAL FOR MONTH	104		
06-64	MECH	.7..............	88	+ .	
		TOTAL FOR MONTH	88		
07-64	MECH	.7..............	72	+ .	
		TOTAL FOR MONTH	72		

FIGURE 2.   Manpower loading report showing estimated man-hours required by time period for a given skill and identifying the performing organizational unit.

have had only limited use, however; the time/cost trade-off approach has seemed more practical. The resource allocation model is considerably less practical than the enumerative and predictive models discussed; indeed, in the widespread practical application of network systems incorporating cost and manpower planning, few managers have tried to refine the system to this degree. It is hard to find a ready correlation between the theoretical approaches proposed and the practical constraints encountered in the actual implementation of the network system. Hence, we will emphasize the enumerative and predictive approach. However, these advanced systems are potentially important and warrant at least the brief discussion which follows.

The Time/Cost Trade-Off Model

In brief, the time/cost trade-off model is based on the assumption that there is a relationship between the dollar cost and the expected completion time for each project activity. This relationship is expressed as a mathematical model used to evaluate variations in resource information. During a typical research and development cycle the technique provides three basic types of decision-oriented information. During the preproposal stage, when they must determine funding levels, the managers get alternate time and cost options for the entire project. They decide on the best schedule, using the time/cost trade-off model. After they have selected the best schedule and established funding levels they use the resource allocation model to allocate resources in detail among the activities of the project. Finally, as work progresses, the system is used to reoptimize total time or cost objectives for the project and then to reallocate resources if the project must be accelerated or slowed. The basic inputs to the time/cost trade-off system model are alternate feasible time/cost points for each activity; these points denote the various levels of effort that could be applied to an activity and the expected duration for each level of effort.

Typically, planners determine a normal time and cost for each activity. In addition, a crash (minimum) time and related crash cost are established for each activity. The procedure then determines the critical path, and begins to shorten this longest path through the network. The relationship established by the normal and crash time and cost estimates allows the planner to select the most economical activity on the critical path: the activity which provides the most time reduction for the fewest dollars. After a given activity has been crashed, the network is reanalyzed to determine the new critical path; this process continues until the minimum project duration is obtained: when all activities on the final critical path have been crashed.

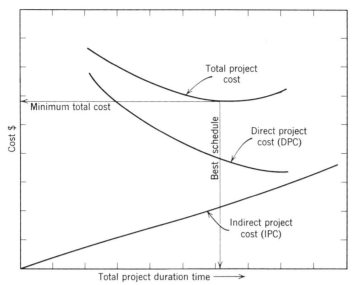

FIGURE 3.   Determination of best project schedule.

The results of this procedure plotted on a time/cost chart produce the Direct Project Cost curve shown in Figure 3. The Indirect Project Cost curve is also plotted (overhead, interest charges, loss of revenue, etc.), and these two curves are then summed to produce the Total Project Cost curve. The best project schedule can be selected and establishd from this final curve. This model is impractical to the extent that it assumes the given network is the best way to plan the project. Actually, revision of the network to reflect improved plans is almost always the way time is saved.

The Resource Allocation Model. Management tends to operate the time/cost trade-off model primarily to determine dollar-funding constraints, without considering sufficiently the real manpower, material, and facilities resources available. This model is theoretically sound, particularly when used to *establish* budgets for optimum schedules, but more-or-less fixed resource constraints call for more rigorous examination of resource allocation.

The resource allocation model deals with this problem by making the best physical allocation of resources among the network activities, within the limits of the existing time/cost constraints. This procedure employs techniques such as manpower leveling, coordination of outside resource constraints, or checks on the feasibility of all resource utilization schemes. This latter helps avoid scheduling more or fewer resources than are actually available at any point during the life of the project. In Chapter

14 we will describe manpower leveling by use of the network plan for a refinery maintenance project.

Resources are allocated by rescheduling activities which have positive slack, because these can be rescheduled at any point between the earliest and latest allowable times without affecting the project completion date. After slack activities are rescheduled the revised manpower schedule is subject to another feasibility check. If the revised schedule still shows peaking of resources, management may decide to divert resources from the critical path, a process opposite to the procedure used in the time/cost trade-off. Such reallocation results in either a later project completion time or an increase in costs.

The resource allocation model theoretically provides for the best resource allocation among the network activities; however, since there is no general model available, programming such a technique would be too time-consuming and inexact to be practical in industry. We have provided this brief summary of these optimization techniques for purposes of completeness, and now return to a discussion of the more practical enumerative and predictive techniques widely used in industry. The bibliography in Appendix E contains a number of references to these topics, to which the interested reader can refer.

OBSTACLES TO THE TIME/COST/MANPOWER APPLICATION

The extension of the basic time-oriented network system to include cost and manpower factors as well as time, increases management interest in, and use of the network system outputs. The integrated time/cost/manpower system can be much more useful than the basic system because it correlates time, manpower, and cost information in the planning and subsequent control of a project. When considering the extension of the system to include cost and manpower, however, we are immediately confronted with certain obstacles which are only incidental to use of the basic time-oriented system. Some of these are:

(a) Basic time-oriented networks are not necessarily all-inclusive. That is, they may tend to center on the technical problem areas most critical from a time-consuming standpoint. When adding cost and manpower resource information it is essential that *all* effort relating to the project be included in the network.

(b) In a typical industrial organization, the project people, i.e., the technical people, are usually the persons who know most of the facts concerning time, manpower, and cost estimates. Paradoxically, they may also be the people most reluctant to divert their attention from technical

Cost and manpower / 127

problems to provide such information. Management must develop input data requirements carefully, and cultivate a responsive environment, if they desire a practical system which will be appreciated and used.

(c) The network-based system with time, cost, and manpower data included, could produce vast amounts of detail in the output reports if the system is poorly planned. Such output would be practical only at the lowest levels of project management, and would require a great deal of manual manipulation to provide summarized interpretations for successively higher levels of management. Hence, the well-planned time/cost/manpower system must contain systematic planning disciplines and associated data summarization methods to provide management at all project levels with valid project information containing the appropriate amount of detail. In addition, these summarized outputs should be produced automatically so that additional manual analysis or expensive chart preparation is not needed.

(d) Conflicts in organizational structures and procedures (discussed in Chapters 2, 8 and 17) must be recognized and correlated with the network planning and resource estimating and accounting; if they are not, the system may become a side loop operation, used for customer reporting if required by contract, but not for internal operating and control. Further, the system costs less and is more readily accepted if existing information channels and data handling systems are used whenever possible. This is particularly true of internal accounting systems: sudden departures from established accounting procedures and routines could result in serious internal control problems.

All these obstacles to effective use of the integrated system are not so much obstacles as pitfalls, which can be avoided if the individuals attempting to design and implement the management system have maturity and good judgment. However, diligent, systematic preplanning is needed to assure the success of the management system.

REVIEW OF SYSTEMATIC PREPLANNING

Chapter 2 discussed the various preplanning elements necessary for network systems, including time, cost, and manpower organization. At this point, it might be useful to review the highlights of that discussion. The basic premise of the network system is that time, cost, and manpower planning and control are done within a common framework.

The planner first prepares the project *Work Breakdown Structure,* identifying all of the work elements and then breaking them down into successively smaller segments until he reaches a level of detail sufficient for realistic planning and control. Next, he identifies the *project organization structure* and correlates it with the Work Breakdown Structure; this involves identifying the organization units or the individuals who will be responsible for, or perform the work for each element and adding these to the identification of the Work Breakdown Structure elements. Then, the manager sets up the essential elements of the time/cost/manpower system; to do so he must decide on the work packages, and must correlate the other planning structures with the Work Breakdown Structure by setting up and coding the chart of accounts. In this chapter we have discussed the enumeration of costs primarily in terms of activities. However, the reader will recall from the previous discussion that the actual enumeration of costs is usually done by groups of activities called work packages. Work packages represent the final subdivision of the Work Breakdown Structure into work elements for which management wishes to make cost and manpower estimates and to accumulate actual expenditures as the project progresses. Each work package is assigned a discrete charge number used to identify estimates and actuals accumulated for it; all activities in a given work package will carry its charge number. The charge number for the work package, as well as all estimated and actual charges to that number constitute the basic elements of the *work package unit record*. The work package unit record also includes the activities in that work package (identified by predecessor and successor event numbers) along with their associated time estimates, and other planning structure information such as organization, contract, etc., coded into the account number. The charge numbers assigned to the work packages constitute the chart of accounts structure for the project.

RESOURCE ESTIMATION

Before continuing the discussion of the cost and manpower extension of the network-based system, it will be useful to discuss briefly some of the general considerations in the estimation and allocation of these resources, and how these considerations affect the plan. The time/cost/manpower technique is useful because it has a unique capability for allocating resource requirements (cost and manpower, as well as time) on the network plan, and correlating these with the Work Breakdown Structure. Time trade-offs to achieve scheduled objectives, basic components of the time-oriented system, are relatively simple compared to the greatly increased visibility the cost system gives. Resource allocation is implied in the time-only plan, but the time/cost/manpower method requires that management consider these resources explicitly and correlate them with the time data. This approach is more powerful, although this specific allocation and manipulation of resources introduces analysis problems of a considerably enlarged order of magnitude, requiring a more sophisticated understanding of the elements and their relationships.

The Estimating Base

A major problem is often encountered in arriving at a realistic project cost plan when the manager attempts to make the available budgeted funds based upon proposal cost data match the cost estimates employed in actual detailed work planning. To overcome such problems, the planner must understand how these cost data are derived.

Estimated costs such as those included in proposals and developed by subsequent contract negotiations, are usually the result of relatively undiscriminating costing;[1] however, when costs are broken down and estimated in detail when the planner is allocating resources to a network plan, estimates are more precise. The hardware item is a typical basis for proposal cost estimating; in such estimates lump sum manpower estimates are factored by average job category rates to obtain dollar equivalents. By contrast, when carried to the finer level of detail required by the time/cost/manpower system, production of the same hardware item is broken down into activities. Manpower estimates may be obtained for each activity or work package involved in hardware acquisition, and the performing organizations can be identified. Unlike the job category rates used in the proposal cost estimates, the appropriate average organizational rate can be used to convert the manpower hours into equivalent dollar values, for the task involved in producing or acquiring the hardware item. Hence, the reader can see that certain inconsistencies may exist between proposed or negotiated budgets for a task, and the actual expected costs for the task, based upon detailed network analysis. This points up the need for vigilance in such matters and suggests that steps be taken to prevent potential overruns from occurring. The planner should consider such discrepancies carefully when evaluating planned resources for a project, and should make an effort to anticipate them in early cost planning, or at the very least, to minimize them when preparing final resource plans.

Utilizing Improved Levels of Visibility

Certain assumptions may lie behind preliminary gross cost estimates; these assumptions may be reflected in both the estimated dollar cost and the manpower requirement for meeting a stated objective. The improved

---

[1]An interesting example of this phenomena is in the aircraft industry, where gross cost estimating for proposal purposes has literally been done "by the pound" for many years.

visibility afforded by the time/cost/manpower system makes these assumptions apparent, and exposes in detail the resource requirements related to the Work Breakdown Structure. With such exposure various discrepancies may be detected and evaluated; these might be improbable manpower loads required to meet scheduled objectives, peak manpower loads which cannot reasonably or economically be provided, or certain budget/cost variances. However, the visibility provided by the network system can provide management with sufficiently detailed information to permit them to identify and solve potential problems during the initial planning process and throughout the life of the project. By clearly defining resource constraints, as well as areas in which there is scheduled slack or a resource advantage planners can consider trade-offs during initial planning, and in many cases, simulate them to assess impact.

Estimating and Pricing

Literature about the cost and manpower extension of the system indicates that manpower and other direct costs can best be estimated and priced by the people who establish the network and make the time estimates. Although this might be the most desirable approach, it is usually not feasible because it assumes that most personnel in an organization possess what is essentially a highly specialized skill: the ability to estimate costs. In reality, resource estimating and pricing are usually done by specialists in various parts of the organizations. In addition, many organizations have developed highly automated formal procedures for this activity. The planner should emphasize these constraints when designing the procedure for acquiring information. Some of the ways he may acquire such data when using the system include:

(a) By receiving completed estimates, with prices, overhead, general and administrative costs, and fee, if applicable, at the work package level. The system would then function primarily to allocate these resources over time (see earlier discussion).

(b) By receiving labor estimates in man-hours (or weeks) with or without skill coding, and other direct costs. The system might then be used to price these estimates, using predetermined rates for man-hours, overhead, general and administrative costs, and fee, if applicable, and to schedule these costs as above.

(c) By receiving manloading estimates (actual anticipated head count) with or without skill coding, and other direct cost estimates. The system would then be used to produce labor cost estimates by factoring time estimates and head count, and then to price, as above.

The first method is usually the most practical to use, at least on an initial time/cost/manpower application, because the estimating does not require detailed competence in the network-based system techniques at a time when experience is minimal. In such cases, the Work Breakdown Structure is established, networks prepared, time analysis conducted (see Chapter 5), and results made available to estimating and pricing personnel for preparation of inputs at the work package level. Even employing this relatively unsophisticated approach, however, responsible management should strive to set up enough estimating guidelines to assure a uniform and valid approach. This is particularly important where there is considerable uncertainty because the company does not have experience with the type of work or product being estimated (for example, in research and development drawing on the latest technology). Cost data developed and estimated must be compatible with the flow of activities in the networks and the work packages identified. Work packages should have been carefully selected, so there is a firm basis for the correlation of time and cost data.

DATA COLLECTION

A useful device for providing clear guidelines to estimating and pricing personnel is the uniform estimating package which includes as a minimum:

(a) A detailed statement of the work to be performed, including a guide to correlation of the Work Breakdown Structure with contract line item requirements.

(b) The Work Breakdown Structure, including a key to summary numbers if used, or to the structuring of work package charge numbers if they are used.[2]

(c) Applicable networks.

(d) Preprinted cost estimating forms on which can be entered the data needed to estimate all cost elements for each work package.

(e) Specific guidelines for providing back-up information to support estimates, particularly for items that may fall under other (or miscellaneous) direct charges.

(f) Specific instructions for estimating the various cost elements to assure their validity and facilitate audit.

Such a carefully designed estimating package can be a valuable aid to the entire estimating process and to use of the network system. If an organization already uses automated or other formal techniques for estimating and pricing, the planner can design the package both to enhance

---

[2]Chapter 8 includes a discussion of summary numbers versus structured charge numbers.

the existing system and to facilitate conversion of these data to time/cost/manpower or Work Breakdown Structure terms. If formal methods are not being used, designing the cost input for network-based systems in the manner described adds a dimension of reliability and order to the estimating process which will benefit the organization generally.

SYSTEM OUTPUTS

Too often in the past, the advantages of completion of a project within or before the planned date have largely been offset by overruns which far exceed planned budgets. Such situations occur in environments in which it is impossible to correlate time, cost, and manpower plans in a meaningful way. Conversely, staying within the budget loses significance if there are serious schedule slippages just as a single victory loses its lustre if the war is lost. Hence, the most important objective of the network-based time/cost/manpower system is the correlation of time, cost, and manpower elements to permit trade-offs among these factors to achieve the best total plan. The planner achieves this objective by integrating plans against the Work Breakdown Structure, and presenting the interrelationships between them in the dynamic form of the network plan. The interplay between the planning elements can be shown by the system outputs in coherent statements which management can use. The network system cost outputs are not accounting reports, as usually defined by finance people; rather they are project-oriented cost reports which give managers the visibility to relate resource expenditures to time schedules for *project* status and control. The network system cost outputs also give a projection of current plans to provide predictions of time/cost/manpower status when the project is finished.

Cost Outputs

The cost output possibilities of the integrated system are:

(a) By automatically comparing actual work cost with the established budgets in force, the output can show dollar overruns or underruns to date.

(b) It can forecast dollar overruns or underruns for selected points *to* completion, and *at* completion; this is done by comparing budgets with actual costs plus current estimates for work not started that leads to selected points and to the end objective.

(c) It can show the effects of cost plan variances on schedules if cost data are projected at the same time as the relevant current time-only outputs.

The cost outputs can also be used to substantiate estimates, because they can present the data in terms of activities, work packages, or cost categories such as labor dollars or material; other formats management requires such as costs by organization, by work breakdown hardware element, or by contract can also be used. The planners can make the data more useful by providing summaries of them with an appropriate amount of detail at various Work Breakdown Structure levels. Using the Work Breakdown Structure as a basis for structuring system outputs, planners can design reports that provide an apparent audit trail; the trail can extend from gross summaries at higher levels down to the pertinent activity or work package account detail which might be the source of a deviation.

Manpower Outputs

Manpower requirements comprise a major element of a project plan, to be considered both in administrative and technical planning and in cost projections and planning. Therefore, manpower data have a special significance in the resources plan and in the display of system outputs. When the planners acquire direct labor estimates for the plan in cost estimating, they can also secure system outputs. These outputs can highlight areas of particular management interest, such as the actual and planned manpower use computed by skill, organization, time increment, or hardware item; labor costs can be computed in the same terms. This information can be used for management planning, or for periodic analysis of performance against budgeted manpower resources. Various types of manpower information can be produced as system outputs:

(a) Labor (expressed in man-hours, -days, or -weeks) overruns or underruns determined by comparing actual expenditures with budgeted manpower for the work completed.

(b) Forecasts of labor overruns or underruns at selected points in the project as well as at completion, computed by comparing allocations with actual expenditures plus current estimates for work not yet started.

(c) Effects of variance in the manpower plan on costs and schedule, produced by simultaneously projecting manpower and cost data with the relevent activity/time data.

(d) Comparison of the anticipated distribution of manpower requirements or rate of use over time (man-loading charts), with the known manpower resources available.

(e) Current measures of the number of men being charged to various account details for comparison with plan.

As with the financial information, manpower data can be presented and correlated in a variety of ways with the relevant network time and cost data to give management the knowledge it needs. Plans can simulate proposed or actual changes in any of the factors (time, manpower, or dollar costs) to assess the impact on project objectives, and to make it easier to consider alternate courses of action.

BASIC SYSTEM ELEMENTS

The basic elements of the network-based time/cost/manpower method are: the systematically developed network plan, against which time, manpower, and dollar cost estimates are entered, certain documentary information required to identify the Work Breakdown Structure, project organization structure, etc., which is added to the estimates; and a basic enumerative and predictive system model for manipulating these data. Collecting, processing, and evaluating these data are the basic steps in the design and operation of the system. In the planning stages of a project, these data can help the planner judge the validity of known schedule or resources constraints, and can generally aid in programming. Because it is designed for use in managing, the system can periodically accept current data such as activity completion dates, actual resources expenditures, or revised estimates; with these data added, the system can be used to present management with current status and outlook reports complete with time, manpower, and dollar costs. A variety of reports can be produced, depending upon information needs:

(a) *Exception Reports* tabulate items identified as potential or current problems in need of particular management attention, according to predetermined criteria built into the system design.

(b) *Work Breakdown Structure Reports* correlate and summarize actual time, manpower, and dollar costs and cost estimates at all levels of the Work Breakdown Structure.

(c) *Event and Milestone Reports* resemble the event-oriented time data or milestone charts which may have been used before the advent of the network-based system; the network event and milestone charts present time, manpower, and cost data in an integrated manner.

(d) *Contract Reports* summarize activities on a contract basis, and show how much the project has cost to date, how far from completion it is, and how much time and money it will take to complete. In addition, by merging such information with the contract data the planner can evolve a comprehensive contract administration report.

(e) *Organization Reports* use organization codes to group data by organization (section, department, division, group, etc.). If the planner uses appropriate formats he can develop manpower loading plans, organizational budgets, and other related types of reports. He can also make provision for subdividing and coding estimated manpower by skill categories, to permit the development of various manpower loading and cost reports by these categories.

IMPORTANT SYSTEM CONSIDERATIONS

The extended network-based system outlined in this chapter includes considerations specifically responsive to needs which, if unsatisfied, might become obstacles to successful implementation. These are:

(a) *Including all related effort in the network.* The systematic project planning or top-down approach consisting of the careful identification and correlation of planning structures ensures that *all* elements contributing to project objectives are included in the network plan. Level-of-effort supporting tasks, such as program management, may even be represented by a single activity, but they *are* identified in the planning structure and included in the project network.

(b) *Realistically designed input data requirements.* If the planner determines what input data are required to complement existing capabilities and data systems, and designs simplified data collection techniques he will largely eliminate confusion and unnecessary transcription of data and provide a service—rather than a burden—to personnel involved.

(c) *User-oriented management reporting.* If the planner designs output reports to organize the data in formats which are the most meaningful to the user organization, and makes provision for report summaries at the various levels of the Work Breakdown Structure, he can help to identify and eliminate the unnecessary detail and volume which otherwise often characterize systems attempting to correlate large amounts and types of data. The upward summarization of data can also directly produce customer reports which present a valid and objective current outlook without compromising proprietary account detail.

(d) *Adaptability to existing cost accumulation and other information systems.* By using the Work Breakdown Structure as a basic frame of reference, and by selectively coding the charge numbers in the project chart of accounts, the planner can use existing accounting formats and procedures for both input and output; by this method he can also produce a network-oriented data analysis as a product of data conversion techniques within the system itself. As a side effect of obtaining PERT-type reports, the validity of data made available to parallel conventional systems is enhanced, and those systems strengthened.

SUMMARY

Chapter 6 has given the reader an understanding of the extension of the network-based system to include dollar costs and manpower factors in addition to time. Armed with this understanding and a similar understanding of the basic time-oriented system, we can consider the operational characteristics of the integrated time/cost/manpower network system in Chapter 7.

7

Design and operation of a model system

In the preceding chapters we have explained the elements of network-based project management systems. In this chapter we will discuss the design of a system which uses these principles, and describe the operation of a model system. In the chapters in Part II we will discuss the conversion of this model system into a management tool operating in a real environment.

DESIGNING THE SYSTEM

It is not possible to design an ultimate general system which will satisfy all requirements for all applications. The planner must design a system using the Work Breakdown Structure, network planning and analysis, integrated time and cost estimating, budgeting and accounting, and electronic data processing, and must tailor it to the situation at hand. Using these same elements, a planner could produce a number of widely differing systems all capable of performing the needed functions. In the following discussion, we will discuss the major questions which must be answered to design an appropriate system for a specific situation.

ESTABLISHING SYSTEM OBJECTIVES

As we have noted before, the network-based management information system will not spring into being with the publication of a directive or a procedure. It must be carefully designed and developed to fit a company's particular needs and requirements. To design such a system, a planner must determine the objectives of the system for the environment in which it will be used, and, by using systems analysis techniques, devise a detailed plan which will achieve those objectives. After designing the system, the planner will develop tools, such as program analysis procedures, input, output and management display formats, flow charts of information to be processed, training aids, a computer program, and a system description. The implementation plan, the basis for the implementation or planning cycle, is also developed in this phase. The final phase is the operational phase or operating cycle, the maintenance and operation of the system once it has been set up. There are four steps involved in achieving an operating system:

(1) Design.
(2) Development.
(3) Implementation (the planning cycle).
(4) Operation (the operating cycle).

As we mentioned earlier, the planner must first identify the system objectives. These should be stated as specifically and unambiguously as possible. In addition to indicating where changes in the existing organization and systems will be required, the system objectives should answer questions such as:

(a) What are the benefits expected from the system?

(b) Upon what sources of information will the system be based? Does the system design provide adequately for data acquisition?

(c) What existing reports, procedures, and methods will complement the system? Which should be replaced or modified?

(d) Who will be responsible for operating the system? Are the objectives of the responsible managers compatible? Will there be sufficient authority where needed?

(e) What is the expected scope of the application? Company-wide? One large project? Several subprojects? etc.

(f) How will other vital management systems be linked to the network system?

(g) What are the desired system performance characteristics (e.g., operating costs, cycle time, manpower required)?

DETAILED DESIGN

Once objectives have been established, the most effective way to begin the detailed design is to start with the end product of the system: output reports and information displays. We can then go backward through the system to establish the requirements for each portion.

Output Reports and Information Display. The purpose of the system is to produce understandable, usable information to assist the manager in decision-making. Thus, the reports and displays produced by the system are its most important element. The planner must know what information is needed by the managers he seeks to serve, and must then establish suitable graphic displays and tabular and narrative reports to provide it. The system design will specify how the displays and reports are to be prepared (by hand or computer), distributed and used, how often they are to be prepared, and what purpose each is to have.

As previous chapters have shown, there is a tendency for designers to use rather standardized network system inputs and forms. The basic input consists of a plan of work, a schedule, and the manpower and dollar budgets related to the work plan. To be meaningful, therefore, the data outputs should indicate progress against these work, schedule, and expenditure plans. However, system outputs may vary widely in format, data arrangement, and even content, because of particular needs in a given application or organization. The basic results of a time analysis of the network plans are almost always the core of the outputs. The planner tailors the outputs to a given situation by varying the types of optional information which are provided, the ways in which it is arranged and made legible, and the ways in which it is sorted and summarized. Factors which influence the design of the system outputs are:

(a) The need to include formats, terminology, contents, and levels of detail which will meet the needs and biases of the intended readership.

(b) The need to devise outputs which can be used to prepare reports for top management or customers.

(c) The need to provide specific feedback information to the operating people who provide the inputs.

(d) The need to tailor the designs to the mechanics of report production, reproduction, and distribution.

The type of output report produced in any situation is also heavily influenced by the type of data processing used, and by the mechanics of the specific application. In any situation, system outputs must be produced and used in a reasonably standardized manner, and actual or predicted problem areas isolated and treated in a predetermined and approved way. Often, computer outputs are bulky stacks of paper containing masses of detailed data; these data must be analyzed and presented to supervisors, managers, and customers in a usable form.

Computer Reports. Machine-prepared reports of various types and degrees of detail will be required, either for management use or for the use of the personnel who prepare management reports and displays. The designer must match the output needs to the capabilities of the computer which the system will use.

Depending upon the degree of automation available, the system outputs can be sorted or summarized in a variety of ways, by any single item of information provided as input. Any one of the various dates, code numbers, slack value rankings, etc., can be used to list the results in a particular sequence. Some of the widely used methods of listing network system outputs are:

(a) By slack path, in which all activities having equal slack time are grouped together, and usually arranged chronologically within each group or slack path.

(b) By event numbers, in which activities are listed in numerical or predecessor-successor sequence.

(c) By expected completion dates, with the most recent dates listed first, so that management attention is directed to the most current action areas.

(d) By organization code number, which facilitates distribution of selected output information to responsible organizations or individuals.

Most current computer programs provide results for each activity in the network; a few provide event reports. Both types of outputs are important and useful, but for different purposes. The early Aerojet-General Corporation IBM 704 PERT program (1959), probably the first operational PERT computer program, provided both types of reports. Many programs, such as the HUGHES-PERT time/cost/manpower program and the IBM 7094 programs prepared by the U. S. Naval Weapons Laboratory and the U. S. Air Force Aeronautical Systems Division, also provide both. Programs developed for the construction and process industries are usually heavily activity-oriented in their outputs.

Computer Operations. The planner must specify the computer program in detail, indicating procedures for deriving each item of information. Procedures must be set for computer run control, with responsibilities specified.

Computer Input Data. The planner must design specific input formats, with procedures for acquiring the data, verifying their accuracy, and pinpointing responsibilities. He must also devise methods for collecting and transmitting the data.

Planning and Progress Information. The system must have procedures for obtaining the required planning information, as well as methods for obtaining progress information on work accomplished and funds and labor expended. The planner must also define rules for systematically planning the Work Breakdown Structure, the networks, and the related estimating, budgeting, and accounting structures.

SYSTEM DEVELOPMENT

To develop the system adequately, the planner must prepare report formats, procedures, etc., to fit the design. Tangible end products of this development are:

Management report and chart formats.
Operating computer program.
Procedures for:
  planning.
  obtaining progress information.
  preparing input data.
  processing data.
  analyzing output.
  preparing reports and charts.
Training aids to be used in explaining the system to all concerned:
  manuals and brochures.
  practice exercises.
  audiovisual aids.

SYSTEM IMPLEMENTATION: THE PLANNING CYCLE

When the planner first begins working to achieve valid plans, he takes the following steps:

(a) He decides to apply the network-based system to the project.

(b) He establishes operating policies and procedures.

(c) He indoctrinates and trains the personnel involved (Chapter 11).

(d) He establishes systematic planning structures for the project (Chapter 2).

(e) He prepares master phasing charts and identifies milestone and interface events as required (Chapter 2).

(f) He makes detailed network plans, and codes and identifies events and activities (Chapter 3).

(g) He establishes time estimates and related input data (Chapter 4).

(h) If a computer is to be used, he prepares input data for machine processing (Chapter 9).

(i) He makes the time analysis (Chapter 5). On small networks the planner may perform this analysis by hand.

(j) He analyzes output results:
    To determine the validity of critical path and slack paths.
    To compare the plan with desired or imposed schedules.
    To identify potential slippages.
    To review and validate estimates.
    To send results to personnel concerned.

(k) He adds dollar cost and manpower estimates to the approved time plan and evaluates the results (Chapter 6):
    To determine the validity of the budget estimate against the schedules.
    To identify potential dollar overruns.

To review and validate cost and labor estimates.

To disseminate results to personnel concerned.

(l) He prepares alternative plans and management and customer reports, which describe specific trouble areas, estimate the criticality of problems, suggest corrective actions, etc.

(m) He resolves problem areas to management and customer satisfaction, and revises plans accordingly.

(n) He repeats steps (a) through (m) until he achieves a satisfactory time/cost/manpower plan.

The upper part of Figure 1 indicates the general flow of the planning cycle. The reader should note that the time plan should be completed before the cost and manpower factors are added. When using the enumerative and predictive costing approach, the planner can often provide a firm basis for cost and manpower estimating in this way, because he can keep work package definition flexible until he has considered the basic schedule.

SYSTEM OPERATION: THE OPERATING CYCLE

In the ongoing operation of the network-based system the status of the project is periodically assessed, and the basis for ongoing direction of the project provided by:

(a) Maintaining the Work Breakdown Structure by continually revising to reflect changes in the overall plan.

(b) Maintaining the network by continually evaluating, and, where appropriate, revising and improving it to reflect current understanding of the project.

(c) Updating activity time estimates, to continually survey and, where appropriate, revise activity time estimates as management learns more about specific activities and gains experience with the project.

(d) Updating resource estimates, to continually survey and, where appropriate, revise cost and manpower estimates as management becomes familiar with specific activities and with the project.

(e) Recording progress, by logging times for activity completion and event occurrence so that a current record of program and status is maintained on the network.

(f) Accumulating actual resource expenditures by conventional cost accounting methods, and providing these in a form which is easily adapted to network system formats.

(g) Periodically performing network analysis and related resource analysis, based on current information; preparing input data sheets and

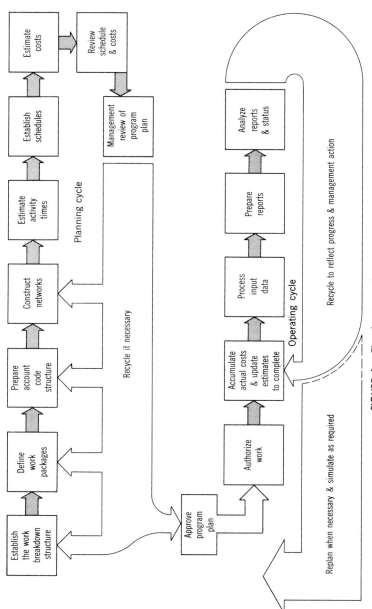

FIGURE 1. The planning and operating cycles.

submitting them for key-punching and integration into existing input data card decks or tape; providing for routing of actual dollar and manpower expenditures from conventional accounting channels, and merging these with the network input. The updating cycle will then include the following added input:

Additions or deletions to Work Breakdown Structure.

Addition or deletion of activities and/or events, and other network corrections.

Revisions in activity time, cost, or manpower estimates.

Revisions in schedule, budgets, or related performance criteria.

Actual activity start or completion dates.

Actual dollar and manpower expenditures to date.

(h) Analyzing output results:

To isolate and rank critical path and slack paths.

To compare estimates and actuals with schedules and budgets.

To identify actual or potential slippages and overruns, or underruns.

To evaluate alternate courses of action.

(i) Preparing management reports:

Summarized versions of output data in the form of charts and reports.

Concise descriptions of specific trouble areas, corrective actions recommended or taken, and management actions required, if any.

Publishing and distributing results to appropriate levels of management.

(j) Repeating steps (a) through (i) on a predetermined periodic basis for the life of the project.

The reader should notice that when the network is to include time only, with cost and manpower factors simply implied, we exclude the points in the cycle that refer to resource inputs, such as the merging of cost actuals with the network data. In other essentials, the cycle remains the same for the time-only application. The lower part of Figure 1 illustrates the operating cycle.

At this point, it may be helpful to look at a model network-based system.

A MODEL NETWORK-BASED PROJECT MANAGEMENT SYSTEM

This example is based on an actual time/cost/manpower system developed in 1961 by the Hughes Aircraft Company for its own use on major weapon system programs.[1] We feel that it amply illustrates the system design and operational characteristics we have been discussing, including the planning and operating cycles. The Work Breakdown Structure and task level planning are key ingredients. The reports structure, format, and content correspond to the generalized considerations we have discussed.

The Planning Cycle

The first step the planner takes is to identify the project objectives. They are specified in terms of the end items (hardware, services, data, and facilities) that are to be delivered to the customer or that the contractor has committed himself to deliver to the company. The subsequent division of each end item into its component parts creates a project Work Breakdown Structure, which then serves as the framework within which the project is planned and controlled. This is pictorially presented in Figure 2.

The development of the Work Breakdown Structure serves to:

(a) Define interfaces at a number of levels within the project (customer, associate contractor, and intra-organizational).

(b) Identify the milestones to be used for summary reporting of progress to both customer and contractor management.

(c) Define the project tasks and establish their relationships to the project end item(s) and project objectives.

(d) Establish a framework within which costs and schedules can be planned and controlled in an integrated manner.

(e) Establish a framework within which the cost and schedule status of the project can be summarized for progressively higher levels of management.

The project Work Breakdown Structure is also the basis for constructing the network of activities and events.

The configuration and content of the Work Breakdown Structure and the specific work packages to be identified vary from project to project, depending upon the size and complexity of the project, the structure of the organizations involved, the manager's preferred way of assigning responsibility for the work, the customer's project management requirements, and the number of contracts involved and their relationship to one another.

These considerations will also determine the number of end-item sub-

---

[1] *Hughes-PERT Guide*, Corporate Industrial Dynamics, Hughes Aircraft Company, Culver City, California, 1962.

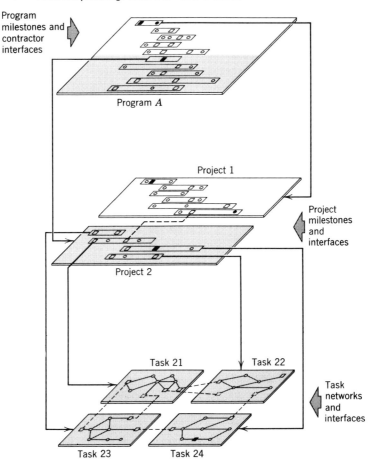

Program milestones and contractor interfaces

Program A

Project 1

Project milestones and interfaces

Project 2

Task 21   Task 22

Task networks and interfaces

Task 23   Task 24

FIGURE 2. Work breakdown structure for model system company program.

divisions included on the Work Breakdown Structure before the work packages are identified and responsibility assigned to operating units in a contractor's organization. The work packages formed at the lowest level of the breakdown then constitute the basic units in the system, for which actual costs are collected and compared with estimates for cost control. The end-item approach to development project planning and control insures that the *total* project is fully planned, and that all derivative plans contribute directly to the desired end objective.

As the work progresses, the project Work Breakdown Structure serves as the framework for "bottom up" summarization of project information.

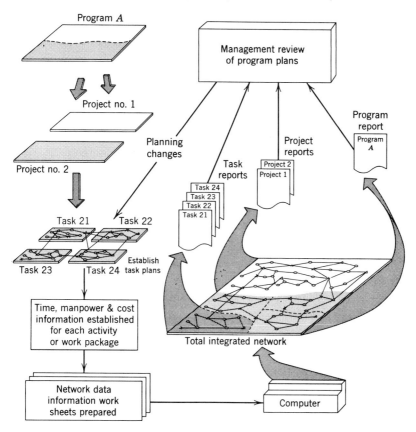

FIGURE 3.   Model system planning cycle.

In this way, the amount of detail presented to management at any level in the project satisfies management's interests and decision-making requirements at that level. When the summary reports indicate problems in a specific project area, the system enables the manager to obtain greater visibility by proceeding to successively lower levels of the Work Breakdown Structure.

The time, manpower, and cost information for each work package is collected and transcribed on to worksheets from which IBM cards are prepared (see Figure 3). These cards, together with a program of computer instructions, are then processed in the computer. During the processing, the computer analyzes each task and joins it to other tasks in an integrated manner so that the whole program network is available within its memory.

Using this work plan reconstruction and the subsequent analysis, the computer prepares a detailed graphic and documentary report for each task. Individual tasks are then merged into the next summary level to create project reports; the projects are merged and summarized to prepare program reports. The planning loop is closed when the manager reviews the reports and the work plans, and authorizes the necessary corrective action. The resulting changes are then incorporated into the network plans, and the cycle is repeated until the plans are acceptable.

The Operating Cycle

Figure 4 shows how the information is updated:

(a) Changes in plans from the preceding cycle are incorporated into the current plan.

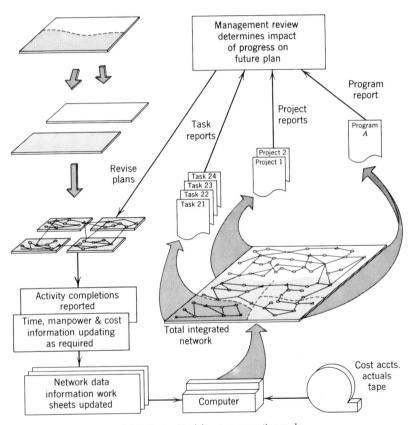

FIGURE 4. Model system operating cycle.

(b) The passage of time is noted and the completion of activities reported.

(c) Actual labor and dollar expenditures are automatically picked up from the accounting system and matched to the appropriate activities.

This cycle provides information needed for performance and progress evaluation related to the Work Breakdown Structure. The computer also uses the information to forecast the effect of attained progress and incurred cost throughout the program. When the manager reviews the reports to determine the impact of progress on the plans and to revise the plans to reflect his corrective action, the system operating cycle is complete.

Management Reports

The system provides management reports at three levels of the Work Breakdown Structure. The reports make available in documentary and graphic form the schedule and cost data project management needs to exercise control. For each segment of the Work Breakdown Structure these reports are prepared:

(a) The current project plan, schedule, and budget.

(b) The time, cost, and manpower performance to date, compared with the plan.

(c) The time, cost, and manpower forecasts to the completion of the project.

The output reports point up potential trouble spots in the project, and make it possible for managers to anticipate schedule and cost deviations. The manager gains enough information about these trouble spots so that he can direct his attention where it is most urgently needed. He can also determine those areas from which he may withdraw resources to assist the more critical phases with the aid of the reports.

The output reports provide current information from the bottom up without distortion to each project management level. This relieves the manager from the need to "wade through" a massive volume of data from subordinate levels to evaluate program status. However, under the network system this detailed information is readily available when any specific area should require his attention.

All the reports are interrelated because they deal with the same basic data, but each one emphasizes a different project element. A set of reports produced by the system will provide the manager with accurate, timely information on the time, cost, and manpower status of the project. In

```
RUN DATE          16 AUG 62      CUST NO AND NAME   738         PROJECT  REPORT  -  TIME OUTLOOK                    RUN TYPE    UD
TIME DATA AS OF   25 JUL 62      PROG NO AND DES    7385    USAF BALLISTIC SYSTEMS DIV.                             RUN ID       2
COST DATA AS OF   29 JUL 62      PREPARED BY - VILLORIA      WEAPON SYSTEM XXX                                      END EVENT 8530400
RESP ORG FOR PROG  DEPT 07
SCOPE OF NETWORK - PROGRAM

PROJECT NO  73858   COMMAND AND CONTROL PKG

RESP ORG   07-90
```

| TASK | DESCRIPTION | | | START | COMPLETION | | DATES | | SLACK | PROB |
| TASK NO | RESP ORG | C | | DATES | | | | | | MEET |
PEN	SEN	N G		EXPECT/ACTUAL	EXPECTED	LATEST	PLANNED	ACTUAL	WEEKS	PLAN
01	BLOWER ASSEMBLY									
0004	07-90			13 JAN 62 A		23 JUN 62		7 JUL 62	-2.0	
	0326									
02	CASE ASSEMBLY									
0005	07-90			10 MAR 62 A	23 FEB 63	02 MAR 63			+1.0	
	0373									
03	HOUSING ASSEMBLY									
0002	07-90	*		16 JUN 62 A	25 MAY 63	04 MAY 63	4 MAY 63		-3.0	
	0395									
04	CONFIG CONTROL PROG									
0007	07-90			19 MAY 62 A	20 APR 63	20 APR 63			+.0	
	0325									
05	ASSEMBLY AND CHECKOUT									
0012	07-90			02 JUN 62 A	11 MAY 63	04 MAY 63			-1.0	
	0368									
PROJECT 00158 TOTALS										
0001	0395	*		13 JAN 62 A	25 MAY 63	04 MAY 63	4 MAY 63		-3.0	

FIGURE 5. Output report—time outlook.

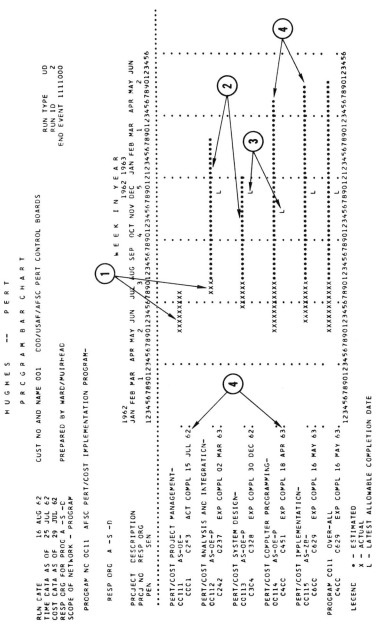

FIGURE 6. Bar chart graphic display. 1 Actual time progress; 2 remaining time to complete; 3 latest allowable completion date; and 4 predicted or actual completion date.

addition, the reports will identify the problem areas and the concerned responsible individuals at each work breakdown level.

Although the terms "program," "project," and "task" are used both in this narrative and in the illustrative reports, these terms are only one means of representing three successive echelons of the Work Breakdown Structure; these levels might also be identified numerically (e.g., level 3, level 4, level 5) or by other names (element, subelement, work package). Users of the system may adopt whatever identification is most meaningful by entering the names on a control card. The computer program will then record the appropriate name on the proper output.

Time Reports

Two types of time reports are included in the system. The first is the product of the basic PERT/TIME computer program which is used within the system. The second type is related to the Work Breakdown Structure and is both activity- and event-oriented. Reports of this second type are provided in both graphic and documentary form.

Basic PERT/TIME Reports

The basic PERT/TIME program provides an unstructured and detailed analysis of activities. The activities are sorted so that the "pacing" activities on the critical path and other slack paths are highlighted. Activity information is also sorted by organizational responsibility, by time sequence, and by event number. The system adds to the basic PERT/TIME reports by structuring time information to insure compatibility with the work breakdown plan; preparing machine-generated bar charts to facilitate rapid graphic analysis of time status, and identification of schedule problems; and preparing event-oriented milestone reports and charts for progress reporting.

Structuring the PERT/TIME information results in reports immediately useful to the manager responsible for each portion of the work breakdown. A sample time outlook report produced by the system is shown in Figure 5. The graphic displays highlight those pacing elements which may delay timely completion of the job. They also give a graphic measure of the degree of the problem. A sample report from the system's bar chart series is shown in Figure 6.

Milestone event reports are prepared to allow management to keep track of accomplishment and to meet customer requirements for milestone reporting. The current status of selected, previously-identified milestones is reported for each level of the Work Breakdown Structure. The milestone event report documents the predicted or actual completion date of an event compared to its scheduled completion date. An example of an event report produced by the system is shown in Figure 7.

Cost Reports

The benefits of meeting a planned completion schedule are considerably offset if budgets were ignored in the process and a large over-expenditure resulted. Conversely, meeting a cost target loses its value if delivery is delayed. The system approach treats time and resources as interdependent factors linked by the network plan. The effect of a change in one factor on the other is displayed by the system outputs.

The cost report series displays information which allows the manager to identify the following problems:

(a) *The cost overruns or underruns to date* by comparing the budget with the actual costs of the work.

(b) *The forecast of cost overrun or underrun at completion,* obtained by comparing the work budget with the actual costs plus the projected (latest revised) cost to complete.

(c) *The amount of schedule slippage* as indicated by the difference betwen the established schedule and the presently expected completion date.

A comparison of the *actual costs accumulated to date* and the *contract estimate for the work performed to date* will show whether the work is being performed at a cost greater or less than planned.

Using the audit-trail approach, the manager can trace down to those activities or groups of activities which are adversely affecting the time and cost targets. This makes it much easier to systematically interrelate the Work Breakdown Structure and the report structure as well as to display the information graphically. A typical graphic cost report is illustrated in Figure 8.

Manpower Reports

Project planners cannot develop realistic work plans and schedules without considering the manpower which will be required during the project life. Also, since manpower is a major cost element in engineering efforts, the manager must have information showing the impact on the total cost of the actual and the planned manpower utilizations. When the project planners have obtained direct labor estimates for the network plan and actual labor charges from the cost-accounting system, they can prepare charts that illustrate manpower utilization and requirements.

A cumulative curve can be prepared to show estimated manpower

T A S K   E V E N T   R E P O R T

RUN DATE 16 AUG 62	CUST NO AND NAME 738 USAF BALLISTIC SYSTEMS DIV.	RUN TYPE UD
TIME DATA AS OF 25 JUL 62	PROG NO AND DES 7385 WEAPON SYSTEM XXX	RUN ID 2
COST DATA AS OF 29 JUL 62	PREPARED BY - VILLORIA	END EVENT 8530400
RESP ORG FOR PROG DEPT 07		
SCOPE OF NETWORK - PROGRAM		

TASK   NO 03   HOUSING ASSY

RESP ORG 07-90

EVENT NO	TYPE *	RESPON ORGAN	COMPLETION EXPECT	COMPLETION LATEST	DATES PLANNED	DATES ACTUAL	SLACK WEEKS	PROB MEET PLAN	EVENT DESCRIPTION
0002		07-90		30 JUN 62	23 JUN 62	16 JUN 62	+2.0		COMPLETE WORK AUTHORIZATIONS
0003	MM	07-90		30 JUN 62	23 JUN 62	16 JUN 62	+2.0		START HOUSING ASSEMBLY FABRICATION
0006	MM	07-90		30 JUN 62	23 JUN 62	30 JUN 62	+.0		START HOUSING DESIGN
0009		07-90		21 JUL 62		14 JUL 62	+1.0		APPROVE HOUSING DESIGN
0072		07-90		04 AUG 62		28 JUL 62	+1.0		START PROCUREMENT OF COMPONENTS
0015	PM	07-90		11 AUG 62	04 AUG 62	11 AUG 62	+.0		ISSUE FINAL SPECIFICATIONS
0021	PM	07-90		18 AUG 62	11 AUG 62	18 AUG 62	+.0		START FAB FIRST ITEM
0030		07-90		18 AUG 62		25 AUG 62	-1.0		PREPARE HANDBOOKS
0055	PM	07-90		01 SEP 62	25 AUG 62	01 SEP 62	+.0		DESIGN HOUSING ELEC CIRCUITRY
0101	PM	07-90	08 DEC 62	01 DEC 62	24 NOV 62		-1.0		ASSEMBLE COMPONENTS OF SCANNER
0109	PM	07-90	15 DEC 62	08 DEC 62			-1.0		CONDUCT PRELIMINARY INSPECTION ROUTINE
0176		07-90	22 DEC 62	15 DEC 62	08 DEC 62		-1.0		START FINAL TESTING SCANNER
0201	PM	07-90-	29 DEC 62	29 DEC 62			+.0		INTEGRATE HOUSING DESIGN
0225	PM	07-90	09 FEB 63	19 JAN 63	05 JAN 63		-3.0		START FINAL TESTING
0275	PM	07-90	13 APR 63	23 MAR 63	09 MAR 63		-3.0		INTEGRATE REWORK
0301	PM	07-90	04 MAY 63	13 APR 63	30 MAR 63		-3.0		FREEZE CONFIGURATION
0361	PM	07-90-	04 MAY 63	04 MAY 63			+.0		DELIVER PRELIMINARY TECH DATA
0395	PM	07-90	25 MAY 63	04 MAY 63	04 MAY 63		-3.0		PACKAGE AND SHIP HOUSING

FIGURE 7.   Ouput report—time outlook—event-oriented.

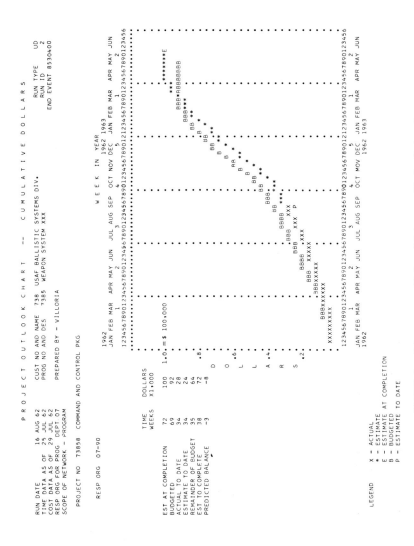

FIGURE 8.   Output report with financial data graphically displayed.

PROJECT MANPOWER CHART

```
RUN DATE            04 OCT          CUST NO AND NAME 670    NASA GODDARD SPACE FLIGT CENTER
TIME DATA AS OF    30 SEP 62        PROG NO AND DES  6700   SYNCOM ADVANCED TECHNICAL DEVELOPMENT     RUN TYPE     NN
CUST DATA AS OF    30 SEP 62                                                                          RUN ID   000001
RESP ORG FOR PROD 22-00-00          PREPARED BY CELLI ACONE GRICE                                  END EVENT 0000172
SCOPE OF NETWORK - PROGRAM

 PROJECT NO 67000   SYNCOM ADVANCED TECHNICAL DEVELOPMENT

   RESP ORG  22-80-00

  MAXIMUM NO MEN PLANNED =  240 = 1.0

ORGANIZATIONS CURRENTLY CHARGING

ORGAN            NO. MEN
21-11-00            19
21-12-00            41
21-14-00            12
21-22-00             3
44-30-00             1
44-41-00            31
50-12-00            23
50-47-00            14
```

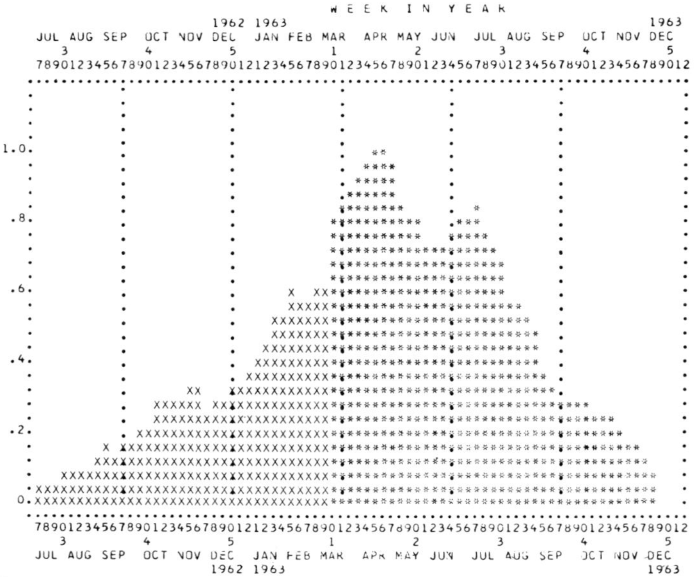

LEGEND - ACTUAL XXXX   ESTIMATE ****

FIGURE 9. Output report with manpower data graphically displayed.

requirements to complete the project, actual manpower used, and the allocated manpower budget, all as functions of time.

A manpower outlook chart can provide the manager with information which enables him to identify the following problems:

(a) *The labor overrun or underrun to date,* found by comparing the budgeted manpower allocation with actual manpower used to perform the work.

(b) *The forecast of labor overrun or underrun at completion,* which is obtained by comparing the manpower budget with the actual manpower used plus the latest estimate of the manpower required to complete the whole project.

(c) *The amount of schedule slippage,* which is indicated by the difference between the established schedule and the present expected completion date.

Visibility is provided in a top-down approach which allows the manager to identify a specific problem on the network rapidly.

A manloading chart (Figure 9) complements the manpower outlook chart by allowing the managers to *compare manpower allocation capacity* (number of men available for assignment) with the manpower requirements of the plan, and by giving *insight into the organization charging* the project and the equivalent number of men "on board" from each organization.

SUMMARY

In this chapter we have introduced consideration of the design and application of the network-based project management system. We have stressed objective analysis of the constraints of the environment in which the system must function, and those relating to the peculiarities of any given application. We have emphasized the need for careful attention to the acquisition of input data; format, structure, content of system outputs; and the information flow in the organization structure. Finally, we have described a model network-based time/cost/manpower system which illustrates the design criteria we consider most important to successful system design and implementation. Our example also illustrates how such a system might operate to control project cost and scheduling. Appendix F presents a summary of the current detailed report formats specified for the Department of Defense (DOD) and National Aeronautics and Space Administration (NASA) PERT Cost System. Review of that appendix will reveal how the DOD/NASA system presents the data we have been discussing.

PART II

Implementing the system

8

Organizing for the integrated system

Many of the difficulties in applying the principles of the integrated, network-based time/cost/manpower system arise over organizational structure. Unless the integrated system is carefully designed to complement the affected organizations and mesh with related information systems, serious problems may show up. Inappropriate persons may gain responsibility, personnel may resist the system, organizations may become involved in unhealthy competition and/or duplication, and the potential effectiveness of the system may be greatly reduced or even destroyed. In this chapter we will analyze the elements to be organized, discuss a major characteristic of the organization problem, trace the evolution of a typical system, and discuss required policies and procedures. Finally we will examine the requirements for personnel who are assigned to design, implement, and operate the system.

ANALYSIS OF ELEMENTS TO BE ORGANIZED

System Objectives. The planner must continually keep the basic objectives of the integrated time/cost/manpower system in mind: the need to provide timely, accurate, meaningful, integrated information relating to schedules, labor and dollar estimates, and expenditure, and the need to supply this information to responsible managers at several levels, to aid in decision-making and action. The system is not intended to produce auditable accounting information, nor does it have the capacity to produce *all* the information needed by any manager.

Functions to be Performed. The first step the planner must take in organizing the company for use of the integrated system is to identify the various functions which must be performed to achieve the system objectives. These are:

(1) *System design, development, improvement and extension:* these

include setting up the initial system and developing and improving it to provide additional capabilities, experimental application to new areas, and so on.

(2) *Indoctrination and training:* all persons involved directly with the system or using its results require indoctrination and training, usually a continuing effort.

(3) *Implementation:* during this process the planner establishes acceptable plans (project work breakdown, network plans, labor and dollar estimates, etc.), prepares input data, obtains satisfactory system outputs, and interprets the system results.

(4) *Operation:* this begins as soon as implementation is complete; the function includes:

> (a) Updating the plans (work breakdown, network, labor and dollar estimate).
> (b) Reporting progress against the network activities and the labor and dollar estimates.
> (c) Preparing input data.
> (d) Obtaining computer processing or performing manual processing.
> (e) Analyzing results.
> (f) Reporting to the managers and the customer.

(5) *Electronic data processing:* this is the actual computer processing of information, including the key punching and editing of input cards involved in machine preparation of input data.

(6) *Management decision-making and direction:* this is directly related to the previously listed functions which are more integral to the network-based management information system. This function is included because it is the primary reason for the existence of the other functions.

System design, indoctrination and training, implementation, operation, and data processing are the functions which, when properly performed, present integrated, useful results to managers to assist in decision-making and control.

Benefits of Proper Organization. If the planner organizes the functions and assigns responsibilities properly, he will benefit in these important ways:

(a) Responsible managers will accept and use the system and its results.

(b) The company will not have to tolerate duplicate and competitive systems.

(c) The required amount of standardization will be possible without overregulation.

(d) All users will have the best procedures, methods, and system advances available.

(e) Skilled manpower at all levels of system design and operation will be used most efficiently.

Need for Advance Planning. Although it may seem obvious to some that advance planning and thoughtful attention to the organization problems are necessary to achieve these benefits, experience indicates that such planning and organizing is rare. In many companies and agencies, network-based system effort springs up almost at random, and the problems of organizing the effort are approached on a hit-or-miss basis. Once functions are entrenched in various parts of a company, it is quite difficult to restructure the responsibilities and create a new, more effective organization. There is a great need for advance planning in this area.

The Elements to be Organized. People are the primary elements to be organized. The planner organizes them by deciding who and what the person is to be responsible for, and what authority he is to have over other people, funds, and policies. *Policies* must also be organized, since it is through policies that individuals are empowered to act. A third element which must be blended into the effort consists of the *methods and procedures* to be used in the system. The *people* use *methods and procedures* to carry out the *policies* which have been established for use of the integrated system.

PROJECT VERSUS FUNCTIONAL ORGANIZATION

In contemporary advanced-systems projects, many technical disciplines must interact; this produces new communication problems. A longer time is now needed for product development because of continual pressure at the threshold of technological progress, and so the target dates for delivery of operational hardware become more and more elusive. Accomplishment of project objectives within specific schedules has become more based upon the manager's ability to organize the job than upon his ability to meet specific technical objectives. In the wake of cold war needs technological advance has been so rapid that traditional organizational disciplines both in government and industry are unrelentingly pressured to change. Technical complexities have increased the need to analyze the impact of any move on program objectives *before* acting in any part of the organization. This need to assess impact requires multilevel analysis and approval prior to each action. So, as communication becomes more difficult, the need to integrate effort and objectives becomes an even more complex problem. Various organizational and management concepts, differing from traditional ideas both in form and in detail, have been and are being developed to meet these needs. One form of organization is most used to accomplish the type of major programs responsive to network-based systems: the project manager or projectized approach. When an organization is structured on a project basis, each project can be virtually self-sufficient, because all its functions are grouped under a single management.

There was no particularly compelling need for the projectized approach as long as an organization was able to concentrate management skills in the product-line type of organization on a few, reasonably small projects. In this traditional management environment, the typical organization structure has functional line managers complemented by a specialized staff. The staff provides the technical and managerial specialization needed to carry on specific functions, and the line organizations coordinate and integrate the whole and perform productive work. As it becomes increasingly necessary for organizations to integrate planning and control on a number of major project efforts, this traditional approach proves considerably less effective; by contrast, the projectized approach possesses significant advantages. Among other things, the projectized approach provides a strong source of authority in the form of the project manager; this makes it easier to integrate project functions efficiently. Project participants work in a framework of common identification and objectives, and the resultant balanced level of effort minimizes the potential conflicts in priorities, objectives, loyalties, etc., which are inherent in the functionally oriented organization.[1]

In an attempt to resolve the real and potential conflicts between the alternatives of strong project control exemplified by the completely projectized approach and the efficiencies of functional organization, companies often employ a compromise form: the "matrix" organization. Basically, the matrix organization provides well-established departments with special skills and capabilities which can be used to perform specific functions on a variety of projects. The projects flow across this essentially functional complex, receiving the services of these specialized departments.[2] Usually, there is a project office to coordinate the efforts of many individuals in various departments on a given project. Although it

---

[1]Fremont Kast and James Rosenzweig, "Science, Technology, and Management," *Proceedings of the National Advanced Technology Management Conference, Seattle, Washington, September 4–7, 1962,* McGraw-Hill Book Co., 1963, pp. 309, 310.
[2]*Ibid.,* p. 312.

is superior to the strictly functional organization, the matrix form has similar inherent weaknesses.

The functional, or matrix organization is often retained for a number of good reasons, however. One, is that the matrix organization is able to retain functional skills. Skills and knowledge gained on one program can be used on all the projects of the organization.[3] When managers are planning a project within an existing organization, they must recognize that although the needs of a given project may be compelling, the project objectives must be evaluated in relation to long-term organization objectives, because the two may conflict. If preplanning is sensible and realistic it will accommodate the existing organization's long-range objectives. Ultimately, as our industrial technology continues to grow in size and sophistication, however, management will have to confront the problems of organizing in an age of massive engineering, and find answers to questions such as those asked by Paul O. Gaddis of the Westinghouse Electric Corporation:

How can we fit complex new products and processes, involving vast capital costs, but also vast potential markets, into existing corporate structures which were never designed to receive them? What principles will make corporate structures flexible and dynamic enough to provide for a large-scale, forced product turnover?[4]

These questions are complex, and the answers are not likely to come easily. Therefore, until quantitative solutions are offered, management must aim at a workable compromise embodying the advantages of the existing matrix structure within the framework of a projectized approach.

Therefore, under present conditions we take two important steps to develop the project organization structure:

(1) We isolate and identify the divisions, departments, groups, or individuals responsible for each element of the Work Breakdown Structure.

(2) We correlate these with the Work Breakdown Structure by adding an organization designation to each element of the structure.

Network-based systems can make matrix organization much more effective. Chapter 2 shows how the project organization structure might be identified in an existing organization structure which uses the functional matrix approach. In the example shown in Figure 4, Chapter 2, a project office has been created to isolate the organization required by the contract within the existing functional organization. This project office will communicate with the customer organization, and direct the

---

[3] *Ibid.*, p. 313.

[4] Paul O. Gaddis, "The Age of Massive Engineering," *Harvard Business Review*, January-February, 1961.

company's project management and technical efforts. The office identifies the functional units which will contribute skills and resources to the project, and assigns them to the project. The planner decides which functions are to be performed within the project office, and which within functional units in the organization, by determining how the alternate resulting total project organization structures will affect project objectives and customer requirements. For example, project-oriented functions such as project administration, systems engineering, or system integration, would be project office responsibilities. Functions which are best executed by existing units because they require certain facilities or other factors (*e.g.*, laboratory, testing, or manufacturing operations), would be designated as functional responsibilities. In the functional areas the project office is supported by the functional organization. During the project people may actually be moved to a single location to facilitate integration of effort; some provision is usually made to transfer such people back to their functional areas once the project is completed, however. In such instances, the functional area is the "home base" for the functional resources diverted to project use, although the project authority directly manages the effort.

On a particularly large program all of the unique functional sections, as well as the contractual and accounting sections, may have to be directly responsible to project management, so that project objectives and/or customer requirements are met. In these cases, the only practical solution may be to establish a separate, autonomous division for the new program. Many persons consider this totally projectized approach ideal; in fact, it has been used quite frequently in organizations contributing to the nation's major weapon system and space programs. This approach does have many advantages for the large-scale, complex program. It minimizes potential interorganizational conflicts, and provides the framework for unambiguous authority and responsibility. In addition, assigning all functions to the project makes it easier to integrate goals and objectives to work towards a common purpose. Another less obvious advantage reported by successful project managers is the *esprit de corps* fostered by the projectized approach, a factor which can significantly contribute to the success of a project.[5]

There is another solution to the problems involved in trying to correlate the functional and project types of organization on a given project:

---

[5]When Vice Admiral W. F. Rayborn, Jr., Deputy Chief of Naval Operations, U.S. Navy, and prime mover behind the eminently successful POLARIS program, spoke before the National Advanced Technology Conference in 1962, he placed great emphasis on this team spirit as a prime factor in the early success of that program. Other managers attending the conference frequently made similar observations.

the planner can change the type of organization during the various phases of the project. In the program definition stage the matrix form which combines functional and project organizations may enable the planner to use the best talents in the organization. Once the program definition phase is past, the projectized approach may be the most efficient one to use to achieve the defined project objectives. Finally, as the project phases out, the company may gradually shift back to the matrix, and then to the functional organization.

EVOLUTION OF THE SYSTEM ORGANIZATION

If the planner wants to introduce the integrated system into the existing organization, he must start *somewhere*. Usually, a single person is assigned to investigate and then recommend whether the organization should adopt the system; if he recommends using it, he is asked to specify the method of implementing it. If a contract requires that the system be used, however, the company must only determine how it will be done. Experience shows that it is unrealistic to establish a rigid organization, expect that it will remain unchanged as the project progresses; instead, the plans should anticipate the necessary evolution of the organization. When a company decides to apply the system, it usually selects a particular project as the pilot application. The company may choose to use an existing system design, or it may decide to develop a proprietary system. When the system has been tested on laboratory data, the company will designate one or more persons as the implementation team. (Later in this chapter we will discuss this team.) The team then implements and operates the system set up for the project according to the methods described in Chapter 7, performing the functions described in this chapter. When this first system is running smoothly, the implementation team may turn it over to an operating team, so that they can begin a second project. In the following paragraphs we will discuss who should be on the team during the design and development, implementation, and operation phases, where the team members should be located in the organization, and how the various functions should be divided among the team members.

The System Design Team. The time/cost/manpower system requires people with knowledge, skills, and experience in the following organizational areas:

(a) Program/project management.
(b) Line (functional department) management.
(c) Contract administration.

(d) Controller/accounting.
(e) Business planning, scheduling, estimating, pricing and proposal preparation.
(f) Data processing.

By assembling a team representing these parts of the organization, the planners can guarantee that the system design characteristics, policies, and procedures will be acceptable to all or that the areas of disagreement and the reasons for them will be apparent, and so that upper management can help to resolve the disagreement.

The Implementation Team. The ideal situation, in which a fully qualified and experienced team designs the system, is rarely encountered. Usually, the company gives the implementation team an essentially complete, operational system for immediate application. In either the ideal or the usual case, the implementation team should include persons with the skills and experience listed in the foregoing paragraph; only the cooperative effort of persons who are aware of the problems, methods, and capabilities of the affected parts of the organization can produce an effective application of the system.

Tailoring the System. The implementation team is usually asked to use a system which is at least somewhat generalized. It must be adapted to the project at hand and to the needs of the project management. The major task the implementation team faces is adapting both the network system and the existing organizational structures to achieve an effective, economical application. Most authorities believe it is better to attain use of 80 percent of the model system capabilities in a manner which guarantees that the results will be *used* for decision-making, than to insist on attaining 100 percent of the capabilities if attaining the last 20 percent results in an exorbitant cost or undue complication. In the latter case, the system would probably not even be fully accepted by all concerned.

The Operating Team. At some stage in the implementation of the system we can identify a routine operating cycle. This would be possible when the system is functioning smoothly to produce useful information, when the procedures have been well established, and when the diverse skills of the implementation team are no longer fully required. At this point one or more individuals can be designated as the operating team; they would be responsible for keeping the plans current, updating the information, obtaining and analyzing results, and so on. The operating team may be drawn from the implementation team, or may be selected from the group of persons who are already involved in planning, scheduling and control.

Staff or Line? Those who think in traditional terms about setting up an organization or finding a place for a new or revised function within an organization will immediately ask, "are these items put in a line or a staff position?" Paul O. Gaddis' statement is a particularly appropriate answer:

It is an understatement to say that the literature, the terminology, the beliefs, and the folklore about "line-staff relationships" from classic management are no longer significant under massive engineering. They are worse than insignificant—in many cases they are misleading and damaging. What is needed is a new understanding of authority and accountability in the organizational environment. And the important first step is to throw out the words "line" and "staff" and replace them with a meaningful terminology.[6]

Authority and control in the network-based system should be given to the individual responsible for overall business management during the design phase; during the implementation and operation phases, responsibility should shift to the program or project manager. Figure 1 gives an example from aerospace experience; it shows a valid organization of the central staff, operating team and data processing service groups. In the following paragraphs we will discuss the general concepts involved, which could be applied in many situations.

Evolution of PERT organizations. Experience shows that definite organizational changes will be required as the system becomes more widely used in the company. The following pattern of change usually shows up:

(1) The first step is to set up a small central staff (perhaps one man) to evaluate the system and recommend to management whether to try it out.

(2) Assuming that management decides on the system, the second step is to enlarge the central staff and to begin to indoctrinate and train necessary and interested people.

(3) The third step is to select a pilot project (perhaps one requiring PERT or CPM in the contract) and to begin to set up the system.

(4) After the system has operated successfully for a while, the fourth step is to split off an operating team to be placed under the project manager, giving him complete control and authority. (The team may be composed of persons who were previously on the project manager's staff, trained and assisted by the central staff.) Thereafter, additional operating teams are set up for other projects or programs.

Centralized versus Decentralized. The most suitable organization is

---

[6]Paul O. Gaddis, "The Age of Massive Engineering," *Harvard Business Review*, January-February, 1961, pg. 138.

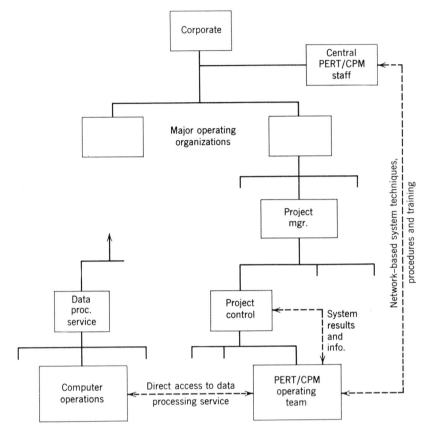

FIGURE 1. Locations of PERT/CPM central staff, operating teams, and data processing.

neither a wholly centralized nor a completely decentralized one; it is one in which there is a strong central authority for system mechanics, procedures, etc., combined with a strongly decentralized operation. In this way, the project manager is responsible for the data and information handled by the system for a given project, but he cannot change the basic system without consent of the central staff. Balancing centralized with decentralized functions properly is a challenge to any manager.

In a decentralized organization system people can be integrated into one organization with those actually performing the work. One advantage of this is that this effort can be charged directly to the project, instead of to an overhead account.

Size of the Central Staff and Operating Teams. The number of people assigned to the network planning and evaluation functions has varied so widely that we can state no general rule about size. In some cases, these techniques can be used effectively by assigning persons to perform some of the duties on a part-time basis. At the other extreme, we find groups of 50 or more persons assigned to PERT organizations which are attempting to apply network planning to large weapon system programs. When we compare the number of people with the results we can see that the smaller the central staff and the smaller the number of people assigned solely to network planning, the more effective the application will be. The responsible managers are forced to plan and evaluate when they do not have large staffs to do it for them. When these managers become directly involved, they develop accurate network plans which will be used for decision-making. Factors influencing organizational size include the scope of the application, the amount of detail in the network plans, the amount of involvement, interest and support the technical people responsible for the effort show, and the extent to which the network-based system is actually to be used for management, rather than for "window dressing," added to give the company a modern look or to comply with a contractual requirement.

The Network System Related to Total Planning and Evaluation Function. We cannot picture the effect of the system on organizational needs accurately unless we look at the total management information scheme, particularly the persons engaged in planning, scheduling, cost estimating, and progress evaluation. If management adds network planning without changing the existing planning procedures, then additional people will be required. However, if management applies the system properly and uses it to replace portions of the older planning and evaluation structures, experience shows that the same number of people or even fewer are required. If a new contract requires more extensive planning than has been done in the past, more people will be required.

Selection of Personnel. Persons with widely different backgrounds have proved very successful in network planning operations. Two characteristics seem essential: interest in planning work, and the ability to deal effectively with people at all levels. Because network systems are sometimes controversial and because they require technical people to give their attention to administrative jobs (which engineers and scientists traditionally dislike) a successful planner cannot add to these problems by creating animosity because of his personality.

The task of preparing valid network plans, or of catalyzing the responsible people into preparing them, calls for a person with the broadest possible knowledge of the work to be done, the organization structure, and the company's operating methods. For these reasons, persons who are experienced with the company, and who are technically qualified in at least some areas of the operation, are the most effective network planners. Persons experienced in scheduling and estimating can best determine specific input data, such as time estimates for the activities. A significant amount of pencil-pushing or clerical work must also be done to get the network data written on the data transmittal work sheets. It is often extremely valuable to have persons on the central staff and operating teams who are experienced with data processing operations and procedures; because they can help set up good procedures, and solve the difficulties which will arise when data are flowing to and from the computer.

System Policies. Before any major effort begins, the planners should issue management policies concerning the integrated project management information system. These statements should make it clear that top management understands and supports the system effort. If they do not have such support, the design, implementation and operating teams will find it very difficult to achieve their objectives. Policy statements should also indicate general organizational responsibility.

System Procedures. Network system implementation and operational procedures should be established for a given part of a company as early as possible, to make the system definite and provide material for indoctrination and training. The procedures are usually set down in a booklet containing a brief explanation of the system; an explanation of network preparation; and an explanation of input data forms preparation (specifically explaining the forms used by the organization); a section on obtaining computer runs; one on interpreting computer outputs; an explanation of the process of updating the network; one explaining how to update the input data forms; and one on management and customer report preparation. Among the publications listed in the selected bibliography in Appendix E are several which give detailed procedures.

SUMMARY

Chapter 8 has explored one of the prime causes of difficulty in using the integrated network-based system: the problem of fitting this system into the existing organization. Persons who design, implement, and operate the integrated system must not be drawn haphazardly from the organization. A proper balance between a central staff group responsible for system development, standardization and implementation assistance, and decentralized operating teams handling day-to-day system operation will produce the best results.

9

The role of the computer[*]

One of the most important characteristics of the network plan is that it can be adapted to electronic data processing. The logical structure of the network makes it possible to have the computer perform the time analysis calculations and relate cost and manpower data to the detailed plan, linking this management information to the calendar scale and predicting when actions must occur and resources must be available. In this chapter we will discuss the function of the computer in the network-based system, and describe the major considerations involved in deciding to use a computer and in selecting the computer and the appropriate network, cost, and manpower analysis program. We will also discuss some of the problems and difficulties involved in using the computer for this application.

There are two basic reasons to use large-scale data processing equipment in any situation. First, the computer may do the job less expensively (or in a shorter time, which results in less expense), and second, it may be the only possible means that can be used on a particular data processing job. In many situations human effort unaided by the computer is too slow and inaccurate to produce useful results, regardless of cost. A nationwide airline reservation system, and the tracking and control of a satellite, are two examples of such situations.

Most businesses decide to use computers because they reduce the cost of a job; in many instances, however, the really significant benefits are almost by-products which cannot be estimated in dollars and cents and so used to justify using the computer. For example, if a computerized network-based system gives managers information weeks before they could otherwise get it, decisions made could save the company many thousands of dollars. Early warning on problems can affect profits, but it

---

[*]The authors are indebted to Joel M. Prostick, Informatics, Inc., for his contributions to this chapter.

is quite difficult to measure the value of such early warnings before they occur.

Factors To Be Considered

Planners must consider several factors when deciding whether or not to use a computer for network analysis. In the following paragraphs we discuss each of these, and then consider similar factors for processing the data derived from the cost and manpower aspects of network-based systems.

Network Size. Although planners mention this criterion most when discussing whether a computer is needed, size must be considered in connection with several other factors. Experience indicates that it is difficult to perform the time analysis by hand on networks with more than a few hundred activities.

Complexity of Network. It is easy to process a large network made up of long independent strings or paths by hand, because the processor has fewer opportunities to make errors than he does in a network containing very involved interdependencies. The nature of the network is, of course, greatly influenced by the type of project being planned.

Complexity and Sophistication of Processing. If three time estimates are used and probabilities must be computed, the computer will justify itself more quickly than if single time estimates are used. Also, if the planners wish to calculate total, free, and independent slack, using manual procedures will be a great burden. When other factors are equal, networks possibly four or five times as large can be manually processed when only one time estimate is used and only total slack computations are desired.

If cost and manpower information is to be added to the network, manual methods are feasible only for the smallest networks. When time/cost trade-offs or manpower leveling is necessary, manual calculation becomes impractical for any network over one or two hundred activities.

Updating Frequency. If a network is to be analyzed only once and never updated to monitor progress, it can be manually analyzed even if it is large, much more easily than if it is to be updated frequently over a period of time. In some cases, preparing input data by hand for computer analysis may take more effort than the actual calculation. In such a case, it is unprofitable to prepare the input data for a single computer analysis.

Timeliness of Reports. After the initial file is established for computer analysis of a network, it can be updated and new reports can be produced in a matter of minutes, although typically hours, a day or two, or even a week or two might be required, because of administrative procedures.

Network analysis results must be made available to the manager in a timely fashion or they may be out of date when he receives them and so unused. Time-consuming manual methods compare unfavorably with the computer except on the smallest networks.

Formality and Format of Reports. If the network processing results are to be used by only one or two persons in an informal manner, the calculations can be placed on the network chart itself. Networks of several hundred activities can be maintained in this manner, without transcribing the data from the network. On the other hand, if the results are to go to many different people, computer processing with multiple copies eliminates the need for typing or transcribing the data. Also, results printed by computer are more accurate, since errors are bound to occur in transcription. (The computer editing and processing program will catch errors in computer input, however.) Computer outputs generally have an air of correctness and impartiality (justified or not) which does not attach itself to handwritten or typed reports.

Alternative Report Formats and Data Sequence. Data sorting is one of the basic uses of computers. A file of information can be rapidly rearranged in another sequence using a data field, a performing-organization code, or a combination of several data fields or codes instead of an event number. This capability is very useful, producing brief, meaningful reports of interest to particular individuals or organizations.

In the case of network processing, the degree of need for various sorts of the data could well be the crucial factor in determining whether or not to use a computer. The basic results from the simple network analysis routine can be rearranged in many ways: for example, when an activity report is sorted by total slack and expected start date, a list of activities in order of criticality, and in chronological sequence within each order of criticality, is produced; when a report is sorted by start or completion dates it presents the activities in the order they should be performed; and a functional sort, or grouping of activities according to responsibility code, provides each organization with a separate list of items concerning that organization only. Many networks can be processed on one master file and the results sorted by organization, to produce a report listing all activities, perhaps by start date, for all the networks separated into organizational categories. Examples of these various arrangements of data are shown later in this chapter and elsewhere in the book.

Processing of Cost and Manpower Data. If cost and manpower data are added to the system, it is necessary to use a computer for networks with more than one or two hundred activities. Manpower loading and dollar expenditure charts can be prepared manually for larger networks, but the computer is easily as economical as such manual operations. Of course, if actual expenditures are fed into the system and a computer is used for accounting, it is quite natural to use the same computer for network processing.

Benefits of Using a Computer

When considering whether to use a computer the planners must consider all the factors we have mentioned; they must also weigh the benefits of applying a computer to the network-based management system. At the beginning of this chapter we mentioned two reasons for using a computer: its capacities to cut costs and to perform otherwise impossible calculation, data manipulation, sorting, merging, or other data processing chores. The benefits of using the computer are discussed in the paragraphs following.

Speed. The computer performs calculations, compares numbers, letters, or words, sorts, merges, correlates and manipulates data and information many thousands of times faster than a person can, even if he uses the latest calculators or slide rules. The manager wants quick answers to questions about the vital plans and schedules in the network-based system. If he has to wait days or weeks for them, either he will not use the system or, if he does, he will find that the system does not fill his needs.

Economy. Beyond a certain volume of data, it is more economical to use a computer than to use manual processing. The cost saving grows rapidly as the volume increases beyond the break even point.

Accuracy. Computer calculation and data manipulation errors are almost nonexistent. The error rate of even the best trained clerk is a source of concern to planners dealing with management information which will be the basis of major decisions. Errors occur when data are prepared manually to be fed into the computer, but the editing and self-checking features built into good computer programs will locate most of these. Faulty but logical information will certainly produce erroneous results, though. Because the computer does not introduce errors in the data managers should not assume that all computer-produced information is correct. Planners must always analyze results, and institute stringent safeguards to assure the validity of input data.

Ability to Handle Large Volume. A great advantage of the computer is its ability to handle large volumes of data economically, quickly, and accurately. This ability makes it possible to maintain very large files of data, to revise the files at will, and to pull out particular information as needed for processing, calculations, and reporting. Networks of up to 75,000 activities can be handled by large computers, and some programs in use can accommodate networks of unlimited sizes.

Partial Interpretation. The computer user can program the machine to apply any number of rules to the information, to at least partially interpret the results and thereby to save many hours of human effort. Examples of such interpretation in the network-based system include:

Sorting activities by slack value (degree of criticality).

Comparing expected or latest allowable dates with the analysis date to identify overdue activities.

Identifying any activity or work package with a cost over a set figure (e.g., $10,000) which has less than 5 days of slack on its most critical activity (if more than one activity is involved).

Printing bar charts and graphs to interpret the digital information in an analog form for quick analysis by the user.

Many other examples can also be cited.

Legible, Uniform Results. When the network system is used to manage a number of projects, the legible and uniform computer reports are advantageous, because they are understandable to anyone in the company, no matter which department is responsible for the project. Too frequently, reports with different formats are prepared by different parts of the organization, making it difficult to compare and correlate the information for top management use. A computer with a well-designed program matching the network-based system introduces discipline and standardization which are quite beneficial to many organizations. The planner must be careful not to use too rigid a standardization, however; variations needed to accommodate real differences in conditions must be allowed.

Problems in Using a Computer

Unfortunately, to enjoy the benefits of the computer the planners must invariably exert more effort than would at first seem necessary. To their dismay, many managers have found that the great expectations the computer salesman encouraged them to develop are painfully slow in becoming realities. Using the computer on the network-based system is much simpler than converting an entire organization to electronic data processing; this is because there are so many kinds of computers and network processing programs, and because many different types have been widely used. Even so, there are several problems often encountered when using the computers to set up an effectively operating system. These are discussed below.

The Aura of Mystery. Many persons, including many managers, feel that a vague aura of mystery surrounds the computer and related electronic data processing hardware and software. Unfortunately, some computer specialists have fostered this feeling. Because they find computer technology mystifying and because they do not know what the computer can and cannot do, managers frequently expect more from the computer than it is possible to get; when they actually use computers, therefore, they are bound to be disappointed.

Discipline and Accuracy. Noncomputer systems such as manually prepared payrolls or manually calculated networks leave lots of room to ignore due dates, fill in forms incorrectly, enter inaccurate data, and so on. If a number is not entered on a form in the manual system the person performing the calculations can fill it in, call someone to get the information, or skip over it and come back later. It is often hard for people to become accustomed to the discipline imposed by the use of a computer: all required data *must* be provided, and must be entered precisely and properly. The computer will detect errors in input data and return them for correction, but this delays the final result.

Administrative Problems. When the planner has prepared the network plan with related time, cost, and manpower information, he is usually very interested in seeing the results of his plan as quickly as possible. He will be rather disenchanted if he finds that he must wait a week or two before seeing an analysis of his plan. Unfortunately, this often happens. Some time is required to transcribe the data from the network to input forms or transmittal sheets for key punching input cards; additional time is required to punch and verify the cards. (Typically, cards can be punched at the rate of 100 to 200 per hour, although this rate varies considerably with the nature of the data, the legibility of the input sheets, etc.) Even when cards are prepared quickly long administrative procedures often delay the actual computer run. Even within one organization, many different people are usually waiting to use the same computer for different purposes, which leads to priority problems: is the payroll more important than the schedule analysis? Too frequently a small bureaucracy grows up around the computer, calling for forms to be filled out, approvals obtained, and so on. Finally, when the data are ready to be fed to the computer, the person working on the project is grudgingly allowed to watch through a window. If an error is found in the data, he is sent an error report; and, even though it may only take five minutes to correct the error, he must then drop back to the end of the line and wait another 12 hours for his next try at the computer. This time-consuming routine can nullify much of the advantage gained from the speed of the computer. Fortunately, new equipment and new time-sharing techniques being developed allow many persons to use the computer simultaneously, providing an answer to this problem.

Access to the Proper Computer with the Proper Program. For any system or application, certain computer types have distinct advantages over others. Therefore, because different applications have different requirements it is impossible to have access to the most desirable type of computer for all purposes. In addition, the best available computer may not have a program available for the application at hand. Any person charged with selecting a computer configuration for an organization faces this dilemma. Such difficulties also confront the potential user who has a particular application, such as the network-based system, in mind; he must locate the best available computer which has a program available to do the job he has to do.

Cost of Developing New or Modifying Programs. The computer must be told in infinite, painstaking detail exactly how to process the information. Developing a new set of instructions, or program, is more costly than most persons unfamiliar with computers realize. Modifying an existing program is also very difficult and expensive. This expense tends to make computer use rather inflexible. You either use existing programs and bend your methods and procedures to conform to them, or you invest a considerable amount of time and money to develop new or modified programs.

Difficulties in Using a Poorly Designed Program. A poorly designed program can be very difficult to use. Poorly laid out input forms, inadequate or misleading error messages, faulty editing procedures, and other items can be very frustrating. Most deficiencies of this type are caused by the programmer setting up the program for his own convenience, not for the user's. Although an experienced user can tell how difficult a given program will be to use by looking over the documentation, only actual experience with the program will reveal many of its faults.

Errors or omissions in the input data produce one kind of problem in this area. A program can be set to make certain assumptions when an error is encountered, to print out the appropriate assumption, and then to proceed with the processing. Too frequently, however, the programmer does not know enough about the application to program such assumptions, so that when such an error shows up he merely prints the error and stops the job. If a series of errors halt the processing, each causing several hours' or a day's delay, the advantages of using a computer quickly disappear.

Another prime source of trouble is inadequate understanding of the various limitations on the computer program; this leads to mistakes in use. For example, if one exceeds the number of items of data allowed in a certain table without receiving proper error messages, it is extremely difficult to figure out why the computer has stopped.

Selecting a Computer and a Program

If we were considering computer selection on the basis of suitability for an entire company, we would weigh certain factors we have not discussed. We are considering selection solely in terms of advantages for the network-based management information systems, however, and so we will discuss only the way a user or potential user might evaluate an existing program for a given computer. (Actually, new programs which will handle the requirements of the network-based system [with various limitations] can be prepared for computers of almost any size, but development of new programs is not within the scope of our present text.) Computer and program selection cannot be treated separately, and so we will discuss the major factors in terms of program characteristics, limitations, and restrictions because they reflect the characteristics and capabilities of the computer for which the program is written.

Computer Programs for Network Analysis

We will first consider the program characteristics which affect the network analysis phase of the network-based system. Later in the chapter we will consider aspects related to cost and manpower plans.

Network Size. Virtually every computer network analysis program limits the number of activities which can be treated as an integrated network. The limit may also be expressed in terms of number of events, or number of activities plus events. Some programs also restrict the largest event number to be used. These limits reflect the size of the computer memory and the calculation procedures used in the program. Network size limits vary from about 200 activities on small computers to 75,000 on the largest commercially used machines. At least one program for a large machine is said to handle networks of "unlimited" size. Occasionally, computer users will encounter other limits, such as the maximum "thickness" of the network, or a maximum number of simultaneous or parallel activities. This last type of limit is often not clearly explained in the operating instructions, and can be completely mystifying to the user who encounters it unknowingly, especially if the program does not provide an appropriate error message.

Random versus Sequential Event Numbering. Knowing whether the program allows random event numbering or requires sequential or i less than j numbering is important. The program planner would not want to use completely random numbers on the network (as discussed in Chapter 4), but having to use sequential or i less than j numbering can be very inconvenient, particularly when the network must be revised. Random numbering allows the user to add sections, delete activities, and so on, without renumbering the entire network, and to use significant coding patterns for the event numbers, if he desires; a program which allows random event numbers is therefore more useful than one which does not. Random numbering requires a larger computer and a more complex program, however. Although most available programs allow random numbering, some do not, particularly on small computers.

The computer user should also know whether or not the program allows him to use letters as well as numbers in the event number field, because this additional coding flexibility is quite useful under some conditions.

Multiple Start and End Events. Many programs limit the number of start and end events, usually requiring one start and one end event. This limitation can cause problems if a project has multiple end objectives. A common solution to this problem is to tie all end events to a single point; this practice will produce erroneous latest allowable dates and slack values, however, unless all project objectives have the same scheduled date, or the computer program is able to accept scheduled dates which override the latest allowable dates. It is therefore more desirable for the program to be able to handle multiple start and end events.

Event as well as Activity Input and Output. To completely define a network on any network analysis program information on all the activities must be included in the input. Input data on events are not required nor even useful in the network calculations, however, although the event numbers for each activity must, of course, be included. Some programs do accept event input data and use them to carry supplemental information such as event description, milestone or interface codes, responsible organization, or scheduled or completion dates. The coding and descriptive information is then used for event reports, and the scheduled event dates are used either to compare against or to override earliest or latest dates calculated for related activities. If the computer program can accept an event completion date it is much easier to keep track of the status of a network, because when an event is reported as complete it implies that all preceding activities are complete.

Using event completion data to determine the status of the project is one of the many time-saving steps the computer user can take; the system designer or programmer will often prevent the manager from using these, however, ostensibly because "we should force the user to report each activity because he needs this discipline and cannot be trusted to remember to check all activities if we make it easy for him to report event completions." Behind this is the fact that using event completions would require the designer and the programmer to work harder and to produce more inventive programming.[1]

---

[1] Another convenience which can be included in the program is the automatic completion of dummies (constraining, zero-time activities). The user will be annoyed if he is asked to fill in completion dates on all dummies when updating a network, and so the program can easily be designed to complete a given dummy automatically as soon as all preceding real activities have been reported complete. This convenience is not included in many presently available programs, however.

The computer can produce output reports presented for either activity or event records, or for both. Descriptive event input enables more useful and understandable event reports to be prepared. Codes placed on the event record can be used to prepare milestone or interface reports and charts.

Schedule Data Input. Activity identification (beginning and ending event numbers) and duration are the only data necessary to define the network, to calculate earliest and latest dates and float, and to identify the critical path. It is necessary to use schedule dates for specific events or activities to depict external constraints and commitments which are not otherwise shown in the network itself, however. Scheduled dates can be used either to compare with calculated ones or to override calculated dates and thereby affect subsequent calculations and results. Some computer programs allow the user to specify if a scheduled date is to be printed on a given report to compare with the calculated earliest expected and latest allowable dates, used in place of the earliest expected start date, or used in place of the latest allowable completion date for a given activity or event.

An outside constraint is a schedule date which replaces the calculated earliest expected start date of an event or activity. It may do so only if it is later than the calculated date, or if the user decides it should be allowed to force an activity to start earlier than calculated, resulting in negative slack. The most important outside constraint is the network start date, which is used to convert calculated network times to calendar dates.

An outside commitment, on the other hand, is a schedule date which will replace the calculated latest allowable date for an event or activity. The user can also decide if this date is to override the calculated latest allowable date if it is earlier (thereby producing negative slack), or whether it is simply to be used for comparison. At least one computer program can override latest allowable dates with schedule dates to produce both the normal set of dates and slack, ignoring the schedule dates, and produce what is called "secondary slack," or slack compared to the imposed schedule dates. This capability is quite useful for some applications, and is programmed into the IBM PERT/COST System available on several IBM computers. A computer program which can impose scheduled dates including multiple start or end events at any point in the network, and can, if desired, override either earliest or latest allowable dates, is very advantageous.

Calendar Conversion. It is most important to be able to submit the activity data to the computer in working days or weeks (assuming a 5-day week, for example), and to have the computer convert all calculated times to the calendar scale, using a specified network start date. However, not all present computer programs can do this, and those that do vary in practicality and flexibility. Some programs have a fixed calendar with established holidays and a 5-day work week, which the user cannot modify. Desirable features in order of priority are:

(1) Control over holidays in the year.
(2) Ability to use variable time units for activity durations:
    (a) hours.
    (b) shifts.
    (c) days.
    (d) weeks.

(3) Ability to vary the number of time units per calendar week (for example, a concrete curing process may take 6 days, but it will continue over a weekend; certain types of jobs will be placed on a 6-day week basis, using overtime labor).

(4) An "8 to 5" schedule rather than a "noon-to-noon" schedule.

Most programs use a "noon-to-noon" schedule which assumes that a one-day activity starts at noon on the first day and ends at noon the next day; the next activity is then scheduled to start at noon on the second day. An "8 to 5" schedule says that a one-day activity starts at 8 and ends at 5 the same day, however; the next activity then starts at 8 the next day. In reality almost any project is scheduled this latter way; so, although the discrepancy is apparent when many very short activities are involved, using a noon-to-noon schedule is confusing and irritating to a manager who does not want to "do things the computer's way"—which in actuality is the computer programmer's way.

Method of Handling Completions. There are several methods of handling activity completions on different programs. The most desirable method uses the last completion date entered on a chain of activities and re-anchors the network to one or more such dates; then the new analysis is performed, showing the impact of actual reported progress on the future schedule. Another way to handle completion dates is to set the duration of completed activities as zero, and use the "run date" (data cutoff or status date) as the start date of the network; this pushes the start of all incomplete activities up to the run date, which may give erroneous results unless precautions are taken against it. To prevent such erroneous results the user must consider which activities are in progress on the run date and revise the time estimate to reflect the time remaining to complete each of these activities.

Each of these methods allows actual calendar dates to be submitted to the computer for specific activities. Less sophisticated programs which do not have calendar conversion routines make the user determine the actual duration of completed activities, revise the original duration if necessary, and reanalyze the network using the original network start date. It is then necessary to code the completed activities so they are easily identified, or to zero out the completed activities, as we described in the preceding paragraph.

The method of handling completions determines in great part the usefulness and practicality of a particular program for monitoring and control purposes.

Computer and Program Efficiency. The cost of running the program is very important to the user. Cost is affected both by the type of computer and by the efficiency of the program. Many of the desirable features discussed earlier increase the computer running time and therefore the cost. The factor which most affects cost is the number of different reports requested or produced by the program. A single schedule report costs from 2¢ to 7¢ per activity for machine time, on typical programs for various widely used computers; this price does not include the cost of labor, data preparation, etc. Re-sorting the data for different reports can consume a lot of computer time, depending on the efficiency of the sort program, which may or may not be integrated into the network analysis system.

The efficiency of network analysis programs varies widely. At least one has turned out to be so inefficient because of the design and the computer language, that few could afford to use it. Using a very large, late-model computer does not guarantee a more efficient or cheaper program than a smaller machine can produce. The only way to determine the cost of running a particular computer program is to test it over networks of varying size under the conditions which the user will actually encounter.

Variety in Types of Reports. Chapter 5 discussed briefly some of the basic ways of arranging network analysis results. The computer can rapidly sort and re-sort masses of data and select particular items to meet various criteria, as well as do calculations rapidly and accurately and convert to calendar dates. These computer capabilities let us obtain a wide variety of types of reports from the basic network information. In addition, the computer can present the data in both digital/word form and analog/graph form, through the use of high-speed printers and plotting devices. If the computer program has been so designed, tabular listings or digital/word reports can be prepared to arrange the network analysis results in one or more of the following ways:

(1) Chronologically, using earliest or latest start or completion dates or schedule dates for events or activities.

(2) By slack value or degree of schedule criticality.

(3) By specific code numbers:

    (a) Event number.

(b) Activity number.

(c) Work Breakdown Structure summary number.

(d) Organization number.

(e) Accounting system charge number, or work package number, if different.

(f) End-use code (subnet or fragnet identification).

(g) Milestone or interface event codes.

(4) By comparing with certain parameters supplied to the computer on a control card, such as:

(a) All activities with completion dates prior to the run date.

(b) All activities due to start between the run date and a given date in the future.

(c) All activities with negative slack, etc.

(5) Any logical combination of the above items.

A number of examples of various types of computer-produced reports appear in other chapters of this book and in Appendix F.

Graphic presentations (analog/graph form) are prepared either on the line printer or with a plotting machine which moves paper and pen to produce the desired graph. The cathode ray tube, essentially the same as a television tube, is also used to present results graphically: to obtain permanent output results the tube must be photographed, however. The line printer reproduces symbols, letters or numbers to form a pattern which, though not continuous, creates a useful picture. The basic chart formats produced by all three graphic methods are the network itself, usually on a fixed or at least a relative time scale; the bar chart display, showing the length of each activity or group of activities over a fixed time scale; and the milestone chart display, which shows the points in time at which an event is expected or scheduled to occur; and the rate or cumulative curves which plot resource expenditure data. When prepared by using the sorting and selecting factors described earlier, these graphs can be very effective for management display purposes. Several examples of graphic output displays appear in Chapter 7 and in Appendix F.

The development of more flexible and economical plotting devices is proceeding rapidly, but they are still too expensive to be substituted for manual or printer-prepared networks or graphs in most situations. The ability to prepare legible, time-based networks rapidly and inexpensively would be most important in bringing more widespread acceptance of network planning systems.

When selecting a particular computer and program, the user should particularly consider the variety of types of reports which each would make available to him.

Input Data Requirements and Editing. The user should study the input data transmittal or key punching forms for a particular program to determine precisely what information is required, and how difficult it will be to fill out the forms. Awkward input forms overshadow the other features of some programs; some require the user to enter the same data more than once on the form or forms, to use many cards for each activity (which increases the key punching cost), and to use elaborate or confusing codes which can be misunderstood by various users.

The ease with which data can be revised depends upon the way the revised data is to be entered. It is most desirable to be able to revise a given field of data (such as activity duration) without having to repeat all data related to the activity in question.

Editing the input data is extremely important in any computer program. Good editing will locate errors as early as possible, before expensive computer time is wasted in processing faulty data. Typical items checked in a network processing program are:

(1) Duplicate event numbers on one activity.
(2) Duplicate activities.
(3) Illogical dates, such as June 35, 1966 or 1 Jan 1866.
(4) Missing items of information, such as activity duration.
(5) Alphabetic information in a field which will always be numerical, such as duration, etc.

The program should edit *all* the data, not stopping when it encounters a particular error which will prevent network processing. Certain types of errors such as gaps in the network (which produce hanging end or start events) will prevent further processing, but other types will not necessarily do so. For example, if an activity is duplicated, either the first or second card can be used to process the network, with the duplicate record listed on the printer; the user can then decide whether to make a correction on a subsequent analysis run. (Such duplication can result from a key punch operator punching a second card to correct an error in the prior card, and then forgetting to discard the prior card.)

Network Integration. For certain situations the user may wish to select a program which can integrate various separately developed networks; this integration capability can be particularly important with a full-blown time/cost/manpower system. (The need for such network integration is discussed in Chapter 12.) This capacity is related to the overall maximum size of network which the program can handle. Any program can integrate networks until the maximum number of activities it can handle is reached, if the bookkeeping on event numbers is kept straight; however, certain programs designed to handle and integrate networks offer certain advantages. One of these, the IBM PERT/COST System, handles up to

100 subnets of not more than 750 activities each; however, this limit on any single subnet can itself be a disadvantage.

Computer Programs for Manpower and Cost Analysis

There are only a limited number of computer programs available for manpower and cost analysis and they have been used far less than the time-only network analysis and scheduling programs. We will discuss the factors to consider when evaluating a program for use with manpower and cost elements.

Manpower Analysis. There are several ways for the computer to handle manpower data. Typically, the number of men and their particular skill code are included in the activity record; then, when the network analysis has determined the earliest expected and latest allowable activity start and completion dates, the manpower requirements can be plotted on the calendar. By summing up all such requirements by day or week, the planner can forecast the total project manpower needs rather accurately. Computer programs differ in the number of skills they can accommodate for one activity (eight or more are fairly practical), and in the capability the user has to impose a resource limitation which may override the critical path calculations.

When the computer program can impose a resource limit, it is necessary to specify on control cards the maximum number of persons or hours of each particular skill which will be available. Some programs allow the user to vary these maxima for different periods of time. The computer then keeps score on each skill as the network is analyzed. When more men of a particular skill are required than the control card specified as the maximum for that skill, the program will delay the start of a new activity until some other activity is completed, freeing sufficient men of the particular skill. Often resource limits will extend the total length of the project beyond that of the critical path. In this case, the critical skill becomes the limiting factor in the project. An activity can be coded so that the computer can specify whether it can be interrupted or not; then, if a more critical activity requiring those resources is scheduled to start while the activity is in progress, the planner can reallocate resources quickly. Such a program is described in Chapter 14.

Other programs simply total the manpower requirements by time period, using either head-count or man-hours. This total is valuable to the user, since it gives him an accurate forecast of his resource requirements. Actually, manpower is simply a special case of resource allocation; dollars, equipment, or facilities can be treated in a similar manner. Most PERT/COST programs handle manpower data in this way.

Cost or Dollar Analysis. Dollars needed to complete an activity (usually called the cost of the activity) can also be related to the activity on the network, and subsequently scheduled over time. Programs differ in the number of types of costs that they can identify (such as estimate, budget, contract, loan value), and in their capacity to allow these costs to be entered at higher levels of the Work Breakdown Structure, rather than requiring that they be entered for each activity. The programs which make time/cost trade-off analyses are considered special applications, and were discussed in earlier chapters.

Generally, programs which accommodate cost data will analyze these data in one or more of the following ways, listed in order of increasing sophistication:

(1) Accumulate dollars of a given cost code, plus project totals.

(2) Accumulate dollars as in (1) but also display the totals by time period (day, week, month) for the past and future.

(3) Accumulate several types of costs (estimate, budget, contract) as in (1) or (2), compare these and show the differences (overruns, underruns) at points in time.

(4) Accumulate data as in (3) and also accept actual expenditure information (either tape to tape, directly from the accounting system, or as manual input) and compare the actual costs with estimates, budgets, loan values, etc.

(5) Sum up any of these types of information in various ways, such as:

    (a) Work Breakdown Structure (summary number) code.

    (b) Organization or subcontractor code.

    (c) Cost-accounting code.

    (d) Other special codes for user identification.

The user must evaluate a particular computer program in terms of its capability in these areas compared with the requirements of the application. The capacity to produce fairly accurate cash-flow analyses of direct costs is a very useful one for construction, research and development, refinery turnaround, and other applications.

PERT/COST Computer Programs

Although a number of computer programs are adapted to resource (manpower) analysis and lump sum cost totaling by activity, few can automate the integrated time/cost/manpower system described in earlier chapters. Two major programs which qualify as operating time/cost/manpower programs are available, however, and several more may be being used by individual aerospace industry contractors behind the proprietary screen. (Actually, these systems combine many individual computer programs in their total operation.) The two systems, which are fully documented and extensively used (or extensively attempted) are the U.S. Air Force PERT/COST System[2] and the IBM PERT/COST System.[3] Each system has the capabilities required to automate the major portion of the integrated time/cost/manpower management information system, although each has particular advantages and disadvantages. The interested reader should obtain information about each of these systems and evaluate them in relation to the various factors discussed earlier and the particular application at hand.

Problems in Using PERT/COST Programs. In addition to the problems discussed earlier in this chapter and elsewhere in the book concerning the development and use of the time/cost/manpower system, there is another very important problem involving the computer: how to tie in the existing cost-accounting system. The problem arises because the cost-accounting system is usually the only one using electronic data processing; this means that cost codes are well-established, usually with requirements which conflict with the network-based system needs; it also frequently means that the controller is the man responsible for all computer operations and programming.

All these potential problems can be resolved by qualified persons who desire to resolve them, if the individuals use the great capabilities of modern computers skillfully.

SUMMARY

In this chapter we have explored the role of the computer in the network-based system, emphasizing its function of not only performing the simple but time-consuming calculations of network analysis, but also performing the complex functions of sorting, re-sorting, comparing, selecting, and presenting particular management information in the form most useful to the manager. Factors to be considered in selecting computer programs for network analysis, as well as manpower and cost analysis were discussed, and problems arising from use of a computer were reviewed.

---

[2]*USAF-PERT* Volumes 1 through 5, Air Force Systems Command, U.S. Air Force, September 1963.
[3]*IBM PERT/COST II Reference Manual,* No. B20-6701, IBM Corp., Data Processing Division, White Plains, N.Y., 1964.

10

The cost versus the value of the network-based system

The basic measurement of the usefulness of anything is made by comparing its cost with its benefit or value. This comparison is valid for management systems, and particularly for the type of management information system which concerns us in this book. In this chapter we will discuss the sources and makeup of the cost of such systems and estimate the magnitude of the cost, to assist in justifying the development and use of the network-based system. Although it is difficult to measure benefit or savings which can be attributed directly to the system (especially in large research and development organizations), we can discuss the areas in which users have saved money.

COST FACTORS

Network-based system costs can be divided basically into labor costs and data processing costs. There are several types of labor costs, as we will show later. Computer or data processing costs can be replaced by labor costs if the analysis is done manually, but we will assume that electronic data processing is more economical than manual processing for most applications. We shall show that the ratio of labor to computer costs is generally about 10 to 1, and so a detailed analysis of the labor costs is of direct interest.

Degree of Acceptance versus Cost

The most important factor directly affecting the cost of installation and use of the network-based system is the degree to which personnel accept and use the concepts and the system itself. If individuals in the organization have adopted the system voluntarily, it is highly probable that the system will cost less than if it has been imposed by contractual requirement or by edict from on high. Imposed use results in higher costs because resistance and resentment lead people to fail to understand the system and its operation; imposed use also results in duplication of existing systems or procedures rather than integration and replacement of them, which inevitably adds expenses. (Chapter 17 discusses the lack of acceptance because of resistance to change and other factors.) These considerations often lead to deliberately expensive network-based system staffs and efforts, especially in the aerospace industry (or in construction work related to it), where the customer (the government) imposes the requirement and then pays the extra cost. Higher costs in many contractual situations mean greater overhead margins and, under some conditions, greater profits. Companies may also have hoped to make the system so expensive that the government would no longer require it. Further, as we remarked earlier, reluctance to discontinue or integrate existing systems creates unrealistic cost burdens which are misrepresented as the cost of the new system.

Those favoring use of the system recognized this problem and have tended over the past several years to disallow this type of direct cost and to force the companies to include it in overhead as a normal management cost. The claim that the government or customer is requiring much more detailed and expensive planning and scheduling than is necessary for the normal operation of the company may be somewhat justified; however, it appears that this has become another cost of doing business with government.

COST ANALYSIS APPROACH

In analyzing the network-based system costs, we will first identify the functions to be performed. We will then estimate the amount of labor and data processing expense for each function in a typical use, and finally estimate the total system costs.

Functions to Be Performed

In Chapter 8 we discussed the various functions involved in developing and applying the system in a given organization. These functions are system design and development, project planning, input data preparation, progress reporting, data processing, analysis, and management reporting.

System Design and Development

The cost of developing a system depends upon whether or not it is to include a computer program of any complexity. Development costs could be as low as a few thousand dollars for a noncomputerized approach primarily involving revision of existing procedures to accomplish the objectives of the network-based system. On the other hand, even an average computer program can cost $100,000 or more by the time it is designed, coded, tested, revised, and successfully used.

The reader should note, however, that a new system invariably is developed in an environment containing other systems of varying complexity and sophistication. Thus the apparent cost of a new system may vary greatly, depending upon how extensively and how long the old systems continue to operate concurrently with the new, duplicating them, how much the new systems can be evolved from the old, and other factors. In assessing the true costs of a new system, it is important to separate the inevitable inefficiencies of the transitional phase from the costs of undue clinging to old and duplicate systems.

In this chapter, we will not attempt to analyze the cost of developing a new system; we will instead concentrate on the operational costs of using an existing system and computer program tailored to fit a particular organization.

Project Planning

Any project, large or small, requires planning. The project must be planned, regardless of what system or procedure is used. The basic planning tasks listed by Baumgartner[1] are:

Task breakdown and definition.
Make or buy.
Scheduling.
Budgeting.
Reliability, performance, and quality.
Organization.
Facilities.
Subcontractor selection.
Fiscal planning.
Personnel.
Reporting.
Termination.
Stop work.
Expanding the contract base.

At present we are attempting to determine the cost of using the network-

---

[1] J. S. Baumgartner, *Project Management*, Richard D. Irwin, Howewood, Ill., 1963.

based system. This is a difficult job, because we must carefully sort out the planning tasks which relate to the capabilities and requirements of the network-based system from those which do not. Looking at the problem another way, we could try to determine the total cost of the network-based system, or the cost (if any) added to the usual planning cost from use of the system. We will take the latter approach here.

The tasks which involve the network-based system are work or task breakdown and definition, scheduling, budgeting of money, labor, and other resources, and fiscal planning. When we have described each of these functions, we will look at the magnitude of the effort each involves.

Input Data Preparation

Acquiring and preparing input data are tasks often overlooked when implementing new management systems; however, overlooking these important jobs often increases costs. Input data preparation includes manipulating both the information originally included in the operational system and the data added to update it as a part of necessary maintenance. The amount of attention the planner devotes to preparation of these data will have great impact on the cost of the system.

Experience shows that a new system such as the network-based one is usually introduced into an environment where planning has been inadequate in the past and where current reporting reflects this inadequacy. Therefore, the new system will often require more detailed and realistic planning than the company has used. In such cases, increased attention to planning, recording of plans in formats communicable to the system, and maintenance of these plans to assure the continuing validity of system output will add a clearly identifiable cost to the new system. The true cost of this increase in planning is difficult to assess, however. It is hard to determine how much of the apparent cost increase can be amortized in terms of reduced overruns and more efficient operations because of increased visibility, since this saving will vary from case to case; a planner can try to evaluate such costs, however, by rigorously examining current efforts to produce project visibility. Such a study may reveal instances in which management tried to gain visibility "after the fact" but did not achieve control. Such costs can be virtually eliminated by designing and implementing an effective system; they can therefore be subtracted from the apparent increase in data input costs. An example of this drawn from an actual case studied by the authors was a situation in which a company conducted an elaborate monthly cost-to-complete evaluation on each project in progress. This monthly exercise called for project, contract administration, accounting, and other operational personnel contributing to the program to spend many hours preparing reports. Organizations within the company were required to coordinate activities extensively to arrive at a consensus on project status, because there were no reliable plans to check progress against. In addition, much of the data produced was more subjective than objective, and so was unreliable and unpredictable from month to month.

Obviously, a network-based project information system would produce current cost-to-complete information inherent in the data output as a function of the system. Costly exercises such as the one just described would be entirely eliminated as personnel gained confidence in the system output. This is only one example of a method for assessing the true cost of data input. The reader can easily envision others in which it is possible to measure the improvement in nonproject-oriented operation information; these improvements are almost always made because management gained current and reliable project expenditure data on cash flow, capital expenditure planning, and personnel and machine scheduling.

Progress Reporting. If the managers were sure the work would progress according to plan, they would have to put forth little additional effort once a project had been planned and initiated. However, we all know that the only certain thing about the plan is that it will change as delays and unforeseen circumstances are encountered. It is therefore vital to report progress and compare it to the revised plan, so that managers can decide what changes (if any) should be made to maintain the present plan or modify it. Management must get progress reports whether or not the network-based system is used; our question is, does the network-based system increase or decrease the cost of this function? Management needs progress reports about physical accomplishment (completed actions), resource expenditure (money, labor, and materials), and technical achievement (quality results of completed actions). The network-based system can give information about the first two.

Data Processing

In its broadest sense, the term "data processing" includes any data manipulation: mathematical, logical, comparative, or, within limits, analytical. Network-based systems generally involve electronic data processing, although there are valid network-based systems which require nothing more than manual data processing. In either case the cost should be determined as precisely as possible. Electronic data processing costs are incurred for key punching and key verifying data submitted in written form onto cards; computer editing of data and error correction; computer analysis of data and report preparation; and printing, reproduction, binding, and distribution of output reports.

Analysis and Management Reporting

As we saw in Chapter 7, the system is not complete without proper analysis and display of results. Evaluating the computer reports, together with network plans and any other pertinent available information to decide on further action takes work. It calls for specialists to study and analyze the computer results and, in many situations, to prepare additional charts, graphs, or displays which will quickly communicate the meaning of the results to higher management. All this takes time and effort, which cost money. Therefore, the question again arises, does the network-based system make this function less expensive than some other system does?

Effects of Organization

We have previously discussed the problems involved in organizing the network-based system functions and integrating these with existing structures. The method used to organize the effort can affect the system cost dramatically. Probably most costly to a large organization is a completely centralized network-based system group which attempts to serve the project managers and their subordinates, no matter how diverse the projects, or what the desires of the managers. On the other hand, in a small organization a central group may be the least costly.

In a typical large, fairly diverse organization, the most efficient organization is one which is decentralized to the lowest level of decision-making, action-taking supervision. In this way duplicate systems are avoided, full-time systems people are eliminated, and cost is minimal, because people intimately familiar with the project do the work with the full support of the project manager involved.

Estimating Costs for a Specific Situation

With these qualifications in mind, let us look at an early attempt to estimate the additional cost resulting from the network-based system. The functions we need to consider are project planning, input data preparation, progress reporting, data processing and analysis, and management reporting.

Table 1 is taken from this early study.[2] Two phases are shown in the table: the implementation phase and the continued operation. A range of manpower and computer time estimates is shown under five headings, related to the five functions. The reader should note the qualifications for

[2]Russell D. Archibald, *PEP Implementation and Operation Cost Analysis*, research report prepared for Management Procedures Division, Hq. ARDC, U. S. Air Force, December, 1960.

TABLE 1

Estimated ranges of PERT manpower & service requirements

	O/ML/P	Implementation phase (Time:[a] O:3 ML:4 P:6 months)					Continued operation (Biweekly reporting)				
		Central PERT Staff[b]	PERT operating group (No. of full-time men)	Engineer's manpower	Data transcription and machine services	Repro. services	Central PERT staff[b]	PERT operating group (No. of full-time men)	Engineer's manpower	Data transcription and machine services	Repro. services
A. Project planning (network preparation)	O	1	1	0.4		As required	0	1	0		As required
	ML	2	2	0.7			1	2	0.1		
	P	4	3	0.8			1	3	0.2		
B. Initial input data preparation[c]	O	0	1	0.2	Hourly: 30 hours per revision (1000 events)						
	ML	0									
C. Progress reporting	O	0					0	1	0	Hourly: 6 hours per report	
D. Electronic data processing and computer runs[d]	O				Salary: 0.5 / 1.0 / 1.0 / 0.85 hrs. comp. time per run/1000 events (AGC 704 PROG.)					Salary: none / 0.85 hrs. comp. time per run/1000 events (AGC 704 PROG.)	
	ML										
	P										
E. Analysis and management reporting	O	0	1	0.1		$200/report	0	1	0.1		$200/report
	ML	1	1	0.2		400	1				400
	P	1	1	0.2		600	1				600
Totals	O	1	3	0.7		Total salary 5.2	0	3	0.1		Total salary 3.1
	ML	4	4	1.3		10.3	1	4	0.4		5.4
	P	7	6	1.5		15.5	1	6	0.6		7.6

[a] O = Optimistic; ML = Most likely; P = Pessimistic.
[b] If central PERT staff is not available, requirements shown must be combined with PERT operating group. [c] Typical for weapon system program at associate contractor.
[d] Machine costs based on PERT network of 1000 activities; initial checkout of existing computer program is included, but not preparation of new program or other development effort.

TABLE 2
Summary of estimated PERT expenses for a major program

	All direct	Central staff on overhead; others direct	All PERT on overhead; others direct
	Implementation (4.12 months)		
Salary°	$81,000	65,170	48,610
Hourly	630	630	630
Computer	1,500	1,500	1,500
Reproduction	400	400	400
Totals	$83,530	$67,700	$51,140
	Continued operation ($ per month)		
Salary°	$10,250	9,324	5,519
Hourly	84	84	84
Computer	600	600	600
Reproduction	800	800	800
Totals	$11,734	$10,805	$ 7,003

°Salary rates used: $5.50/hr. on overhead, $11.00/hr. direct.

the type of project and network size.

Table 2 shows the results of extending Table 1 into dollars. The ranges in Table 1 were averaged, using the PERT formula of

$$t_e = \frac{a + 4m + b}{6}$$

Depending upon which functions are considered overhead, the cost for the first year of operation varied from 112,000 to $177,000, and the operating cost thereafter ranged between $84,000 and $141,000 a year. In discussing these results, the study says:

In actual practice, this cost could vary considerably one way or the other, depending on the contractor's attitudes, organization, acceptance of PERT, etc. It is emphasized that a 1,000 event network may cover a widely varying scope of work depending on the level of detail included. This study presumes a rather gross level of detail. It is interesting to note that, using the Navy's estimate of 0.1% of contract price, this level of PERT effort could handle a program costing $84,000,000 to $140,000,000 annually. This compares favorably with actual experience on certain current programs.

The ratio of labor to computer costs is at least 10 to 1 on this project; the widespread conception that computers are fantastically expensive is therefore not valid for this type of application.

Available Experience

Companies using the network-based system have made widely differing statements regarding its cost. A survey conducted by a prominent management consultant firm produced the following conclusion:

The variety of conditions present in companies makes it difficult to come to any precise conclusions on what it may cost to apply PERT. However, considerable insight is provided by the recent survey respondents. Some 47% regarded the cost of applying PERT as minimal, some 45% as moderate, and 8% as high. In more specific terms, the Air Force indicates that PERT runs 0.5% of the total cost of an R & D project, and 0.1% of the cost of other major programs. Companies in the survey had the following typical comments.

"Technique is more expensive than bar charts, yet benefits, thoroughness, and detail are well worth the effort."

"Too insignificant to measure."

"1% of total program costs."

"PERT/TIME, 0.2% to 1.0% of project cost; PERT/COST, 1% to 3% of project cost."

"0.5% to 0.75% of total program cost."

. . . . the general consensus of companies using PERT is that cost is not a major deterrent to its use.[3]

The U.S. Navy has stated that the original PERT effort for the POLARIS Fleet Ballistic Missile Program cost approximately 0.1% of the total contract price for those parts of the program to which PERT (time only) was applied. At Aerojet-General Corporation, the PERT effort (time only) for the POLARIS solid rocket motors was handled by a team of three people, on the average, for R & D contracts totaling some $40 to $50 million annually.

In the construction industry, the range seems to be 0.1% to 0.5% of construction costs. The larger the total cost, the smaller the percentage cost for the network-based system. System cost also varies with the complexity of the project and frequency of monitoring and updating. The residential construction industry has had considerable experience in using network-based systems for daily scheduling (see Chapter 15); the cost for such service ranges from $45 to $75 per house on tracts, and $30 to $45 per apartment unit for multiple housing projects. This cost includes the cost of updating every two weeks for the life of the tract plus the cost of preparing additional production control reports. Thus, with an average of $50 on a $20,000 house (construction cost), the system cost is 0.25%.

A major engineering and construction firm has found that the costs of

---

[3]Booz-Allen Applied Research, Inc., *New Uses and Management Implications of PERT*, 1964.

developing networks for four heavy engineering construction projects vary from 12 to 65 manhours per 100 activities. On another project the system cost was 0.5% of engineering billings.

Effect of Extension to Include Cost and Manpower

Most of the experience available to date has been with network-based system applications using the time dimension only. Extending the system to include cost and manpower will naturally increase the expense; it is difficult to measure how great this effect actually is, however. Once again, it is important to measure the cost the system *adds* to the project and to refrain from charging to the system the entire cost of functions necessary for effecive management of any project. Early experience with PERT/COST in government R & D programs produced statements that adding cost and manpower to the system triples its cost, but this statement is not easy to substantiate.

Cost of Obtaining Equivalent Information With Other Systems

Perhaps the most effective way for the manager to judge whether the network-based system is worth its cost is for him to determine whether he can obtain equivalent information by a less expensive method, presupposing that the information provided is useful and necessary for project management. If another method does provide timely, equivalent information at lower cost that method should be adopted.

Value of Savings

We cannot predict with certainty the value of savings in a given situation. In addition to direct project cost savings due to better planning, scheduling, and control, there are often also administrative or indirect cost savings. Company acceptance, management philosophy, and the adequacy of the existing administration are the factors determining whether network planning increases or decreases administrative costs. System users have noticed significant improvement in planning, management control, progress reporting, identification of problem areas, communications, management of resources, decision-making, and time saving. All of these can be translated into cost savings. A survey[4] of 44 PERT users produced the following summary of time saving:

Percent of project time saved	Number of companies
1% through 5%	7
6% through 10%	12
11% through 15%	9

---

[4] *Ibid.*

16% through 20%	9
21% through 25%	2
26% through 30%	3
31% and over	2
Total	44

Over two-thirds of the companies in the survey estimated time savings of 6 percent to 20 percent.

Time is money. We can assume that money is saved when such large amounts of time are saved. For example, construction firms and others have reported cost reductions of from 5 percent to 30 percent of the total project. Figures of such magnitude may seem outstanding or even ridiculous, but they actually do represent fairly typical results of the effective use of the network-based system.

SUMMARY

In this chapter we have given some insight into the cost of using the network-based system. Basic costs under the system come from labor and data processing. Labor costs may be divided into project planning, input data preparation, progress reporting, data processing, data analysis, and management reporting costs. System cost will be affected by the degree to which the system is accepted, the manner in which the system is organized, and the extent to which cost and manpower are added to the basic time system. Experience shows that the cost ranges from 0.1 percent to 0.5 percent of the total project costs for time-only applications, to as much as 3 percent for time/cost/manpower applications. Computer or other data processing usually accounts for about one-tenth of the total cost for the system.

The value of savings and other system benefits cannot be measured with any precision although managers do see improvements in many areas. The majority of one group of users report time savings of between 6 percent and 20 percent of original project duration; other users have reported cost savings of from 5 percent to 30 percent. The benefits obtained seem to have justified the additional costs involved for many companies.

11

Indoctrination and training

As we discussed earlier, if managers do not recognize that the network-based system is not a substitute for capable management they may have difficulty using the system successfully. The system is simply a *tool* which can help managers to acquire and process data as an aid to decision-making. Top management, middle management, and operating personnel must all understand, accept, be proficient in the system; in addition, because it is a relatively new technique in most situations, the network-based system may initially even require more management attention than traditional techniques. Hence, although top managers need not become technically proficient in networking, they should understand the approach and be aware of the benefits and performance they can expect from their subordinates. At middle management levels, introducing the network-based system disciplines tends to threaten established functional "empires," and even to push into the spotlight a nonperformer who had been previously hidden from evaluation. The middle management man may also feel he has the most to lose from improved performance evaluation techniques; unfortunately, he is usually in the most advantageous position to sabotage implementation of the system by using his direct organizational power. Hence, middle management must be thoroughly indoctrinated with an understanding of the system, to eliminate irrational attitudes and foster enthusiasm for the technique as a way to improve performance. Finally, the apparent simplicity of the technique can lead management to underestimate the amount of specialized training required to apply the system successfully at the operating level. For example, a requirement for the system will often crop up unplanned, and either be added to an individual's duties, or assigned to available personnel who are not able to meet the demands of the system. We cannot overemphasize that personnel at the operating level should be selected carefully and trained adequately, to insure successful implementation of the system. Many users have found on-the-job training under the direction of an experienced specialist in the field to be an ideal method. This is a particularly advantageous approach when personnel are also required to attend one of the many excellent full-semester courses increasingly available in colleges and universities. Conversely, most of the widely available "quickie" two- to five-day seminars, which ostensibly make personnel competent in the technique have often been found inadequate for anything other than general familiarization purposes. An individual can learn by himself, particularly where good programmed texts are available for this purpose; no text can provide the necessary insights and experiences as effectively as on-the-job training, however.

Most initial attempts at implementing the network-based system will require a thoughtfully planned and executed training program. In many cases, in-house training by a skilled staff will be best; this chapter is therefore primarily concerned with developing such a program,[1] which includes the ingredients necessary to provide each level of personnel with the understanding they need.

Some Guiding Principles

Although many companies now require an operational network-based system, managers must realize that network system technology is still evolving. In the current state of development, network technology can be used in company organizations only by using an orderly training process; this training must be done within the limitations imposed by a developing technology and must include a capacity for informing personnel about subsequent technical advances in an orderly way. Industry experience in introducing a new technological advance such as the network-based system shows that three things are required: reliable and consistent *information* about the advance, *enthusiasm* for it, and an official *policy or authority* to govern use of it. If any of these three is not adequately transferred to the users, the system will lose efficiency. Transferring enthusiasm, information, and policy or authority from one group to another requires both groups to change their activity patterns; because of this, many people will resist the changes, and if development and training efforts in the company duplicate one another, this resistance can grow. Consequently, we can give certain rules of thumb for communicating information and attitudes about the new technology to operational groups.[2]

---

[1] The authors have included as Appendix D an outline for an eighteen-session, full-semester course in network-based techniques, to be used with this text. This outline will also lend itself readily to in-house use.

[2] See James Quinn and James Mueller, "Transferring Research Results to Operations," *Harvard Business Review*, January-February, 1963.

Rule 1. Keep the number of points of communication in the information flow to a minimum.

This implies that there is a centralized source for information and training in the new advance, to avoid conflicting interpretations and undue intramural competition, and to provide a clearly defined implementation program.

Rule 2. Keep the same personnel involved as much as possible.

This applies both to the trainers and the trained. Personnel policies must, of course, encourage the key people communicating the information and attitudes to remain with the organization. There is also a need for transfer personnel (i.e., instructors, staff specialists, and consultants) to keep in touch with the various sources of technological advance so that they do not acquire rigid patterns of thought and opinion which do not allow progressive technological changes. Responsibility for the current "last word" in the technique should be centered in an individual or group with common objectives and source of authority, so that conflicts are avoided, and uniform training for the individuals in the projects or groups who will use the technique. This last is vital. If training is not uniform, the conflicts and contradictions inherent in decentralized communication will multiply, particularly in a large organization.

Rule 3. Give the people involved in the change enough information about the consequences of the new technique to keep them from having irrational reactions, and to foster their enthusiasm, understanding, and acceptance.

This rule calls for management to explain the new techniques uniformly and consistently, paying particular attention to limitations, transitory adaptations to meet customer requirements, and potentials. It also implies that the company will establish a central source of reliable information to fill specific individual needs in a manner consistent with planned, overall policy.

Another critical aspect of training concerns the technical aspects of the techniques themselves. Current or future development work in a fluid technology may profoundly affect the technology in general, as well as some of the fundamental mechanics of current practice. In the network-based system, these changes may result from redefining and updating customer requirements as easily as from technological innovation. In such circumstances, there must be a centralized source of information and authority which directs the training program. Particularly with the network-based system, the technology may have far-reaching implications in other areas of the company's operations (i.e., finance, contract administration, etc.) which simply cannot be subjected to conflicting points of view if they are to function efficiently. For internal control alone, the new techniques must be implemented uniformly in a planned, orderly fashion, if they are to succeed.

Managers must realize that real implementation problems exist, to avoid working these same problems through on each new project. For example, accounting and project cost accumulation objectives are correlated by identifying work packages and establishing the chart-of-accounts structure for cost versions of the network-based system. Experience shows that this process involves similar problems wherever encountered; they must be resolved by the same complex, and sometimes painful, process. Nevertheless, in some organizations the problems are resolved anew each time the system is used. This is almost always due to a poorly planned training and implementation program. Conversely, a centralized, companywide communication and implementation plan provides the framework to develop uniform procedures for this and other problem areas; this means that the time and cost of setting up the system are cut, and acceptance is greater.

When planning an in-house training program for the network-based system, it may also be advisable to:

(a) Plan the introduction of the new techniques into the company operations.

(b) Design an information program to serve the company's subcontractors who may be supplying vital inputs to the company's larger programs.

(c) Specially train marketing personnel in network-based system techniques, particularly those concerned with preparing proposals in the defense and space market.

(d) Set up a system of feedback from operating groups using the technique, to gain the benefits of experience and assure compliance with customer requirements.

(e) Set up a uniform program of integrating the new management tools with existing company systems, procedures, and policies.

SUGGESTIONS FOR AN IN-HOUSE PROGRAM

Integrated Network-Based System In-House Training

As we suggested earlier, all training must be integrated into a coherent, continuous program for implementation to succeed. Management must review and integrate the time and cost aspects to eliminate potential implementation conflicts. The time-only system must be explained to the operational groups in a manner which provides a basis for using the system to plan time and costs. As the integrated time/cost/manpower application becomes the most commonly used type of network-based system, it is becoming increasingly evident that the time-only and integrated versions are not separate systems, but simply two aspects of an evolving technique. Training which includes explanations of the time, cost, and manpower aspects will guarantee that personnel are capable of expanding the basic system to include cost and manpower, and that they are flexible enough to do so.

In addition, as we observed earlier, management must give attention to the *levels* of training. Personnel at all levels do not need to absorb an equal amount of detail. At least one basic division into *top management indoctrination* and *operating manager training* should be made, to save the time of the former and assure proper attention to detail for the latter. Project planners may also consider differentiated training for engineering management, project management, networkers, and analysts, according to the needs dictated by individual situations in the company. This training should stress the integrated management of time, cost, and manpower, provide adequate training materials and exercises, and follow up training with programmed instruction that can be carried back to the job. Manager/instructors drawn from the operational groups should be trained to assume the role of network-based system experts in their areas. This will make it certain that the system is effectively established, with little distortion and few irrational reactions at the operating levels.

Using the Network-Based System in Contract Administration

Particularly in the defense and space companies, there are two general types of contract administrators to whom the network-based system is immediately and increasingly important: the administrator who oversees prime contracts or associate contracts from military or NASA customers and the administrator in charge of procurements from subcontractors and suppliers. The specific activities of these administrators affected by the network-based system seem to be almost the same; these are proposal and preproposal activities, contract award and definitization activities, and implementation.

In the proposal and preproposal phase, both kinds of administrators can significantly enhance the company's capability if they understand and use the network-based system effectively, as a part of important proposals to fulfill customer requirements. If they use the tool in the preproposal and proposal effort the company can "buy" valuable proposal time, and reduce the cost of the proposal effort. In addition, important systems preproposal and proposal efforts often involve contact with many companies. The procurement subcontract administrator can play a major role with suppliers and can reinforce the validity of proposals if he networks the subcontractor and/or associate contractor programs jointly *during the proposal activity*. In many cases, this means educating suppliers and *potential* suppliers, a time-consuming process. Even if a company plans a system it cannot "guesstimate" the important segment of its business carried on by subcontractors, however, and expect to measure up in the highly competitive defense and space market.

To be brief, we will not discuss in detail the areas of contract award, definitization, and implementation; it should be obvious, however, that it is very desirable to use the network-based system effectively in these areas. Effective and acceptable time, cost, and manpower plans and controls are most valuable, both in dealing with customers and in dealing with suppliers. Therefore, marketing and contract administration are increasingly important areas in which personnel should be familiarized with the network system; training courses should be especially designed for them. Such training should be separate from the regular network-based system training, and should be designed to meet the particular needs of marketing and contracts personnel.

In addition, project planners should consider conducting special sessions of the regular course to indoctrinate suppliers and potential suppliers in an impartial way. This program increases the subcontractors' capability and their motivation to support the company; industry experience clearly shows that, when trained in this way, suppliers feel they are benefiting and respond enthusiastically.

User-Oriented Training

In addition to the types of training we have discussed, supplementary user-oriented training may be an effective means of communicating complete information about the systems and techniques to the operational organizations that must use them. Such supplementary training would be in the form of special in-house seminars and workshops, designed to emphasize the special requirements and problems of the particular user and to provide controlled adaptation of the technique to the user's particular problems. These activities would also provide the desired direct feedback to the development and training staff.

User-oriented training can be concerned with project implementations or with special applications. User groups needing project implementation training include the specific project groups who are required by contract to use the network-based system, and who must meet customer requirements and the needs of the particular organization involved by special accommodations. Those needing training in special applications include the more-or-less functional groups, such as finance, contract administration, marketing, and procurement, which are involved with specialized application of the tool, or (as in the case of finance) implications of its use elsewhere in the company; in these situations use of the system may affect existing procedures, and so implementation warrants special handling.

Preparing a Training Program

The in-house training program should have high priority in any implementation plan when a network-based system is first installed. It should be planned very carefully and well in advance; in this way the training will be an effective and important part of the overall implementation. Some of the factors project planners must consider when preparing for a network-based system training program are:

(a) The development and preparation of various training aids and related materials.

(b) The provision of adequate facilities, i.e., space, furniture, and related gear.

(c) The physical processing of the personnel to be trained.

(d) Scheduling of the training sessions to be compatible with work schedules and to fit neatly into the overall implementation effort.

(e) The selection of instructors to conduct the training sessions.

(f) The tailoring of the training program to overall management objectives.

Training Aids and Related Materials

One of the most important training aids, as well as one of the vital components of overall implementation, is a manual or handbook containing a detailed description of the network-based system as it is to be used in the particular situation, and equally detailed procedures for implementation and use. This manual not only can provide an excellent classroom device to add continuity and illustration to the proceedings, but can subsequently be employed as "the book" for disseminating approved policies, methods, and procedures for implementation and use of the system. This handbook is one of the most important implementation and training elements, and should be prepared very carefully. It should be reviewed regularly after publication, and should be updated to reflect current experience, current requirements, and management objectives at all times. The system manual will be a basic reference at all levels of management, starting with the indoctrination and training sessions, and continuing through all phases of implementation.

The system manual is not valuable simply as a means of communication; it has other values, as well. Experience shows that it is a useful psychological aid to establishing the new techniques in day-to-day situations. When set down in this form, the system is not simply understood in the general way fostered by briefings; it has become tangibly part of company policy. The manual becomes a part of the manager's reference equipment. Some managements have fostered this use by including a management policy statement right in the manual. The manual itself can also be an incentive to acquiring knowledge of the system. By "dressing up" the manual, restricting its distribution to the training sessions, and by associating a certain measure of importance to these, some companies have actually made possession of the manual a status symbol; this status creates initiative and enthusiasm for participation in the training and implementation program. While this may seem somewhat whimsical to some (and no one would suggest it as a primary motivation for preparing such a document), many individuals involved in implementations of this sort, including the authors, have observed this phenomenon.

Other valuable aids in training programs of this type are:

Slide Presentations of the Entire Training Program. Such presentations can be used as portable briefing materials for customers or for busy executives who do not have time to attend the regularly scheduled sessions.

Film Presentations of the System Description and Operation. Short (10 to 20 minute) presentations of this type are excellent media for disseminating the system, because they can be taken almost anywhere at anytime, and have high audience appeal. A film presentation is also a welcome interlude in a regular training session, providing a change of pace and reinforcement of learning.

Case Problems Involving Simulated Operating Conditions. These can range all the way from simple exercises using paper and pencil to complex and sophisticated management games which closely approximate reality. Such devices allow personnel to "learn by doing" and so encourage all personnel to participate and learn. Gaining actual practice in using the system is also the best-known way to eliminate lack of confidence and fear of the unknown; the closer such exercises come to real operational problems, the more successful they tend to be.

Programmed Instruction. It takes considerable skill to prepare programmed instruction texts, but they are among the fastest and most effective teaching devices available. They are ideal when large numbers of persons must be trained in a short time. In addition, they permit the student to take the course home with him so he can use it at his own discretion and pace to reinforce his learning and skills. A number of programmed texts covering fundamentals of the network-based techniques are now available.

Space and Facilities

Facilities for conducting a network-based system training program need not be elaborate, but they should be adequate to handle groups of up to twenty persons at a time (larger groups are not recommended). Acoustics should be such that a normal conversational tone can be easily heard anywhere in the room. Where these conditions do not exist, the sound capability should be increased with small microphones. The room should be well lit, and equipped with a blackboard, plenty of wall space for hanging networks and charts, work tables, and comfortable chairs. When the trainers plan to use slides or films, a projection screen should be available in a position visible from any part of the room.

Usually, a fair-sized conference room will serve as the training room. When the room is to be shared by other users, charts, networks, and other paraphernalia should be transportable, so the room can be cleared after use. Keeping large charts on rolls or portable easels facilitates this. Often it is desirable to set aside a special room as the training room, so that it can be used later as an analysis and chart room for system operation. This is particularly desirable in situations in which a central staff implements and subsequently administers the system. In such cases, it is also highly desirable that the room be near the quarters of the staff group, and the data processing facilities. When data processing facilities are available nearby, networks can often be run as part of the training exercises. In the later stages this proximity speeds cycle time and permits faster analysis of system outputs.

If the sessions run several hours or more, provision should be made to serve coffee, and, ideally, to allow some area for participants to get up and "stretch their legs" at breaks. Adequate ventilation or air conditioning for the comfort of instructors and participants increases the attentiveness of the group assembled and insures the overall success of the sessions.

Scheduling the Training Program

Training sessions should be carefully scheduled to be as convenient as possible for participants, and to phase into the overall implementation program at an appropriate time. If possible, training sessions should be scheduled during normal working hours. Participants often consider evening or Saturday session an extra burden, and so do not participate in them very enthusiastically. However, if sessions must be scheduled after normal working hours, planners should be most certain to use time efficiently, so that employees do not resent the extra demand. When sessions run several days, it is best to avoid halfday sessions and to spread the days over a week or two (perhaps including one Saturday session), so that participants are not away from their desks for too long a period at any one time.

Personnel should be carefully scheduled for attendance at the training sessions. Planners should arrange the sessions with the individuals and their supervisors well in advance. They should also remind participants just prior to the session, so that participants are not absent because they forgot. A certain number of participants inevitably drop out; arrangements should therefore be made to replace dropouts quickly, and to include the dropout in a later session. Alternate instructors should always be available so meetings do not have to be canceled unexpectedly. It is often good practice for teams to conduct training sessions, so that instructors are interchangeable; then, if necessary, each can carry the entire session alone. This practice has other benefits as well, which will be discussed.

It is often a good idea to conduct top management sessions early, to build support for the program and heighten the interest of lower management. Conducting these sessions first lets upper-level managers know what their subordinates will be involved in, so they are receptive when the time comes for them to release the subordinates to attend sessions; early sessions also keep upper management levels confidently "on top" of the implementation process at all times.

Interfacing the Training Program with Overall Management Objectives

Often, if project planners give some advance thought to integrating overall company objectives into the training program, they can improve the way the training program is conducted. Coordinating network-based system training with the company's industrial relations or other regular training programs can often produce forms and procedures for handling personnel as well as other administrative aids, for handling details, which would otherwise fall completely on the shoulders of the systems people conducting the sessions. These regular procedures may include giving recognition or credits to the participants, which will make the course much more appealing. Equipment, facilities, publicity, and even instructors and professional help from the company's regular training operations can help project planners and trainers to build a successful and effective program. The training effort should be coordinated with all affected areas of the organization, if for no other reason than to assure that company practices and policies are observed.

Training Techniques

In addition to the usual classroom situations involving lectures and supporting displays (i.e., flip-charts, slides, etc.), the in-house, network-based system training program offers special problems and opportunities. The problems lie in the type of material being transmitted and in the special kind of student encountered; these factors also give rise to special opportunities, however. At this point, it is useful to explore these problems briefly, and to discuss some of the techniques which have been useful in training sessions.

A good example of problems leading to significant opportunities comes from the fact that effective network-based system training must include financial, marketing, contracts, purchasing and other personnel, as well as technical and administrative people. This kind of traditionally heterogeneous group can be an instructor's nightmare (at the lower operator levels it is often best to schedule these people separately and to tailor the presentation to the particular group); the heterogeneity can also be a unique management opportunity, however, particularly at intermediate and higher management levels. The in-house training sessions conducted to implement the network-based system frequently bring together for the first time technical and administrative people, financial and marketing people, etc., to discuss common objectives in cost control and performance improvement. Whereas in the past the diverse backgrounds of the managers prevented understanding and sometimes even created hostility, the interdisciplinary approach fostered by the network-based system often increases their mutual awareness of information and control problems from a *company* standpoint, and leads to real progress and cooperation in acquiring and using new techniques. The instructor in this situation should be alert to such opportunities; in fact, he should try to create them, because exploiting these unique aspects of network system training can add to the success of the implementation, and benefit the company and the individuals involved.

Another problem concerned with the type of people involved in such training is the presence of irrelevant biases against the unknown. The network-based system is often associated with esoteric operations research and/or computer technology. Therefore, if the instructors take too high-toned or esoteric an approach, they can make the students feel inadequate and obsolete, and so produce irrational resistance. The managers may also fear the new and untried. Many otherwise capable managers may let their confidence in their wealth of practical experience obscure their objectivity when faced with change. Another problem may arise: the instructor may have difficulty establishing rapport with the individuals attending the sessions, particularly if he is younger than they, and has been with the company a shorter time. Older, experienced people tend to be impatient with what they see as a theoretical approach to a practical business situation which they feel can best be handled by depending upon experience such as their own. The "outside" intervention of staff people is also feared by some managers. In systems work, it is common for managers to believe that if they are asked to consider methods of improving operations, the company is implying that either they are presently handling things in a way which is basically wrong, or that they have not been progressive enough to develop new techniques themselves. This is commonly called the "NIH" (not invented here) factor, and can be a formidable problem if the trainers do not recognize and deal with it.

These problems will vary in intensity and frequency from situation to situation, depending on the company, the people, the recognition of the need for the system, the familiarity of the instructor to the participants, and other factors. When such problems do exist, however, the instructor will need maturity and good judgment to minimize or eliminate them. There are some guidelines which can be drawn, however.

The instructor must not attempt to impress his group, and must avoid even giving the impression of trying to do so. He must present the material in the most familiar, practical, and relevant terms possible, and must avoid "computerese" when explaining the data processing functions. He should not mention as background material disciplines and techniques which are likely to arouse visions of "bright young men" with their mathematical models and automation taking over. He should not try to "razzle-dazzle" the participants with "totally-integrated-on-line-management-information-systems." He is trying to successfully implement a system to improve operations. He should therefore be familiar with both the strengths and weaknesses of present techniques. He should be receptive to questions and should recognize that any technique must be adapted to various management situations. He should always acknowledge the seniority and responsibilities of the participants, by planning sessions so that the allotted time is used efficiently. He should promote discussion and avoid a lecture atmosphere. It is important to keep a high level of interest to foster maximum participation. A good way to do this is to use absorbing problem situations. Posing a problem to be solved evokes response and aids in "selling" the technique, particularly if the problem is structured to point up the advantages of the network-based system as a problem-solving tool. Exercises and case studies which require active participation are best for this purpose. If they are well done, they encourage objective analysis, discussion (particularly valuable among divergent disciplines), and retention of the material.

When it is possible and practical, it is an excellent idea to combine training with actual implementation on a pilot operation. This makes the problem immediate and the activities of the participants very meaningful. When one major weapon system was being set up, the participants developed in the training program the networks they would be actually working with. In addition, they had an opportunity to run them on the computer (next door, in this case) and to perform initial analyses. Hence, when they went back to their jobs (which now included maintaining the networks they had developed) they did not face the problem of transferring classroom knowledge to operations, but simply continued the work that had started as a training exercise. A good substitute for this more or less ideal approach is the computerized project management "game" we mentioned earlier; the programmed instruction technique is another device which provides continuity from the training program to the actual operation.

Finally, we have some suggestions about effective ways to conduct the sessions. It is good practice to work in teams. This prevents administrative details from using a great deal of class time, and also permits instructors to observe participants' attitudes and reactions while the sessions are in progress. By making such observations the instructors can anticipate problems and redirect the approach while the sessions are in progress, as well as in subsequent contacts. When instructors use exercises and problems, they should subsequently "debrief" the participants to make sure that they saw and understood all that transpired. It is helpful if the instructor develops relationships with the students which foster subsequent contacts. Informal bull sessions and other informal follow-up assistance and counseling can help break down resistance and increase confidence in the system. To avoid making participants feel that something is being thrust upon them, the instructor should emphasize the objectives of the system in terms of their particular problems and objectives. All in all, an in-house training program in network-based system techniques requires an instructor with a good deal of maturity, judgment, flexibility, and responsiveness, as well as skill in the technique.

SUMMARY

This chapter has emphasized the importance of indoctrination and training in obtaining effective use of network-based systems. Various training methods were suggested, and the preparation of an in-house training program was discussed. Finally, comments were offered concerning the instructors' qualifications, training techniques and attitude. Indoctrination and training is a vital part of the overall implementation of network-based systems.

12

Multiprojekt integration

Few, if any, actual projects are carried on in complete isolation; most are constrained by external factors, such as receipt of parts or common use of men, machines, or facilities. One great benefit of the network-based system is its ability to integrate the plans of related projects to show the effect of these interdependencies. Our aim in this chapter is to illustrate the need for such integration, and to indicate how the network-based system can bring it about. We will discuss integration of projects which are all related to a larger program, as well as integration of those related only because they are carried out by the same organization. In the latter case, the system enables a company to sum up all project business, using the common denominator of time to provide relatively accurate status reports and forecasts of business.

Types of Interdependence Between Projects

Projects can be interdependent or related in several ways. We will discuss these in the paragraphs following.

"Result-of-Action" Dependency. This is a condition in which an event must occur or an activity be completed in one project before another activity in a second project can begin, because the result of the action in the first must be made available to the second. An example of this is a process in which a component part is completed and shipped to another company for final assembly, when the part fabrication is treated as one project and the assembly as another.

"Common Resource Unit" Dependency. In this relationship an event must occur or an activity be completed in one project before another activity can be started in a second project because they both use the same skilled man, piece of equipment, or facility. For example, concrete work requiring a large crane must be completed on one building before

214

an activity requiring the same crane can be begun on another building or project.

"Common Resource Rate-of-Use" Dependency. Activities in different projects using the same common resource (certain labor skills, for example) can proceed simultaneously unless the rate of use of the common resource exceeds the available supply. For example, design work in one project requires 3 mechanical engineers; in a second, similar work requires 4. There are 9 mechanical engineers available, and so these activities can proceed at the same time. However, if a third project requires 5 mechanical engineers, that activity will be constrained because only 2 more are available. The mechanical engineering work in these three projects is dependent because the rate of use of a common resource is restricted. Limitations in working capital or specified types of funds can often create such a dependency between projects.

Separate Planning of Projects

Because it is difficult to determine these interproject dependencies, each project is usually planned and scheduled by itself, without specific consideration of the dependencies. Much of the difficulty comes from the fact that the dependencies are all caused by time, and so the planner does not know if a constraining or delaying dependency exists until all the projects are scheduled. Even when scheduling is complete the conflict is frequently not identified until trouble crops up on the job; by then, any solution is expensive. By use of the network-based system, the planner can overcome, or at least anticipate problems of this nature which are not identified when projects are planned separately.

When different organizations work on parts of the same total effort, a broad overall plan should be initially established, to correlate the various parts and provide a starting point from which each group can develop its plan. After all the detailed individual plans have been prepared, they must be tied together so that the total effects of the interactions and interdependencies can be determined; each individual plan may be sound and valid by itself, and yet when all are integrated into an overall scheme serious conflicts may be revealed. The network plan with interface events gives the planner a means of evaluating these interrelationships simply and effectively.

INTEGRATING PROJECTS THROUGH COMMON EVENTS

The first two types of dependencies described above, the result-of-action and the common resource unit dependencies, can be represented by using common events to link the project network plans together.

Thus, the event "complete concrete work" in the Project A network can also be identified as "start moving crane to Project B." The term "network integration" designates the process of linking together previously separate network plans by common or interface events. An "interface event" is the point at which responsibility for a given action changes hands. Integration thus means meshing plans by joining them at well-identified points to denote transfer of responsibility. These ideas have previously been discussed in Chapters 2 and 3.

It is important for the reader to note that we are talking about interface events, not interface activities. Occasionally, one sees references to "interface activities," since some persons find it easier to understand the concept of interfaces if a line is drawn from one network (or subnet) to another. If a planner considers this carefully, however, he will realize that the activity must be included in one network or the other, not both. Sometimes a zero-time constraining activity is used in one network to show that the preceding event is an interface with the succeeding event in a second network. This introduces a sequential relationship between the interfaces which may not be valid under some conditions, however.

Prior Identification of Interfaces

Before networks can be integrated the interface events must be identified: the same event must be present on each of two networks if the two occurrences are to be joined. It is desirable to identify such events before preparing the network, although it is often not possible to do so. When the planner identifies an interface while preparing a network, he must notify the planners preparing the other affected networks of the exact identity of the event, so that all the networks will contain all the interface events.

Importance of Systematic Planning

The importance of systematic planning (described in Chapter 2) becomes evident when we attempt to integrate networks by interfacing common events. It is essential to break down the overall program elements and to identify milestones and interfaces for each element at every level of management interest. The planner must study the information carefully, to determine the number and scope of networks necessary, so he can prepare a minimal number of individual plans; using as few networks as possible cuts down on the analysis and integration problems.

Problems Involved in Network Integration

Two basic problems are encountered when integrating networks by the use of interface events: difficulties involved in identification and control of interfaces, and those involved in processing very large networks.

Identification and Control of Interfaces. We have previously discussed the importance of identifying interfaces before drawing networks, if possible, and, on the other hand, have pointed out that interfaces are often not evident until the networks are fairly well developed. It is a real problem to maintain interface-event control as the networks change and evolve; this problem has led planners to use rather formal methods and procedures.[1] In major weapon system programs, an event log is used, listing all interface and milestone events. In a typical procedure for handling interfaces the originator (one in whose network the interface is an outgoing event) is responsible for identifying, describing, and numbering the interface events to be included in the event log:

(a) To establish an interface event, he sends a TWX to the user(s) (contractor/government agencies) of the event.

(b) He sends the IACC (integration, assembly, and checkout contractor) interface-event data on punched cards.

(c) The interface data he transmits are the assigned event number, the event description, the contractors and/or government agencies (users) involved, the schedule date, the effectivity date,[2] and a clear narrative description of the event.

Within one week after receiving the TWX from the originator, the user(s) (the contractor/agency in whose network the interface is an incoming event) of the event will be responsible for notifying the originator by TWX to indicate agreement or disagreement with the interface data. If the user disagrees with the interface data, his TWX to the originator must specify his reasons for disagreement. Disagreements will be resolved between the affected contractors/agencies, and conflicts will be subject to individual resolution by the SPO (System Project Office). The originator will notify the IACC by TWX of final agreements.

A contractor/agency which identifies an input requirement (incoming interface event) will review existing event lists (prepared and distributed by the IACC) to try to identify the event as a previously documented interface event.

(a) If the event has not been listed, the using contractor/agency sends the originator a TWX indicating the user's suggested description and

---

[1]The following paragraphs are taken from the *MMRBM Program PERT Cost System*, Hq. Ballistic Systems Division, U.S. Air Force Systems Command, BSD Exhibit 62–9, 15 April 1963.

[2]The effectivity date is the date when the event must appear in the integrated network, if it is approved by all concerned.

schedule date for the event. The originator will then respond. (The identity of the originator will be obtained from the IACC, if the user does not know who the originator is.)

(b) If the event appears on an event list, the contractor/agency notifies the originator by TWX.

The SPO must give final approval on event data and schedule dates for all interface events after agreement has been reached between contractors/agencies.

(a) The IACC will send the SPO event lists showing all interface events which the affected parties have agreed upon.

(b) All concerned will assume these events have been approved by the SPO unless the SPO notifies them by TWX within one week after receiving the list.

Interface-Event Network Construction Rules

The rules guiding construction of interface event networks are:

(a) The originating or performing organization will include in its network as activities prior to the interface event, all activities leading to and including delivery at the user's door or dock.

(b) Interface events will appear as the same event (have the same number) in the detail network of the involved contractors/agencies.

(c) All incoming interface events will have no immediate preceding activity in the user's network.

Milestone Events

Milestones are key (noninterface) events which occur in a contractor's/agency's network. The SPO, the contractors, and the agencies may select milestones to be included in the Integrated Weapon System Network.

(a) To establish a milestone event in the Integrated Weapon System Network, the SPO will send a TWX to the contractor/agency responsible for the event.

(b) The contractor/agency will respond within one week by forwarding to the IACC milestone data on punched cards. These data will include the assigned event number, the event description, the contractor/agency responsible for the event, the schedule date, the effectivity date, and a clear narrative description of the event.

(c) The IACC will add these milestone data to the approved event list.

(d) A contractor/agency wanting to establish a milestone in the Integrated Weapon System Network will transmit punched cards containing the information to the IACC.

(e) The IACC will prepare a list of new weapon system network milestones received from the contractors/agencies for SPO approval. All concerned will assume these data have been approved by the SPO unless the SPO notifies them by TWX within one week after receiving the list.

Event Collection

The IACC is responsible for maintaining and updating an event-data file which contains all integrated network milestone and interface events, both approved and not approved. From this file, the IACC will produce event lists and an event log. There are two major types of event lists: those containing milestone and interface events which require coordination or approval action, and the approved event list, containing milestone and interface events which have been completely approved. Lists of events requiring action will be prepared and distributed no less frequently than once a week. The approved event list will be produced once a month, concurrently with the event log. These lists will show all the requested data for each milestone and interface event. The IACC will prepare and distribute the following special event lists:

(a) The *interface coordination list* contains interface events on which agreement has not been reached among the responsible contractors/agencies. Each contractor/agency will receive a complete list of such events.

(b) The *interface approval lists* contains interface events which have been coordinated and agreed upon by the responsible contractors/agencies, but have not been approved by the SPO. Each contractor/agency will receive the complete listing for information. The SPO will receive three copies of the complete list.

(c) The *milestone approval list* contains milestone events which have been selected for inclusion in the Integrated Weapon System Network, but have not been approved by the SPO. Each contractor/agency will receive the complete list for information. The SPO will receive three copies of the complete list.

(d) The *event log revision list* contains those events which have been added, deleted, or changed since the previous issue of the event log. It is sorted by transaction code major and event number minor, and will be a part of each event log.

(e) The *approved event list* contains all approved interfaces and milestones from the event-data file. The SPO and each contractor/agency will receive the complete list. This list will segregate interface events and milestone events into distinct groups. Changes to this approved event list are subject to approval by the SPO.

On the second from the last Friday of each month, the IACC will prepare and distribute an event log, containing (from the approved event list) those interface and milestone events with effectivity dates which occur on or prior to the cutoff date (last Friday of the month). Each contractor/agency will receive the complete log, and the SPO will receive two copies of the complete log. Changes to the log are subject to approval by the SPO.

Even on somewhat smaller projects the planner can avoid difficult problems caused by ommissions, gaps, loops, and other errors in network logic and validity, by controlling interfaces carefully. Of course, there is no interface problem if the network is small enough to put on one sheet of paper.

Difficulties in Processing Very Large Networks

Because the planner has the ability to integrate a number of networks, and will use this ability, he will encounter the difficulties involved in processing or analysing the resulting integrated network. This may be very large.

The major difficulty is caused by the total number of activities in the network. Most network analysis computer programs can only handle a limited number of activities, and these usually are a function of the size of the computer memory. The limits range from 200 to 5,000 for many computers, although one widely used program will handle 75,000 (with some restrictions on the network structure); some programs now available even have "unlimited" network-size capability. The final limit is, of course, one of economy.

Even if the planner is using a computer program which handles "unlimited" size networks, he quickly learns that human ability to prepare error-free data falls far short of the machine's capacity to process them. Preparing huge networks and trying to keep them up-to-date is an expensive operation. In one weapon system program, networks were prepared with a total well over 50,000 events, but because of the volume of data they could not really be used, and so were discarded in favor of less-detailed networks which could be handled more easily.

The Subnet Concept

Many semantic difficulties are encountered when discussing network plans in relation to the Work Breakdown Structure. Use of two different approaches has particularly produced confusion and misunderstanding.

One approach says that the network should be developed from the top by exploding *selected* portions of the network downward at a given level. In this case, to get the total network the system planner has to total all networks vertically. The second approach, which the authors favor for its practicality, says that the network should be developed from the top by exploding the *entire network* downward at a given level: every event in the level 2 network will be carried down to the level 3 network, where additional events and activities will be identified. In practice, it is not feasible to maintain networks at several levels, and the most detailed network is usually the one which is maintained and analyzed. Reports on events (milestones) or elements of the Work Breakdown Structure can be easily derived from the detailed network. Subnets are simply convenient pieces of the detailed network produced so that those responsible for a given area can work with their portion of the network only. Subnets (also called "fragnets") generally are networks of elements of the Work Breakdown Structure, and are linked at common interface events.

Analyzing Large Networks

In this discussion, we will deal with computer network analysis although very large networks can be manually analyzed with great expenditure of labor and a high error rate. A total integrated network can be analyzed by computer if it is within the network size limitations imposed by the particular computer and program. If the network size exceeds the capacity of the machine, several alternatives are possible:

(1) The planner can manually reduce the network size by eliminating or consolidating activities and events.

(2) He can cut the network into blocks or treat the subnets separately.

(3) He can prepare a summary network showing major events (milestones), determine the time span between milestones from the detailed networks, and then analyze the summary network.

(4) He can condense the subnets while retaining all interfaces and milestones, and can then integrate the condensed subnets into an integrated network, analyze this, and feed the results into the detailed subnets.

Each of these alternatives is discussed in turn in the paragraphs following.

Reduce Network Size

Reducing network size can result in more practical networks if it is done properly. If the planner combines detailed activities in series, for example, he can improve the network by eliminating excess detail, provided no interfaces are adversely affected and only one authority has responsibility for the activities.

Separate Analysis of Subnets

If a network can be cut in half, let us say, at one interface event, then the planner can easily analyze the first portion by making the forward pass, use the resulting earliest expected completion date as the start of the second portion, and then go through the forward and backward pass in the second part; he would then have to feed the resulting latest allowable time for the connecting interface into the first portion to calculate the latest allowable times and the critical path for it. However, this procedure is impractical if several subnets are involved, or if a number of interfaces tie the subnets together at different points in time. To illustrate the difficulties, it is suggested that the reader perform this analysis on three simple subnets of his own creation.

Use of a Summary Network

Use of summary networks has proved to be very confusing to many planners, because lines drawn between milestone events or even major interface events on these networks do not represent actual activities; such a network may have some value when used to display the results of a detailed analysis, however. Although it is theoretically possible to transfer span times derived from subnet analyses to a summary network to make an analysis, it is very difficult. One must be sure that *all* interfaces appear on the summary network, and that all interfaces changed on the detailed networks are reflected in the summary one.

Example of Network Condensation and Integration[3]

Out of the need to solve these problems, a method was developed which made it possible to handle very large networks economically on the computer by using subnets.[4] In this method, a computer logic is used to reduce a network to an absolute minimum. The user defines this absolute minimum by specifying the interface and milestone events he wishes to include, and the computer then determines the relationships between these events to form the condensed network. Therefore, an event table must be included in the computer input, in addition to the detailed activity records fed into the computer during a normal network analysis run. The condensing program pairs these listed events in every possible combination to determine whether or not there is a topological relation-

[3]Joel M. Prostick, *Network Integration,* a paper presented to the Twenty-second National Meeting, Operations Research Society of America, Philadelphia, Pa., November 7–9, 1962.
[4]The first computer program to do this was designed and programmed by Joel M. Prostick at Hughes Aircraft Co. in 1962, within the Management Information Systems Department managed by Russell D. Archibald. Mr. Prostick's program was incorporated into the U.S. Air Force PERT/COST System. The IBM PERT/COST System uses the same basic method.

ship between any two (i.e., at least one path which passes through both). If such a relationship exists, then the condensed network shows a constraint between the two events, and the critical path between the two events becomes the expected duration of that constraint.

The constraints obtained in this manner describe the full effect of the detail activities; however, there may be several constraints included among them which are redundant. For example, if the user specifies that events X, Y, and Z be included on the condensed network, and it is determined that X restrains Y and Y restrains Z, it is not necessary to show the restraint of X upon Z if Y lies on the critical path between X and Z; in this case, $X:Z$ is a redundant constraint. However, if the critical path between X and Z does *not* pass through Y, the restraint $X:Z$ and its time duration must be included on the condensed network, since the sum of the time durations of restraints $X:Y$ and $Y:Z$ is less than the duration of $X:Z$. By eliminating all redundant restraints the condensing program reduces the network to an absolute minimum.

Figure 1 shows a typical detailed network with the events that are to be included on the condensed network indicated by capital letters; both interfaces and major milestones are included. Figure 2 shows the same network in condensed form. The reader should notice that constraint $A:D$ is shown on the condensed network, because the critical path between the two events does not pass through C. Constraint $C:G$, on the other

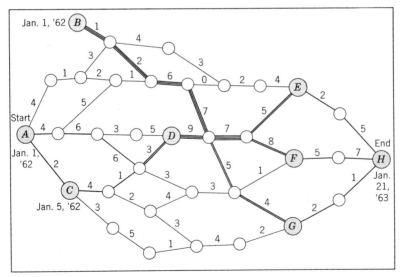

FIGURE 1.   Detailed network.

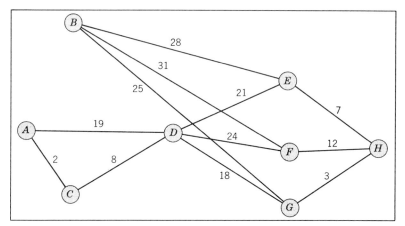

FIGURE 2. Condensed network.

hand, is redundant, because event D lies on the critical path between C and G. We can see the importance of eliminating redundant restraints when we realize that there is a total of ten in this sample network ($A{:}E$, $A{:}F$, $A{:}G$, $A{:}H$, $B{:}H$, $C{:}E$, $C{:}F$, $C{:}G$, $C{:}H$, $D{:}H$). The reader should also note that the condensing program does not calculate expected or latest dates for events, or any of the other standard network values; it simply inputs one network and outputs another.

Integration Process

Using the condensing technique, the planner can link networks through interface events rather than dummy activities. This makes networking more flexible, because the interfaces need not "dangle" from the network (i.e., interface events may be constrained by activities from more than one network).

All interface events must be included on the condensed networks which are to be used for integration. Usually all interfaces for the networks which are to be condensed and integrated are included in one table: the condensed network event table. Major milestones may also be included. When the condensed networks have been produced, using the condensed event and, if desired, the major milestone tables, these condensed networks (not the original, detailed ones) are integrated. The computer network size limitation is then imposed on the number of constraints allowed on one integrated, condensed network instead of being imposed on the number of detailed activities; in effect this increases the computer network-size capacity by the degree of condensation, usually on the order of 10 or 20:1. Figure 3 shows the original condensed network integrated with other condensed networks. The critical path and the new expected program completion date are particularly interesting: event F, which would have been on the critical path if the original network had not been integrated, now has 18 weeks of positive slack; event G, which was thought to be on a slack path, now lies on the critical path.

Processing the integrated condensed networks produces an expected total program duration as well as expected and latest dates for each interface event. Although this information is very valuable, the problem of determining valid expected and latest dates for every event and activity on the detailed networks remains.

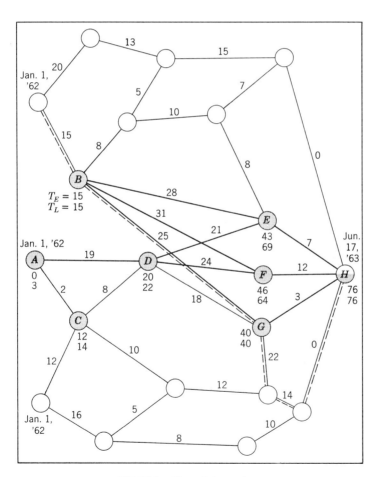

FIGURE 3.   Network integration.

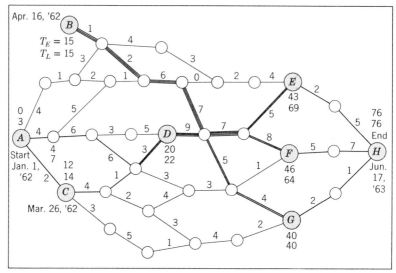

FIGURE 4. Updating.

Updating the Detailed Networks

To obtain the valid expected and latest dates for every event on the detailed networks, the network analysis program must now accept the expected and latest dates for the interface events as input. These dates are produced as output from the integration run, and reflect the effect of integration of all the networks. For an incoming interface the critical value is the expected date, since the interface event may be constrained by activities in other networks; the latest date for an outgoing interface may be similarly constrained, and so it is the critical one. When the interface event dates have been fed into the computer, the detailed networks are processed with the expected and latest dates computed in the normal manner, except that for interface events the integration run dates are accepted and the succeeding calculations based on these values, if they are more constraining than the calculated dates resulting from the subnet analysis. Thus, valid dates reflecting the results of the integration are computed for every event and activity on the detailed networks. Figure 4 shows the original detailed network after it has been updated, using the results of the integration.

Preparing Summary Networks from Condensed, Integrated Networks

When preparing summary networks from the condensed, integrated one for presentation purposes, the user specifies the milestone and interface events which are to appear by listing them on a summary network event table (list of milestones). Even interface events must appear if they are to be included. If more than one summary network is desired, the user submits a separate summary network event table for each. Since each of these networks is to be a summary of the integrated network, to compute each one the integrated condensed network is simply condensed again, using the appropriate summary network event table. It is therefore necessary that every summary network event table be used in the first condensation, so that appropriate milestones can be included on each summary network after the first condensation. (As we stated earlier, the interfaces are input into the first condensation on the condensed network event table.)

Using the condensing technique, the planner can produce many different types as well as levels of networks. For example, one may want a function-oriented network, even though the original detailed network may have been hardware-oriented. By including all the activity start and complete events for a given function in one of the tables, the planner can see to it that all activities in the different functions are condensed, and that all the detailed activities involved in the function would appear on the network. Computers can now be used to produce any level of summary network while requiring the maintenance of the detailed task networks only.

INTEGRATING PROJECTS WITH "COMMON RESOURCE RATE-OF-USE" DEPENDENCY

There are often several projects in progress within an organization at one time. These projects may differ in size, nature, and importance, but they will make demands on the same manpower, materials, machines, tools, buildings, facilities, land, or capital. If we can subdivide each project into activities, prepare network plans for each, and obtain reasonably accurate estimates of the type, combination, and amount of resources required per unit of time to perform each activity, then we can integrate these projects.

To do so, we first develop network plans for each project and prove them valid. We then add to the networks information showing the quantities and specific nature of each resource required. By proper coding we can identify all the manpower skills, machines, materials, facilities, and funds needed. Attaching these resource data to each activity will, of course, lead to more accurate results. However, it is usually impractical and expensive to obtain such a detailed breakdown of resource estimates; estimating resources by work packages or groups of activities produces useful results at less cost.

As the planner schedules each project against the calendar, he notes the resources required during each time period. He could assume that the resources would be expended at a constant rate (or linear rate on a cumulative scale) for each activity or work package; however, he might use a more sophisticated approach, in which he would allow the rate to vary in certain preestablished patterns for each activity or work package. By summing up the various resources required for all activities or work packages within each project and then summing each resource for all projects, the planner can project all resource requirements over time.

If resources were unlimited, the management task would be simple: merely to be sure the resources were at the right place when required. However, most resources are severely limited, due either to cost or to insufficient supply (such as highly specialized skills). What do we do when two projects are competing for the same resource at the same time?

First, let us limit our consideration to the critical path activities in each project. If there is a conflict in resource requirements we will have to decide which project has priority. Resources given to one critical activity and withheld from another in a different project will delay completion of the second project, unless other management action can recover the lost time in later critical activities. Generally, all required resources should be provided for critical path activities.

Let us next examine the remaining activities or work packages, those having slack time. We are able to delay the start or completion of the activity or work package by the amount of slack available. Thus, when we have exceeded the limit of available resources on critical activities, we can delay noncritical activities to reduce the peak demands on the resource in question. Delay beyond the amount of available slack will, of course, cause the activities involved to become critical. This can become an exceedingly complex problem, because use of common resources makes interactions within a project and between projects necessary. Using the network plan as a base and the computer as an aid, the planner can obtain solutions to this problem, although he will still have to make certain simplifying assumptions.

One computer-based solution to this problem is RAMPS (Resource Allocation and Multi-Project Scheduling), developed by du Pont and C-E-I-R, Inc. Although this system has not been widely used, probably because of the difficulty in setting up the required input data, it is one of the most advanced systems available. Input to the program, which uses an IBM 7090 (or 7094) computer, consists of:

The *resource card* for each resource, which notes the code, the description, the time period available, the maximum number of resource units available per period without overtime, and the normal cost per resource unit per period. The resource priority is indicated by the card sequence.

The *overtime card* for each resource, which records the code, the description, the time period for which the data is valid, the maximum amount of overtime per period (by actual number or percent over normal), and the premium unit cost per period.

The *project card* for each project, which has on it the code, the description, the calendar due date, the time period covered, the earliest start date of the project, and the penalty for each period of delay past the due date.

The *activity card* for each activity in the project network, which lists the activity identification, the event numbers, the amount of work, the type of continuity factor (if the activity can be interrupted, etc.), the continuity penalty factor, and the maximum and minimum number of time periods.

The *activity resource card* for each activity, which notes the resource code and identification, and the normal and crash assignable and efficiency rates.

The *control data*, which consists of a series of priority values and weights which are assigned to give certain activities precedence in using available resources, etc.

The mathematical algorithm used by RAMPS to process these data is proprietary and therefore not available. It is mathematically complex, involving a number of simplifying assumptions, used to make it possible to produce certain types of answers. RAMPS output reports contain schedules listing each project activity, and showing their resource requirements over time, based on the derived overall schedule; resource summaries, which show the type of resource (laborers, for example) required by each project over time; and total resource summaries over time.

The Integrated Network-Based System for Weapon Systems

A general schematic picture of an integrated system applied to a major weapon system is shown in Figure 5. Under this system each contractor establishes and analyses his own networks, based on the work breakdown of the entire weapon system. In addition to summarized reports for his own use, the contractor prepares work package and interface event data which are fed into the central government-agency computer, together with similar data from all other contractors. Both overall summaries for the entire weapon system and revised data relating to interface event schedules are prepared and fed back into each contractor's computer, to allow him to revise his plans.

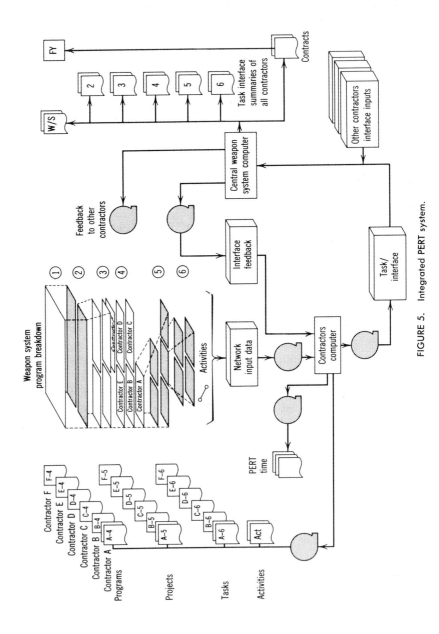

FIGURE 5. Integrated PERT system.

Multiproject Corporate Operations

By using the schematic representation in Figure 5 with minor revisions, a multiproject company can integrate all schedules and resource requirements to derive appropriate summary reports for the organization as well as for the projects themselves. ,

SUMMARY

Projects are interrelated because one depends on the "result of action" of another, or shows a "common resource unit" or "common resource rate of use" relationship with it. By using interface events the planner can integrate the first two types. However, the third requires him to total the required resources over time, to determine whether demands on available resources will exceed supply; if he finds that they will, he knows that delays or increased costs will result. It is vital for the planner to identify and control interface events. Integration of detailed subnets often produces large networks which may be difficult to process; computer condensation is one practical way to make it possible to process larger networks on a given computer. A computerized, network-based information system allows the planner to draw up fully integrated plans and schedules for weapon systems and multiproject corporate operations.

PART III

Case studies

13

Applications of network planning

In this and the next two chapters we will illustrate the varied uses for the network-based management system. In some cases we will describe applications in detail, and in others give only general outlines. In many cases, we will present the illustrations in the words of the management people who used the system in the application described. In this chapter we briefly describe some varied applications of the technique gleaned from published sources; in chapters 14 and 15 we will describe cases about which a great deal of documentation was available. Chapter 16 discusses an approach of potentially great significance.

Although network techniques were developed in the highly project-oriented environment of the aerospace industry, American businessmen were quick to explore many other possible applications. Since network planning developed, it has been used in virtually every business function. Networks have been particularly successful in such diverse areas as construction, research and development, new product planning, automotive model changeover, and advertising. Network planning has even been applied to the production of a Broadway play; the producer said that its use allowed the play to open on time, and allowed the backers to save several thousand dollars in preproduction costs.[1]

CASE STUDIES

The following group of various applications reported in the March, 1964 issue of *Factory Magazine* will give some idea of the range of present uses.

---

[1]Unfortunately, quality of the play did not equal its efficiency in production. It was panned by the critics and closed out of town.

Automotive Model Changeover

Ford Motor Company (Detroit) used network planning on a model changeover at a transmission plant. It was a big job, involving layout, machine installation, and facility tryout, as well as the purchase of 200 pieces of machinery and tooling, 1,500 gauges, and 60 types of raw material and finished parts. The company states that using the network system cut project time 20 percent and saved a "substantial" amount of money. Management ran the conventional reporting system parallel to the network-based system until everyone gained confidence in it. The company also provided for use of network planning on future projects; eight master networks were prepared, covering "buy new machine," "rebuild and retool on-hand machine," "set up quality-control and inspection operations," "buy raw materials," etc. The networks are now selected and interfaced as needed for each new operation.

Isolating Problem Areas

PERT frequently alerts managers to impending trouble; Kelsey-Hayes Company (Romulus, Mich.) used it to isolate and resolve a trouble spot. A welding operation in the production of rims for automobile wheels was producing a large percentage of rims which were defective and had to be scrapped. The research team assigned to investigate this problem had a tiny budget and an urgent timetable. They decided that for their purposes the sequence of events carrying the highest risk of technical failure, not the longest sequence, was the "critical path." The team discovered the production problem—inadequate "upset pressure" on pieces being butt-welded—and then designed and tested a device to measure the pressure. Scrap fell from 3 percent to under 1 percent.

Maintenance Shutdowns

Kendall J. Gibson, of the Casper (Wyo.) Refinery of The American Oil Company told the Fifteenth Plant Engineering & Maintenance Conference that network planning consistently cuts project manpower from 10 percent to 25 percent. American Oil recovered the cost of teaching the technique to some 50 supervisors in less than 3 months, without a major application.

Inspired by the success of networks for maintenance planning, operating people diagrammed the start up and shutdown of facilities as well as certain administrative and emergency procedures. They regularly use arrow diagrams to train equipment operators now, and have even used them in a discipline case.

All this is not to sneer at maintenance applications, of course. These were and are the principal operations on which networks are used at Casper. Networks are employed on all projects having major effects on plant operations, and some complex repetitive or routine jobs have been set up as more-or-less standard networks. A diagram, the job data, and a bar chart are filed for each job and can be pulled out each time the work is scheduled. The most dramatic applications of networks are in routine turnaround maintenance projects, in which entire facilities are shut down, overhauled, and put back into production. In one case, the critical path based on day-shift work only exceeded the four weeks the plant could afford. By using network analysis to selectively allocate overtime where it was needed, management was able to shorten the job. (The predicted amount of overtime proved correct within a few hours.) In another case, a preliminary project study fixed the most economical period of downtime; the problem then was to schedule the turnaround into this fixed period. The project was first diagrammed without regard for downtime, and fell short. Planners then went through the critical path, altering elements to extend the elapsed time. To do so, they reduced the work force on critical jobs, to give demands on scarce crafts special attention. In this way the desired downtime was neatly bracketed, and the planners hit their objective on the next cast. A four-week downtime was once planned, but, shortly before shutdown, the operating budget was changed to require an extra week of production. By postponing some noncritical items and reducing manpower, the planners were able to come up with a three-week schedule.

Casper applied network planning most extensively on a 40,000-man-hour turnaround in spring, 1963. The turnaround involved contracting for a supplemental maintenance force, and took over three months to plan. Both work forces used the same work list, diagram, and bar chart, and got assignments in the same manner. The contractor participated in diagramming and manpower leveling. The two work forces often worked together, once "back-to-back" on a single job, but clear distinctions between owner and contractor responsibility produced a grievance-free job, completed within a percent or two of the estimate. A refinery turnaround is discussed in more detail in Chapter 15.

Production Control

The Brown Engineering Company (Huntsville, Ala.) found that its conventional production control system which depended upon expediters, planners, liaison men, etc., was unable to cope with the pressures of NASA contracting; it therefore set up a central Project Management Office (PMO) with authority to cut across organization lines when necessary. The "cognizant engineers" in the PMO "represent the customer,"

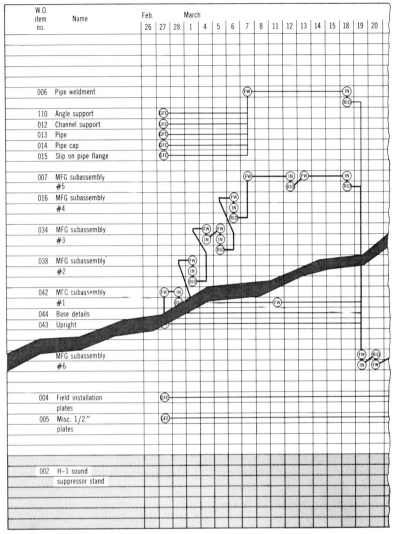

FIGURE 1. Graphic display of network planning data in production of rocket engine sound suppressor.

and concentrate on getting jobs out on time, within specifications and estimated costs. Network planning is their key tool. All items are arranged chronologically and connected by a simple tie line showing the relationship between them. These are drawn against a calendar background, such as the one shown in Figure 1. The dates for each event correspond to the dates the scheduling office put on the route sheets. On this particular job networks saved two full weeks on an already compressed six-week schedule. The critical path is depicted by the heavy line, and represents the greatest man-hour total from the beginning to the end of a job. When there is more than one critical path or when there is a second path almost as long as the first, Brown flags and watches it closely. These charts make it possible for the PMO to recognize the danger of slippage in an item *not* on the critical path. The delivery date for a purchased part, for example, might place the event to the *right* of the critical path, indicating that it will not arrive on time.

Multiproject Integration

The East Alton (Ill.) brass operations of Olin Metals had three massive projects under way at once. The three, which involved two outside engineering firms as well as Olin employees, included a new primary brass mill, expansion and modernization of two finishing mills about 500 miles apart, and the consolidation of scattered fabricating departments. All three projects maintained their own critical path schedules, and fed information into a central control room at East Alton. Olin has been able to roll with several hard punches in the course of the four-year undertaking, and credits network planning with enabling it to respond swiftly, on target, and at minimum cost whenever problems show up.

New Product Introduction

Chrysler Corporation used network planning when embarking on the manufacture of a complete new line of commercial air conditioners, consisting of eight models, ranging from 20- to 100-ton capacities. Chrysler management set the project deadline to coincide with The All-Industry Air Conditioning Show scheduled for February 11–15, 1964. In less than a year's time, therefore, all segments had to be complete. This project involved not only basic designing, but all the tooling, purchasing, costing, plant rearrangement, and marketing. Thirty-one hundred parts were involved, 40 percent of which were to be fabricated as well as designed in the plant. PERT was used to keep the entire undertaking, including management approvals, under control and on schedule. All was not rosy. PERT was new, and on-the-job application left some doubt as to its benefits. At one crucial stage, a cost reduction committee had seven different critical paths at the same time; careful analysis revealed that there were problems in engineering design and in manufacturing, and concentrated effort was then able to break these bottlenecks. Using the method, the PMO set up a two-pronged reporting system: a weekly report was sent to each responsible manager outlining his past-due activities, and a PERT weekly reminder was sent to each person on the project, underlining all of his activities that were coming due. This "fore" and

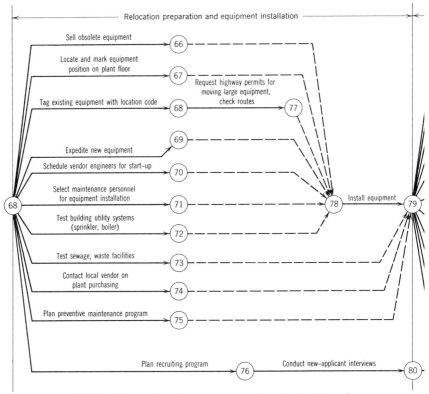

FIGURE 2. New plant start-up-cross country relocation of existing plant.

"aft" system kept everybody on his toes, by forcing good planning and needed teamwork.

Heavy Machinery Installation

A major castings company used PERT to plan and control purchase and installation of a sand slinger in one of its foundries. The sand slinger is a major piece of equipment; purchase of it involves considerable site preparation, engineering, and associated equipment and facilities. By adroit use of resources (especially manpower, which was color-coded by skill right on the network) the job was cut from 15 months to eight.

"Standardized" Networks

In the same article, *Factory* reported on other uses. Most interesting was the reported development of "standardized" networks to demon-

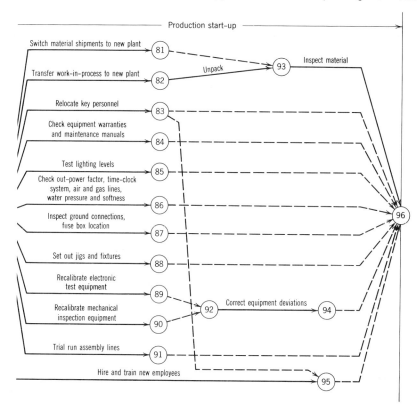

strate the variety of applications possible, and to provide guides for the development of adequate networks. The networks illustrate introduction of a new product, maintenance painting, management succession, new plant start-up (Figure 2), plant consolidation on existing site, and short and long distance relocation. For the sake of space and clarity, these networks omit a substantial amount of data and subnetworks. The networks were prepared originally for *Factory* by Booz-Allen-Hamilton, Inc., consultants.

NETWORK TECHNIQUES IN GENERAL ADMINISTRATION

Managers are paying increasing attention to the many routine administrative and operating procedures which, although different from the more-or-less discrete project efforts usually networked, are adaptable to network analysis. Usually, use of the network-based system is justified on the basis that the project to be networked is a long, unusually complex, one-time-only project so that the system will be used primarily for planning and control of resources and schedule. Although the routine administrative and operating procedures in most business organizations often do not possess these characteristics, an increasing number of observers believe that such procedures could benefit from network analysis, evaluation, and possible restructuring, if management were willing to justify use of network techniques in a more flexible manner. G. B. Davis, writing in the *NAA Bulletin*, suggests these additional justifications:[2]

(a) There is a definite objective to be achieved.

(b) There is a completion-date requirement for this objective event.

(c) There are many identifiable and interdependent tasks (activities) which must be performed in proper sequence before the objective event can be achieved.

(d) Time estimates (and associated cost estimates) can be made for the tasks.

(e) Resources may be shifted from one job to another in order to affect the intermediate completion dates of tasks.

Monthly Closing—Accounting Application[3]

Granville R. Gargiulo gives one example of the application of network techniques to analysis and evaluation of administrative procedures, when he describes the network analysis of the closing procedures used in preparing monthly operating statements. The objective event in this procedure is to issue a set of financial and operating statements to management. These reports are to be completed by a certain day following the close of the accounting period, thereby meeting the completion-date criterion.

To identify a specific task in a procedure the planner must know when the task begins and ends. Although maintenance of an account ledger is a continuous operation, summarization of the ledger and preparation of a journal entry are identifiable tasks which are necessary preliminaries to the closing procedure. Interdependence and required sequences among individual tasks are also evident in the closing procedure: consolidated profit-and-loss statements must wait for completion

---

[2]G. B. Davis, "Network Techniques and Accounting—With an Illustration," *NAA Bulletin*, May 1963.

[3]The following case is an abstract of an article by Granville R. Gargiulo, "Network Techniques: A Means for Evaluating Existing Systems and Procedures," *Systems and Procedures Journal*, November-December 1964.

of individual plant statements; plant statements cannot be completed until the cost of sales is determined; the cost of sales cannot be calculated until all appropriate journal entries are prepared; and so on, back through the entire procedure. When the planner is able to identify individual tasks in the closing operation, he can assign time estimates for the accomplishment of each task. Managers can obtain highly accurate time estimates for routine and repetitive procedures such as the monthly closing much more easily than for other projects to which network techniques have been applied, because long experience with the procedure is available for statistical use.

Case Background. The company involved has a relatively sophisticated budget program and an elaborate hierarchy of responsibility for accounting and reporting. Its investment in developing and maintaining these programs has paid off in improved management control and more effective management guides for taking action to reduce costs or improve profits. Management relies heavily on detailed monthly operating statements as a basis for adjusting plans and modifying emphasis on existing or new control programs; consequently, the timeliness of these statements is a key issue.

The company operates four, geographically separated manufacturing facilities. Persons at the various locations have recognized the importance of a tight closing and over the years have made independent efforts to improve the closing procedure; the result was that the statements were issued to management by noon of the seventh working day after the close of the previous month's operations. For a company of this size, a six-and-one-half day closing time is exceptionally tight. Although management appreciated the results of these independent efforts, it was not sure what it was costing to attain them. The managers felt compelled to look at the entire closing procedure closely enough to see how the tight closing was being accomplished, and to try to recommend shortcuts or other methods which would obtain the same results with less cost. Network techniques were selected as the means of studying the problem.

The Network Diagram. Each company manufacturing facility performs essentially the same operations in the closing procedure. Since the final output of the procedure is a consolidated statement, the facility with the greatest volume, and consequently the largest ledgers, will determine what overall time is required for the closing; therefore the most productive facility was selected for the analysis. The first step was to document the required operations in the closing process. These operations represented an intermediate level of detail, since it was initially only desirable to identify the points where specific interrelationships occurred and to define significant segments of work. At this point no attempt was made

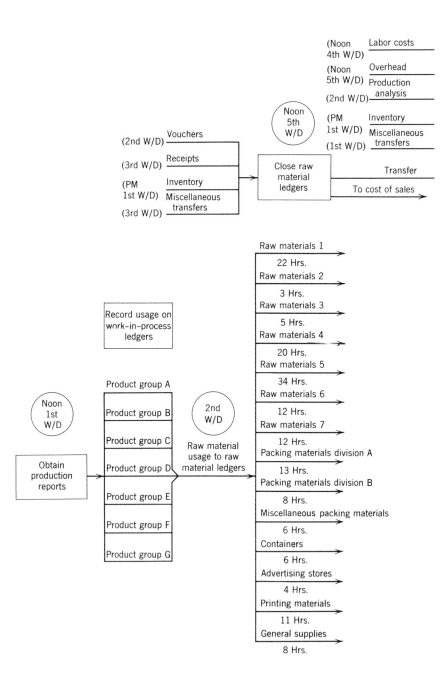

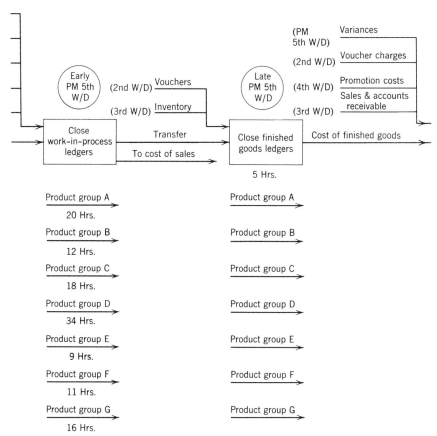

FIGURE 3. Flow of major activities—monthly closing of the books. The first step in the study was to document the operations in the closing process. A flow chart was prepared depicting these operations in general sequence. No attempt was made to diagram the operations as a network. The work day on which each operation was normally completed was noted on the chart; so were the total estimated elapsed times (in hours) where the time required to complete the work varied widely in a given operation. (Figure continued on p. 246.)

to adhere to the conventions of diagrammatically representing these operations as a network. Instead, a flow chart (Figure 3) was prepared, depicting the general sequence of these operations in the closing process. To relate this sequence to the present timing of the closing, the managers noted on the flow chart the workday on which each operation was normally completed. Also, where the time required to complete the work varied widely among the ledgers within a given operation, the total elapsed times (in hours) were estimated and inserted on the chart.

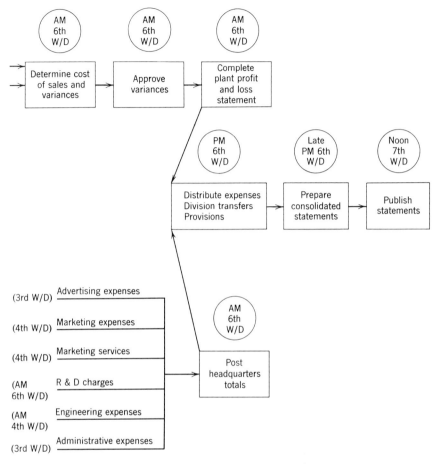

FIGURE 3 (Continued).

A review of this chart led management to conclude that only a limited number of tasks in the procedure were potential bottlenecks: the work required on the most time-consuming ledgers, Raw Materials 5 and Work-in-Process Product Group D. Since every ledger within a given operation requires the same tasks, management decided to prepare a network diagram and to conduct the analysis based on these bottleneck ledgers and their tie-in with the headquarters portion of the closing. A list was made of the required tasks on the selected ledgers and on the headquarters and general accounting portions of the closing (Figure 4). The task list was turned into a network by answering three basic questions as each task was entered on a diagram worksheet:

Plant tasks	Headquarters tasks
Raw Material Ledger	
Record plant inventory	Check and post media charges
Record usage from work in process	Close out voucher register
Extend usage dollars	Voucher payable set-up
Record receipt pounds	Distribute general office services
Foot and extend total debits	expense
Record receipt dollars	Distribute central engineering
Calculate price variance	expense
Calculate loss and gain	Post and summarize other
Summarize ledger and prepare	miscellaneous journal entries
journal entries	Record general ledger transfer
	Summarize co-op advertising, division A
Work-in-process ledger	Summarize co-op advertising, division B
Record raw material usage	Summarize jobber promotions
Record lab analysis	Prepare provision for redemption of
Calculate formulation, yield and	coupons
degrading variance	Prepare corporate preliminary
Calculate pounds made	controls and tabulations
Record production analysis	Post R & D charges from corporate
Record plant inventory	Calculate cash discount and advertising
Value inventory	expense transfer from division A
Prepare to balance accounts	to division B
Calculate loss and gain where possible	Calculate provisions for new product
Record over and under filling	research and future advertising
Record transfer of reconditioned material	Distribute corporate expense to
Balance remaining accounts and	divisions
complete loss and gain calculations	Obtain plant variances
Record direct labor charges	Prepare consolidated profit and loss
Record factory overhead charge	Compare statements with budget
Reconcile with cost ledger	estimates
Prepare journal entries	Publish consolidated statements
Finished goods ledger	
Record book or physical inventory	
Record voucher register charges	
Prepare journal entries	
Other	
Record other voucher register charges	
Record sales and accounts receivable	
Record promotion cost	
Determine cost of sales	
Approve variance	
Prepare profit and loss and transmit	
to headquarters	

FIGURE 4.   Task list. A list was made of the required tasks on the selected ledgers of the separate corporate facilities and the headquarters and general accounting portions of the closing. This task list was later represented as a network.

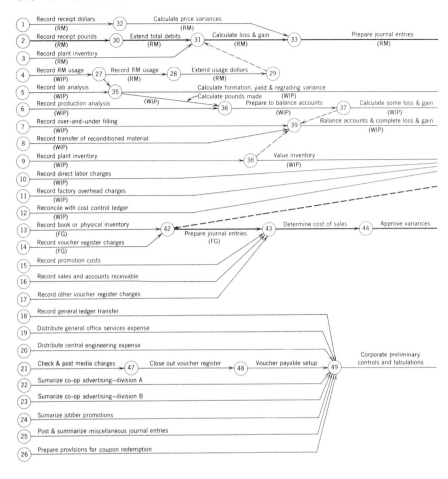

(1) What task or tasks immediately follow this task?
(2) What task or tasks must be completed to make this task possible?
(3) What tasks may be performed concurrently with this task?

The resulting network or arrow diagram (shown in Figure 5) clearly identified the sequence in which the tasks had to be completed and their interrelationships.

The Critical Path Analysis. The next step was to estimate times for the individual tasks, considering each task independently to avoid any bias in estimating. These estimates were the "normal" elapsed time required to perform the task. Normal time was defined as the duration associated with the lowest direct cost to do the job when performed by a

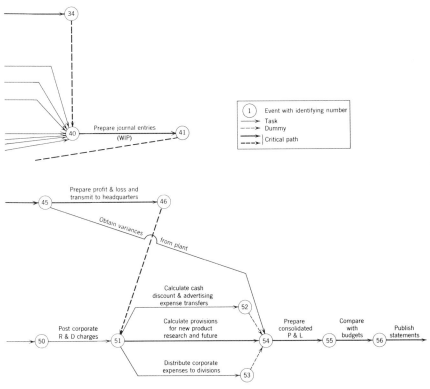

FIGURE 5. Monthly closing network diagram. After each task listed in Figure 4 was entered on a diagram worksheet, the resulting network diagram clearly identified the sequence in which the tasks had to be performed and their interrelationships.

single qualified individual on straight time. Using such time estimates, planners determined the earliest time in which the closing could be completed with the current assignment of one person on any one task and the elimination of overtime which resulted in the existing six-and-one-half day closing period. This was determined by calculating the earliest finish time for each task in the network. The information was entered on the network in tabular form as described in Chapter 5. The planners determined the earliest finish for the closing process as $62\frac{1}{2}$ working hours after the month-end. Using an eight-hour work day, the company could publish consolidated statements by the end of the eighth working day without overtime effort. The company's existing closing schedule of six and one-half days reflected a gain of approximately one and three-tenths working days; this gain, however, required 41 overtime hours (15.5 percent of the

Task	Total elapsed hours	Overtime hours
Record receipt dollars (RM)	9	3
Record receipt pounds (RM)	8	2
Record raw material usage (WIP)	6	2
Calculate loss and gain (RM)	12	4
Prepare journal entries (RM)	12	4
Record lab analysis (WIP)	5	1
Calculate formulation, etc. (WIP)	14	4
Calculate pounds made (WIP)	9	2
Record production analysis (WIP)	6	2
Calculate some loss and gain (WIP)	5	1
Record transfer of reconditioned mat'l (WIP)	8	2
Record plant inventory (WIP)	9	1
Balance accounts & complete loss and gain (WIP)	5	1
Record factory overhead charges (WIP)	7	1
Reconcile with cost control ledger (WIP)	5	1
Summarize co-op advertising, div. A	16	4
Prepare provision for coupon redemption	10	2
Post and summarize misc. journal entries	24	4
All others	94½	0
Total	264½	41

FIGURE 6. Distribution of overtime hours. It was determined that, based on an eight-hour work day, consolidated statements could be published on the eighth working day without overtime effort. This was one and three-tenths working days more than the company had taken to complete its closing, because of the 41 overtime hours it had applied.

total hours worked) applied to the tasks listed in Figure 6. Management felt the reduced closing time justified the overtime premium. They then raised two questions: Could the six-and-one-half day closing schedule be accomplished at a lower cost? Could the closing be further reduced with no additional overtime? To answer these questions the analysis was first extended to evaluate the effectiveness of the present 41 hours of overtime, to consider whether the overtime was allocated to those tasks which governed the overall duration of the closing procedure. To determine the key or critical tasks, the planners calculated the latest finish time for each task.

A comparison of the critical tasks with those utilizing overtime (indicated in Figure 6) showed that a total of ten hours overtime was being applied to only three tasks on the critical path: "record receipt pounds (RM)," "calculate loss and gain (RM)," and "prepare journal entries (RM)." In effect, these ten hours alone made it possible for the statements to be issued in slightly more than six and one-half days after the previous month-end; the balance of overtime (31 hours) did little or nothing to reduce the total closing time.

This simple appraisal answered with qualifications both questions raised by management. First, it was obvious the six-and-one-half day closing schedule could be maintained at a lower direct cost if overtime hours applied to tasks which had no impact on the overall closing time were eliminated. Second, the closing time could probably be reduced further if the total 41 hours of overtime were allocated differently and if the available manpower and skills could be exchanged among the tasks.

Further Analysis. A key to evaluating alternative allocations of overtime is the identification of the critical tasks. The information gained from calculating slack is as valuable as that gained from identifying the critical path. Slack is a quantitive measure of the seriousness of delay; it provides the basis for directing action where required and most effective. Network analysis reveals that four tasks each have eight hours of slack. These tasks had ten hours' duration, and there were 18 hours available to accomplish them. Thus, when we take the four tasks as a single series, we can see that there were only eight hours of total slack available. If any one of these tasks took up all eight hours of available slack, the remaining tasks would become critical. This example shows how total slack occurs on a path that may include one or more tasks; these tasks may flow in and out of the critical path or be completely independent of it. The total slack is equal for each task on that path; if one is delayed or prolonged, the amount available for the succeeding tasks is reduced by that amount.

On the other hand free slack is associated with a single task: it is the time available to accomplish the task, less the time required for the task, assuming all preceding steps have been completed by their earliest finish times and all succeeding steps start at their earliest start times. Free slack is equal to the difference between the earliest start of the task immediately following that task and the earliest finish of the task for which free slack is being calculated.

Final Results. Figure 7 represents the revised closing schedule as a network, with a slight departure from the usual diagraming format. Each task, represented by an arrow, is made to correspond to a more familiar time scale of working days. The scale does not represent elapsed time, since the last part of each day may include overtime hours. The solid portion of each arrow designates the period from the time the task is

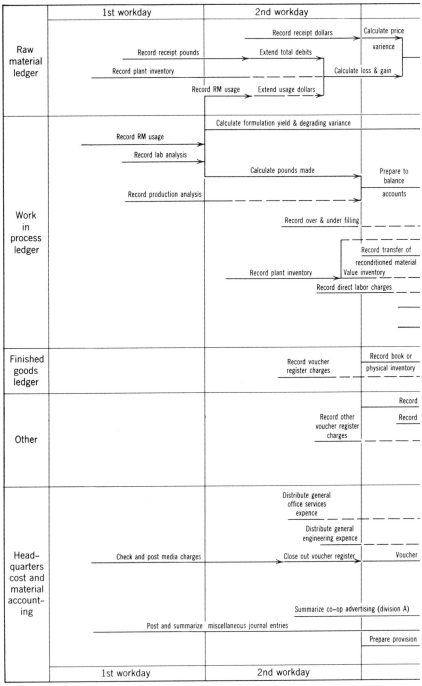

FIGURE 7. Revised monthly closing schedule. This is the revised closing schedule represented as a network, but with each task, designed by an arrow, shown in such a manner as to make it correspond to a time-scale of true working days.

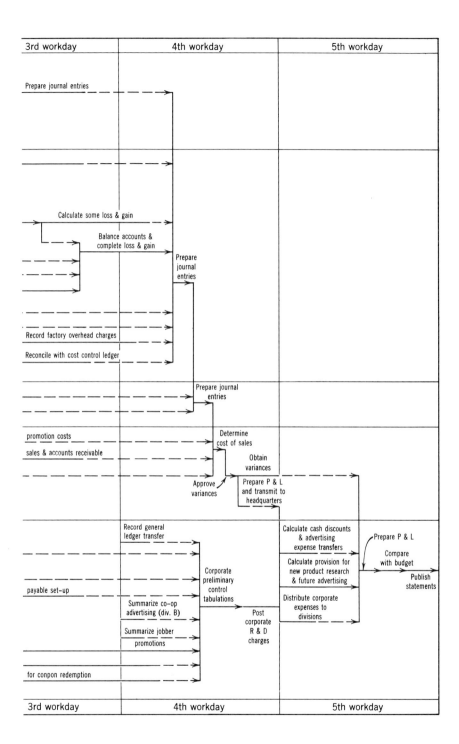

started to the point where it is completed. The dashed part of each arrow indicates the total leeway available before the task output is to be used to perform the succeeding task(s).

The revised schedule represents a closing time of five days (a reduction of one and one-half days); it incorporates the following actions, which were based on network analysis of the old closing procedure:

(1) The straight time required on critical tasks was shortened by assigning personnel from noncritical tasks (and utilizing part of the available slack) to assist individuals on the critical tasks.

(2) Overtime, restricted by management to 41 hours in total, was allocated to critical tasks only, using the total and free slack to assure that the reduced flexibility in tasks with positive slack did not produce additional critical tasks, or near-critical situations in areas that had been troublesome. (Only 12 hours of overtime were needed, a 70-percent reduction from the previous schedule.)

(3) The flow of input data, upon which the start of critical and near critical activities depend, was improved, by setting up a schedule of realistic due dates and responsibility for follow-up. Departments and individuals preparing more than one input received priority schedules with due dates staggered on the basis of the criticality of the starting tasks and the minimum amount of data needed to get each task underway.

The case study described in the preceding paragraphs suggests that similar administrative systems or procedures are amenable to network analyses. The prime consideration in conducting such analyses is that management must foresee some payoff in dollars or other tangible or intangible benefits; such payoffs are not unlikely when the procedure has been in operation for a long time, and is normally repetitive. Long experience with a procedure improves the accuracy and reliability of the data required for a network analysis. Familiarity with the procedure also enhances the opportunity for investigating it at a low cost. Any improvements gained from the analysis are recurring, because cumulative benefits rapidly offset the initial investment in the study.

NEW PRODUCT-PLANNING APPLICATION

Manufacturers have reported numerous instances in which they have used network planning to plan for a new product. The following is an abstract of a brief report on such use prepared by Yung Wong; it appeared in the October 1964 *Journal of Marketing*.

A hypothetical new product venture network is shown in Figure 8. The objective of the network is to bring a new product to the market by a specified target date. The relevant project activities are scheduled as the figure shows. The network represents the logical flow of progress for the project; the planner can determine the direct dependencies and independencies among the activities at a glance. For illustration purposes, events in Figure 8 are numbered successively, 1 through 63; in actual practice, some skip-numbering system would facilitate network revisions. The network covers the actual jobs in the product-planning process comprehensively; the degree of detail can be increased, however. For example, the activity "Motivation Research" (2, 7) may be broken down into sequential activities, "Hire Social Psychologists," "Select Interviewers," "Conduct Interviews," and "Evaluate Findings." On the other hand, the network may be simplified by combining several activities into one. For example, "Estimate Market Characteristics" (7, 10), "Estimate Competitive Behavior" (10, 11), and "Estimate Market Potential" (11, 12) can be collectively termed "Preliminary Market Research." The level of detail can be varied to gain the amount of control desired.

This network is not an ideal or even completely realistic representation of any particular product introduction. The company organization, policies, and environment determine the relevant activities in a project for that company, and any list of activities must be altered to adapt to existing conditions. A nonvertically integrated company without its own sales force must investigate the availability of jobbers or manufacturers' representatives, for example; in such a case it would not be necessary to select and train a sales force. Planning can be carefully and effectively performed by using network techniques to integrate it by cutting across organizational lines; the network system thus fills a need in the new product-planning process under the marketing management concept.

Figure 8 shows the critical paths in heavy lines. The paths have been determined from the computation procedure by using hypothetical time estimates. The activity durations are not shown here to avoid the display of lengthy computations. Two paths are critical: the one between events 7 and 18, and the one between events 19 and 26. Whereas expediting (39, 42) shortens the project time, expediting (19, 26) without similarly expediting an activity on the other critical path between nodes 19 and 26; that is, either (19, 20) or 20, 26) only makes (19, 26) a noncritical activity, because the longest duration path between events 19 and 26 remains the same, and it passes through event 20; the project duration remains unchanged.

To prepare data for a network the planner must investigate the project requirements thoroughly. Operating personnel must thoroughly understand the project, and have clear designations of responsibility from the beginning.

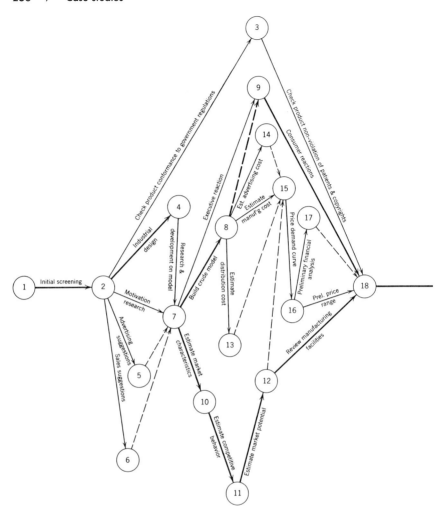

EXPERIENCE WITH A CONSUMER PRODUCT

The Van Camp Sea Food Company used network planning in introducing a new pet food. In a presentation to the fall 1965 meeting of the Southern California Chapter, American Marketing Association, Robert L. Eskridge, Vice President, Marketing, Van Camp Sea Food Company, reported that "we saved 169 days in the process of bringing the Purina Tuna for Cats product to the market place, through using the critical path method. The product actually was on the market 78 working days from the decision that a new product was needed. . . . "We found the use of a

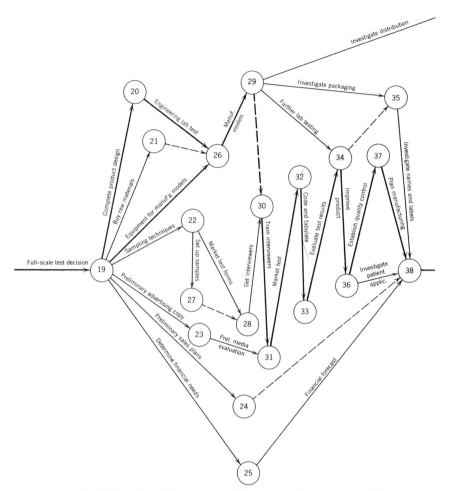

FIGURE 8.   Network for new product introduction. (Continued on p. 258.)

network planning consultant, Mr. Russell D. Archibald of Informatics, Inc., was necessary and desirable to gain the benefits which were realized."

A MILITARY PROJECT APPLICATION

As noted elsewhere in this book, network planning or PERT has been widely used in government contracting. The following paragraphs de-

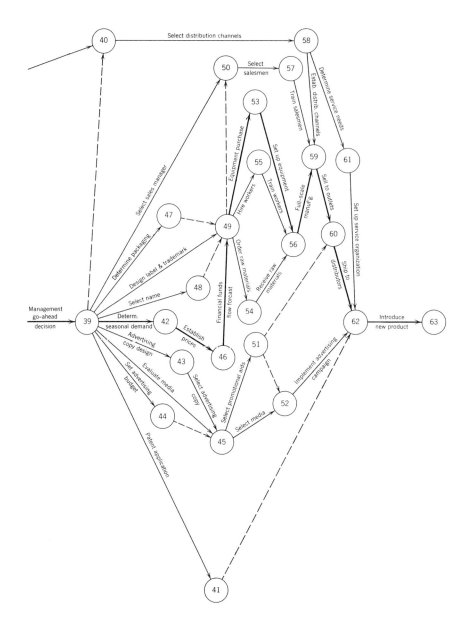

FIGURE 8 (Continued).

scribe a typical application on a major aerospace project, reported by T. L. Senecal and R. M. Sadow.[4]

In August 1960, the Dyna-Soar System program director decided that PERT with electronic data processing would be used to plan and control the program. In September, 1960 the commander of the Wright Air Development Division sent personal letters to the commanders of participating Air Force agencies and the presidents of contractor organizations, telling them that PERT was being set up in the Dyna-Soar efforts. He received replies assuring full cooperation and support. In September the System program director also established a PERT working group to assist the SPO (System Program Office) in implementing and operating PERT. The working group, composed of representatives from the principal contractors and government agencies participating in the Dyna-Soar program, met two or three times a year to review the status and progress of the PERT operation, to discuss new procedures and techniques, and to promote PERT standardization both inside and outside of the Dyna-Soar program. During the fall of 1960, the SPO developed an approach and some principles of application, and then proceeded to establish PERT in Dyna-Soar. The Program Evaluation Division of the SPO administered and operated the system. The division was named with five development system managers who were responsible for applying and developing PERT as a management technique for Dyna-Soar personnel, both government and contractor, with particular attention to program evaluation.

The Approach

The SPO employed what they called the "reporting network approach," consisting of a system of discrete detailed networks which collectively cover the total program. These networks, called "reporting networks," are developed by contractors, government agencies, or both, using the Dyna-Soar Program Elements (Figure 9) as the "road map" for delineating areas of effort and integrating these efforts into the total program. Reporting networks are generally prepared at subsystem or component level for hardware items, and at prescribed levels for other segments of the program, such as facilities construction and site activation.

The Dyna-Soar program elements are themselves unique management tools for the program. The elements divide the system into its areas, subareas, and subsystems, and provide a fundation for the application and

---

[4]T. L. Senecal and R. M. Sadow, "PERT In Dyna-Soar," presented at Institute of the Aerospace Sciences National Summer Meeting, Los Angeles, Calif., June 1962.

DYNA-SOAR PROGRAM ELEMENTS

SEPTEMBER '61

LEVEL 1	LEVEL 2	LEVEL 3	LEVEL 4
SYSTEM	AREA	SUB - AREA	SUBSYSTEM OR PROJECT
1. MILITARY TEST SYSTEM	1. AIR VEHICLE	1. GLIDER/TRANSITION SECTION	1. AIRFRAME 2. PROPULSION 3. SECONDARY POWER 4. ENVIRONMENTAL CONTROL 5. SAFETY SUBSYSTEM 6. CREW STATION 7. CYRYOGENICS SUPPLY 8. PERSONNEL PROTECTION 9. FLIGHT CONTROLS 10. PRIMARY GUIDANCE 11. FLIGHT INSTRUMENTS 12. COMMUNICATION & TRACKING EQUIPMENT 13. TEST INSTRUMENTATION 14. UNMANNED REMOTE CONTROL RECOVERY EQUIP 15. ANTENNAS, WINDOWS & FEEDLINES 16. GLIDER/TRANSITION SECTION - GENERAL
		2. BOOSTER	1. PROPULSION 2. FLIGHT CONTROLS 3. HYDRAULIC 4. RADIO GUIDANCE 5. TEST INSTRUMENTATION 6. RANGE SAFETY SYSTEM 7. ELECTRICAL 8. BOOSTER - GENERAL 9. MALFUNCTION DETECTION SYSTEM 10. AIRFRAME (STRUCTURE) 11. ANTENNAS, WINDOWS AND FEEDLINES
		3. AIR VEHICLE - GENERAL	1. AIR VEHICLE DESIGN 2. AIR VEHICLE INTEGRATION 3. AIR VEHICLE INTEGRATION TESTS 4. FLIGHT SIMULATION
	2. AEROSPACE GROUND EQUIPMENT (AGE)	1. GLIDER/TRANSITION SECTION AGE	1. HANDLING & TRANSPORTATION EQUIP (OGE) 2. SERVICING & ENVIRONMENTAL EQUIP (OGE) 3. MAINTENANCE & TEST EQUIPMENT (MGE) 4. GROUND CHECKOUT EQUIPMENT (OGE) 5. GLIDER/TRANSITION SECTION AGE-GENERAL
		2. BOOSTER AGE	1. HANDLING & TRANSPORTATION EQUIP (OGE) 2. SERVICING & ENVIRONMENTAL EQUIP (OGE) 3. MAINTENANCE & TEST EQUIPMENT (MGE) 4. GROUND CHECKOUT EQUIPMENT (OGE) 5. BLOCKHOUSE INSTRUMENTATION EQUIP(OGE) 6. BOOSTER AGE - GENERAL
		3. BASE & RANGE AGE	1. CARRIER/GLIDER AGE 2. LAUNCH SUPPORT AGE 3. TEST CONTROL CENTER 4. RANGE SAFETY AGE 5. BASE SUPPORT EQUIPMENT 6. SEARCH RECOVERY & RESCUE EQUIP. 7. BOOSTER RANGE INSTRUMENTATION EQUIP 8. GLIDER RANGE INSTRUMENTATION 9. BASE & RANGE AGE - GENERAL
		4. AGE - GENERAL	1. AVIONICS AGE INTEGRATION

FIGURE 9.   DYNA-SOAR program elements.

operation of management procedures and techniques. The program elements are used by all Dyna-Soar contractors and participating government agencies, for consistency and continuity in planning, controlling, and reporting the program. The contractual work statements, program documentation, budget estimates, cost reporting, and SPO records administration are all based on these same program elements.

Each reporting network must contain enough information to be a significant management tool for the personnel responsible for the area of effort it covers. The networks, generally containing from 150 to 500 events, must each present valid pictures of the planning to accomplish established objectives; both the contractors and Air Force personnel must

LEVEL 2	LEVEL 3	LEVEL 4
AREA	SUB-AREA	SUBSYSTEM OR PROJECT
3. FACILITIES AND ACTIVATION	1. TEST SITE FACILITIES	1. AFFTC (EDWARDS AFB) FACILITIES 2. AFMTC (AMR) FACILITIES 3. TEST SITE FACILITIES – GENERAL
	2. TEST SITE ACTIVATION	1. AFFTC (EDWARDS AFB) ACTIVATION 2. AFMTC (AMR) ACTIVATION 3. SITE ACTIVATION – GENERAL
4. SYSTEM DEVELOPMENT TEST	1. TEST AIRCRAFT	1. AIR LAUNCH CARRIER 2. FERRY AIRCRAFT 3. CHASE AIRCRAFT 4. TRAINING AIRCRAFT
	2. AIR LAUNCH TEST	1. AIR LAUNCH TEST PLANNING 2. AIR LAUNCH TEST CONDUCT 3. AIR LAUNCH TEST SUPPORT
	3. GROUND LAUNCH TEST	1. GROUND LAUNCH TEST PLANNING 2. GROUND LAUNCH TEST CONDUCT 3. GROUND LAUNCH TEST SUPPORT
	4. MATERIEL SUPPORT	1. MATERIEL SUPPORT MANAGEMENT 2. SUPPLY SUPPORT 3. TRANSPORTATION & PACKAGING 4. MAINTENANCE 5. OPERATING & MAINTENANCE MANUALS 6. TECHNICAL SERVICES
	5. SYSTEM DEVELOPMENT TEST - GENERAL	1. TEST FORCE ORGANIZATION 2. AIR CREW SELECTION & TRAINING 3. GROUND PERSONNEL SELECTION & TRNG. 4. INTEGRATED CREW TRAINING
5. PROGRAM MANAGEMENT	1. REVIEW & EVALUATION	1. ENGINEERING REVIEWS 2. MOCKUP INSPECTIONS 3. C.T.C.'s 4. SAFETY OF FLIGHT INSPECTION 5. SYMPOSIA
	2. PLANNING & CONTROL	1. PROGRAMMING 2. STATUS & PROGRESS REPORTING 3. PROGRAM EVAL. & REVIEW TECHNIQUE (PERT)
	3. PROGRAM MANAGEMENT-GENERAL	1. ADMINISTRATIVE OPERATIONS 2. INTERNAL LIAISON (NOTE: THESE ARE CONTINUOUS INTERNAL SUPPORT ITEMS)
6. RESEARCH & DEVELOPMENT PRODUCTION	1. PLANNING	1. MANUFACTURING PLANS 2. MFG. METHODS DEVELOPMENT PLAN 3. FACILITIES PLAN (IN PLANT) 4. TOOLING PLAN 5. MAKE OR BUY
	2. FACILITIES	1. TOOLING 2. MANUFACTURING FACILITIES 3. GLIDER CAT. I TEST FACILITIES 4. BOOSTER CAT. I TEST FACILITIES 5. GLIDER/BOOSTER CAT I TEST FACILITIES
	3. FABRICATION	1. SPECIFICATIONS 2. MATERIALS & PROCESSES 3. METHODS 4. QUALITY CONTROL 5. REFURBISHMENT
	4. DELIVERIES	1. ACCEPTANCE TESTS 2. DELIVERY SCHEDULE
7. SYSTEM INTEGRATION	1. SYSTEM INTEGRATION	1. SYSTEM TRADE STUDIES 2. SYSTEMS REQUIREMENTS 3. SYSTEM ANALYSIS 4. MAINTAINABILITY PROGRAM 5. RELIABILITY PROGRAM 6. SAFETY PROGRAM 7. HUMAN ENGINEERING 8. MILITARY EQUIPMENT APPLICATIONS 9. TECHNICAL INTEGRATION

be able to use them. The reporting networks contain interface events (see Figure 10) which link the networks together to form a total program network just as a map of the United States can be constructed by using individual state maps connected by national and interstate highways. Through the use of data processing techniques, all the reporting network data are integrated into a total program PERT output. For example, there were separate reporting networks for each of the Level 4 subsystems (airframe, propulsion, etc.), comprising the Level 3 "Glider/ Transition Section." If fed into the computer simultaneously, all the PERT data for these Level 4 networks will produce a computer printout covering the glider/transition section and its interfaces with other parts of the program.

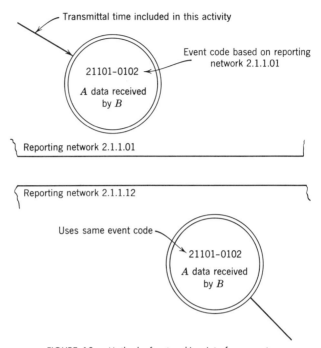

Transmittal time included in this activity

Event code based on reporting
network 2.1.1.01

21101–0102

A data received
by B

Reporting network 2.1.1.01

Reporting network 2.1.1.12

Uses same event code

21101–0102

A data received
by B

FIGURE 10. Method of networking interface events.

Application Principles

In applying PERT to Dyna-Soar, planners established basic principles
covering both the psychological and the operational aspects. Our experience has proven the soundness of the following principles:

(a) PERT is a management technique developed for SPO engineers
and specialists and for other government and industry managers and
engineers directly engaged in the Dyna-Soar program. The application
and use of PERT was not restricted to the staff planners, programmers,
and schedulers; it was provided as a working tool for all persons working
on Dyna-Soar. PERT was used as the basis for engineers and managers
to discuss events and activities, sequencing of events, scheduling, etc.,
just as circuit or wiring diagrams are used as a language or mode of expression by electrical and electronic engineers.

(b) PERT was the basis for integrated reporting to all levels of management. Starting from the reporting network data, planners prepared sum-
mary-type reports encompassing larger functional areas with fewer de-

tails, for different levels of management. Each management level received reports appropriate to its needs, and yet all reports were based on the same detailed information.

(c) PERT was used as a mutual management tool for both industry and government. Expected dates forecast through use of PERT were not considered as contractual commitments, but contractors were expected to take reasonable and acceptable corrective action to alleviate conditions or delays forecast through PERT. We must emphasize that PERT was applied to Dyna-Soar as a management technique to be used both by the contractors and the Air Force.

(d) PERT is nondirective; that is, it is not a means management can use to direct or approve contractor approaches, plans, actions, etc. PERT reflects what has been directed or approved, or the contractor's understanding and planning for tasks not firmly projected or defined. Under PERT, changes in approach or planning are accomplished through normal management channels, and simply indicated on reporting networks. Authorized redirection or change is indicated by revising and updating the applicable reporting networks.

The SPO Engineer's Role in PERT

Another feature of Dyna-Soar PERT was the unique role played by SPO engineers. In the SPO operation these engineers were the key figures: the depth of detail on reporting networks was based on their judgment, and they tried to have the responsible contractor or government agency develop reporting networks with information sufficient for SPO planning and evaluation needs. In effect, they sought for SPO use a "corridor of information" for each reporting network lying between the unreliable, relatively meaningless gross network and the overly complex and detailed, hard-to-handle network. However, a contractor is permitted to develop whatever detailed networks he may require for his own internal operations. The SPO engineers would review and accept the appropriate reporting network for their area of effort.

The SPO engineers' PERT effort was not intended to support a centralized SPO PERT operation, but rather to act as a mechanism for increasing their capability to perform their own job functions. The SPO authorized and encouraged the SPO engineers and the contractors to correspond about and discuss the network content. The reporting networks further supported the SPO engineers by giving them a basis for preparing work statements, for establishing technical reporting requirements, and for coordinating interface events. In addition, PERT automatically created a better understanding between the SPO and the contractor engineers.

SPO and Contractor Responsibilities

PERT management was not delegated to a contractor or management consulting firm: the in-house SPO directed and managed the entire Dyna-Soar PERT operation. Specific tasks and responsibilities performed by the SPO were:

(a) Establishing criteria for development of reporting networks; guiding responsible people to use the appropriate depth of detail for networks; establishing the schedule for data submission, processing, and analysis; setting up the analysis format; and managing other PERT operations pertaining specifically and solely to Dyna-Soar PERT.

(b) Receiving input data from all participating organizations—the system contractor, associate contractors, and other government agencies.

(c) Providing computer processing for Dyna-Soar PERT data and forward computer print-outs to the responsible contractors and government agencies for review and analysis. (Data were processed on the IBM 7090 at the Aeronautical Systems Division, Wright-Patterson Air Force Base, Ohio.)

(d) Summarizing and/or integrating reporting networks into a total program network.

(e) Analyzing integrated PERT outputs for impact on the program. Reporting networks were developed by the organization having functional responsibility for a particular subsystem or portion of the program. On parts of the program which had more than one responsible organization participating, all participants worked together to develop the network. The SPO did not suggest or tell a responsible contractor what planned effort, work flow, etc., to include on a reporting network, because the SPO wanted the network to present the doers' honest and true appraisal of the job and the time required to do it.

The SPO used the same approach to the PERT output. The organization that developed the reporting network was responsible for reviewing and analyzing the computer printout, and for taking remedial action within its own management authority and prerogatives. Situations and problems beyond their authority were referred to the SPO for action. The responsibilities of the contractor (and other participating government agencies) were to develop and/or participate in the development of reporting networks covering their areas of program responsibility; to report on, and update reporting networks on a biweekly basis; and to analyze computer printouts received from the SPO and to submit to the SPO for each printout an analysis report identifying problem areas and the corrective action being taken or recommended.

Procedures

The development of a reporting network was scheduled, and the network was prepared and submitted to the SPO. The SPO reviewed and accepted the network and initiated biweekly update and analysis reporting against it. The organizations responsible for reporting networks simultaneously submitted to the SPO update reports for the immediate reporting period and analysis reports covering the previous reporting period. Subcontractors participated in PERT through their respective prime contractors, to the extent determined by those contractors. All participating organizations used a common reporting period so that all data had the same "as of" date. The SPO processed the PERT input data and forwarded the resultant computer printouts back to the submitting organizations for review and analysis. This data flow from responsible organizations to SPO and back took place on a regular two-week cycle.

Reporting networks, related printouts, and analysis reports, were distributed within the SPO to all persons requiring them. PERT administrators actively supported the SPO engineers in their use of PERT and then, satisfied that the reporting networks were valid and significant, integrated them into more comprehensive reports for program evaluation purposes. In 1962, the SPO was operating 27 reporting networks consisting of some 4,300 events and about 5,500 activities. A few months earlier, the Dyna-Soar program had been redirected from a suborbital mission using a Titan II booster to an orbital mission using a Titan III booster. Existing reporting networks covering hardware items were revised to reflect the new requirements (new contractual work statements) and new reporting networks for other program areas were prepared, as soon as the basic management decisions and plans for these areas were completed.

The SPO realized to some degree all the benefits visualized at the start of the PERT implementation. Overall program planning was improved, cooperation was better, and efforts were coordinated earlier than usual through mutual consideration and review of networks; parts of the program were expedited or replanned to avoid forecasted delays or slippages, and simulation exercises were used to aid conversion to the redirected Dyna-Soar program. Contractors replaced their conventional in-plant progress reporting methods by PERT, and the SPO replaced former contractor progress-reporting requirements by PERT reporting, and dropped contractual requirements for any detailed schedules.

Areas of Emphasis

Valid Networks. Reporting networks must be valid reflections of the work. The planner must be confident that each network pictures the effort honestly. If responsible persons faced with a PERT computer print-out or analysis say "we're not really doing the job that way," or, "that isn't really a problem, the network is all wrong," PERT may fail before it has a chance to prove its worth. Networks are the heart of PERT, and so PERT is only as good as the networks.

Interface Events. In a program as complex as Dyna-Soar, the identification and coordination of interfaces among contractors and government agencies, and even among contractors, is an enormous but mandatory task. Successful integration of reporting networks depends upon including coordinated interface events on the networks.

MAJOR CONSTRUCTION PROJECT

In the fall of 1964, the B & R Construction Co., San Francisco, was awarded the contract to build the Oakland City Museum for about $6 million. The contract required that the project be completed in 24 months, and included a $500 per day penalty for late completion. B & R Construction decided to use the Critical Path Method and retained a consultant from the CPM Systems Division of Informatics Inc. Mr. J. D. Poulsen was the principal consultant on the project, assisting in the planning, scheduling and monitoring.

The structure is reinforced concrete with 3 primary levels, much of which is below grade. Eero Saarinen Associates designed it, and the specifications contained very tight requirements for the concrete work.

The network plan was developed in a four-week period while the site preparation and excavation work was under way, and contained approximately 800 activities. It was completely activity-oriented. Most of the plan was prepared in the construction shack in consultation with the field superintendant and the project manager, Mr. Leo Buckstaber; the rough field network was cleaned up and straightened out in the consultant's office. The network was processed on the IBM 1440 Project Control System, and after 3 edit runs to locate and correct errors, the first computed results were obtained.

This first report indicated that 36 months would be required to complete the job, if the plan shown on the network was followed. Obviously, this was unacceptable, in view of the $500 per day penalty over 24 months, and so the project manager, field superintendent, and planning consultant reviewed the network in detail and made numerous changes. Some of these removed inadvertant restrictions; some changed the sequence of operations from "desirable if there is time" to "acceptable if we have to." The planners overlapped more activities and made time reductions on some critical path activities. The next computer run, which took into account all these revisions, showed that the critical path was 29 months long—still unacceptable. At this point, B & R Construction realized that it faced serious difficulties in meeting the contract completion date. The project manager would have found this out without network planner, but he would not have learned it until six months had gone by. Network planning told him this when he was still excavating—and had time to do something about it.

To get the schedule down to the required 24 months, several major decisions were made relating to the methods of forming and placing the concrete. Additional form sets were purchased, and concrete pumps were used instead of more conventional concrete placement procedures. These actions, plus a number of others relating to architects' approvals and specifications interpretation and requirements, enabled the company to be assured that it would meet the contract completion date.

The master schedule for the job was prepared in time-based network fashion and hung on the wall of the construction shack. (Actually, it was over 20 feet long and so had to be folded to display the portion of the job currently in progress.) The field superintendant marked his progress daily, and was able to see exactly how he was doing against the plan. Once a month, the planning consultant collected all completion data and made a computer run to determine the collective impact of the progress; he then prepared a summary report on the job which he submitted to the field superintendent and the project manager. Since many items were ahead and behind the master schedule, this showed the overall effect on the critical path, and the field superintendent could then take required action to recover to schedule if necessary.

During this project, unprecedented rainfall delayed the job about 40 days. When work was resumed, a new master schedule was prepared within a few days, and all subcontractors on the job had their new schedules by the time they had mobilized their crews and equipment. An extension of time to the contract was granted because of the weather delays.

The contract for the network planning consulting effort was $12,600, which included the monitoring of labor and materials expenditures against the plan. The contractor, B & R Construction, credits the use of network planning with preventing major losses which could have been experienced by the $500 per day penalty. At the time this report was being prepared, the job was eight months from completion, and gave positive indication that the revised completion date would be met.

SUMMARY

In this chapter we have described a number of applications of the network-based system, to illustrate the flexibility and adaptability of the basic concepts. Although the military weapon system and aerospace programs have been the spawning ground for much of this effort, we have not emphasized these programs in the case studies in this book, concentrating instead on other applications. We have done so because national security requirements create difficulties in discussing specific applications, because those responsible are often reluctant to discuss situations which seem to reflect adversely on their organization, and because the massive size of most of these applications would make it necessary for many people to work together to properly document a case study. (The result would probably be a volume as large as this present work itself.) Network planning in government-sponsored projects has produced a large number of bad experiences; the many reasons for this are discussed in Chapter 17. However, since PERT has now become a standard contractual requirement, and there is no sign of this requirement being removed for a considerable number of years, we can conclude that the network system has resulted in more and more effective applications, and that government and industry alike agree that the results are worth the effort.

14

Commercial and industrial case studies

In this chapter we will continue our survey of network planning with three commercial/industrial case studies: a major refinery-unit turnaround, a machine-tool rebuilding operation, and a major industrial construction project. In Chapter 15 we will study a rather large, multiple-unit, residential construction project.

MAJOR REFINERY UNIT TURNAROUND[1]

An oil refinery contains many different types of units, each of which has to be repaired and overhauled at periodic intervals. Network planning and scheduling have been effectively used in turnaround work on many different kinds of units, including combination units, delayed cokers, reforming units, gas plants, ethylene units, crude units, fluid-cracking units, and fluid-coking units. We can illustrate the use of network planning in any of these turnarounds by a specific example of the planning and scheduling of a turnaround of a Thermafor Catalytic Cracker (TCC). The TCC unit produces high-quality gasoline and increases the yield of distillate fuel fractions with attendant reduction in the yield of residential fuels. There are a number of these units in use in the world; they generally consist of a separator, a surge tank, a reactor, a kiln, heat exchangers, piping, pumps, and air compressors. All of these parts are stacked in a structure to allow air lift to raise the catalyst to the top of the structure. In the case under discussion, the structure is 275 feet tall. The catalyst, a powder-like substance, flows through the reactor where it mingles with the charge stock, then flows to the kiln where it is regenerated, and is finally lifted through the lift pipe back to the separator surge tank for reuse. We will not attempt to discuss the intricacies of the opera-

---

[1]This case study was prepared for this book by Elmo F. Pace, Vice President, Pierose Maintenance Corporation, Los Angeles, California.

tion or the reaction within the operating unit, but we give this general description to illustrate the wear due to heating and cooling, as well as the erosion caused by the moving catalyst, and by corrosion within each of the units. Because of the wear the owner company must periodically shut down the unit and repair or replace portions which have corroded or eroded. It is always most important to remove the accumulated coke within the reactor and kiln to bring the entire unit up to full capacity and efficiency. This operation is known as a "turnaround."

Preliminary Planning

During the operating period (which may be from 12 to 24 months) the owner company must prepare a list of items in the unit which will probably need repair, replacement, or alteration. Operating personnel are normally responsible for accumulating the data needed to prepare this work list; the data consist of measurements of corrosion and erosion rates, heat transfer rates, pressure losses, and so on. Nondestructive testing, using methods such as x-ray and Sonizon, is often used to measure metal thickness while the unit is operating. Such data, compared with the years of experience with the unit which the owner company has, make it possible for the operating personnel to compile an accurate list of work items.

Besides establishing the work to be done, the company must balance several critical factors to determine exactly when the turnaround should take place. Special materials such as alloy piping, and replacement parts requiring fabrication such as tube bundles and piping, must be ordered far in advance. The company must also determine the best time to shut down the unit so that markets can be supplied from storage facilities without interruption. To do this, the company must be able to predict how long the unit can continue to operate efficiently and economically without danger of failure in any major component due to excessive corrosion or erosion. The prime objective, of course, is to operate the unit as long as possible, and to schedule the turnaround and repair work at the most economical time to shut the unit down.

The owner company must also decide who will perform the repair work. In the past ten years the petroleum industry has found it increasingly economical to contract for this work, rather than to maintain a large labor force on a year-round basis to do it. If the company plans to use an outside contractor, the responsible manager must select a contractor with completely qualified supervision who has had experience in planning and scheduling as well as in executing the overhaul or turnaround work necessary. He must also be sure that the contractor is able to recruit qualified labor in the quantity required and to supply tools and equipment necessary to effect the turnaround efficiently and economically. The company must also choose a contractor who is thoroughly acquainted with the development and use of a planning and scheduling procedure which will completely define the operating method for the work, to enable the owner company to follow progress on a day-to-day or even an hour-to-hour basis; in this way, the owner company is assured that the work will be completed in the time allotted for the shutdown.

The best way for the owner company to meet its need for complete planning and scheduling, to keep fully informed of progress at all times, is for it to use network planning; use of the network, combined with periodic progress reporting and analysis, allows management to identify the current critical path and keeps them informed of the work status.

Turnaround Duration

A major refinery unit processes 30,000 barrels per day or more; it is therefore imperative that the turnaround be as short as possible, that it start on schedule, and that it be completed on schedule. Each day of the turnaround brings a production loss worth thousands of dollars; each day of delay entails a similar loss and endangers the company's share of the market. The turnaround contractor is normally given a specific start date on which the unit will be sufficiently cooled so that work can begin, and a specific date on which the unit must be ready to start up again.

When he knows the number of days allowed for the shutdown, the contractor can assist in planning the work to meet this schedule and can recommend the hours to be considered as one workday, the days per week, and whether a two-shift, three-shift, or staggered-shift operation will be used. Use of the Critical Path Method will assist him in determining this.

Preparation of the Network Plan and Schedule

When the contractor was awarded the contract covering the planning and scheduling of the TCC unit turnaround, he appointed a superintendent and made him responsible for developing all necessary information to produce a suitable network plan; the plan had to contain the proper descriptions and event numbers so that the data could be processed through the computer to identify the critical path.

In the preliminary planning stages of this kind of operation, the company must use the services of a man who is qualified to act, or will act as the work superintendent. He must be responsible for reviewing the entire work list and realigning it in sequential order so that it can be further divided to develop the information in network-plan form. The superintendent assigned to the TCC unit turnaround described here visited the job site with the owner company's engineer, inspected the entire unit, and accumulated all the information he needed to supplement the data on the work list. After studying the operation and investigating the step-by-step requirements in detail, the superintendent prepared an overall list, dividing the work into physical areas to expedite manpower allocation and network plan development. He also had to prepare a complete tool and equipment list to meet the project requirements, because the availability of such tools and equipment would directly affect manpower needs and activity time durations. Prior to leaving the job site, the superintendent also investigated the labor market in the area and checked with various labor organizations to ascertain the qualifications and availability of local labor; with this information he could further support the contractor's claim that he could secure manpower in the required quantities with the required skills to perform the work in the assigned time.

As soon as the basic information had been obtained and the sequential operations developed, a consultant skilled in network planning and associated computer operations was retained (CPM Systems Division of Informations Incorporated, located in Sherman Oaks, California). The consultant was assigned to work with the superintendent to develop the data necessary for processing. The superintendent and the planning consultant were assigned to develop the complete network plan, showing the sequence of operations, the duration of the work, and the critical path. Using the preliminary information secured by the superintendent, the two men developed manhour requirements by crafts for each activity in the network and prepared the data for machine processing. To do so, they made a preliminary draft of the network plan diagram and wrote a preliminary description of each operation on the plan. The superintendent assigned duration times in hours by crafts, including the numbers of each craft for each particular operation or activity.

After completing the rough network, they placed the information on input data sheets. To prepare the data in proper form for processing through an IBM 1620 computer, they had punched cards prepared from the sheets. The first run was made to determine the total project duration, based on the first preliminary network and the time estimates. The run indicated a total of 316 hours' work time against an allowable 240 hours', so that a second run was necessary to adjust duration requirements. In the preliminary stages the owner had decided that the job was to run twelve days; to meet this deadline, the contractor decided it was necessary to work two ten-hour shifts a day for the total twelve-day period, to get in 240 work hours. The other four hours each day were left open for overtime spot work which might become necessary if unanticipated problems were encountered in any particular area. The second computer

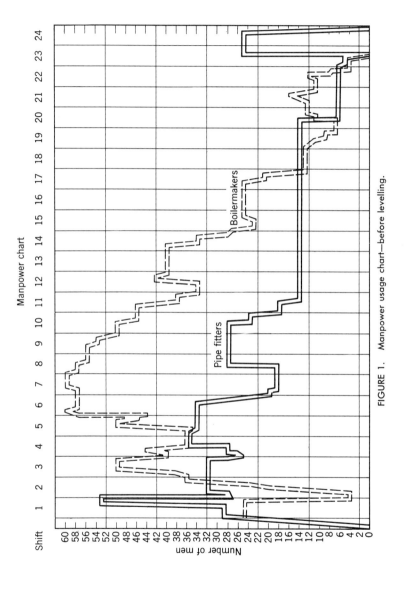

FIGURE 1. Manpower usage chart—before levelling.

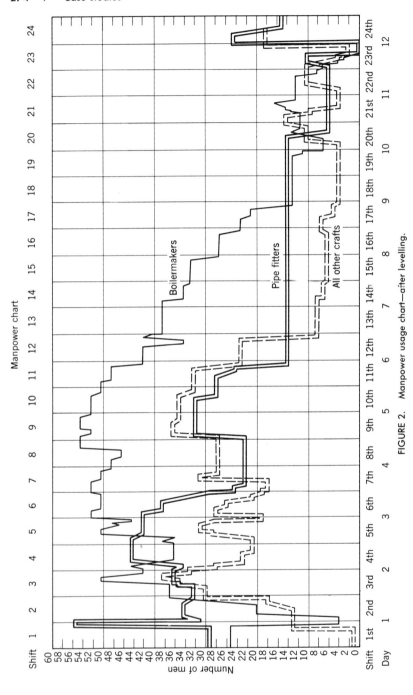

FIGURE 2. Manpower usage chart—after levelling.

run developed the proper elapsed time by changing the activity time estimates, which involved increasing manpower requirements; it also indicated the critical path, with second and third possible critical paths which might develop due to unanticipated troubles on the most critical path.

Manpower Requirements

It was then necessary to process the data through the computer to determine manpower requirements. The proposed network plan indicated too great a fluctuation in manpower over short time periods; it was therefore necessary for the superintendent to adjust the numbers of craftsmen required in certain crafts, to lower the peaks and level out the manpower requirement curve. This adjustment also made it necessary to change the time estimate for the affected activities, but the changes were limited to those not on the critical path. The manpower requirements were also leveled by delaying the start of some activities which had available slack.

The first run to determine manpower requirements (the third computer run) indicated a requirement of 50 boilermakers for the first four hours, 20 for the next six hours, and 35 for the following period. A similar unrealistic fluctuation in pipe-fitter manpower indicated that it would be necessary to level out the requirements. After adjustments in numbers of personnel and time for operation were completed, a fourth run was made to compute manpower on a more realistic and economical level. Figure 1 indicates the fluctuation resulting from the first manpower-leveling run through the computer. Figure 2 shows the final manpower chart, indicating a smoother curve for the operation. (In this application, these charts were hand-prepared. Figure 3 is an illustration of a computer-prepared chart of the same kind, for a subsequent refinery turnaround.) The superintendent and the consultant decided that the fourth run provided an acceptable schedule with realistic manpower requirements, and that the operation could be performed in the time allocated, using these basic data. All network planning, scheduling, and manpower leveling encounters fluctuations not readily definable in the program; the superintendent in charge of the work must therefore be prepared to make adjustments in manpower requirements and to change the assignments on a day-to-day or even hour-to-hour basis, to pick up lost time caused by unforeseen difficulties. The final graph of manpower requirements, while not completely smooth, was symmetrical enough to indicate that day-to-day operations would permit the most efficient manpower utilization, and that miscellaneous emergency adjustments would smooth out the curve when work was in process.

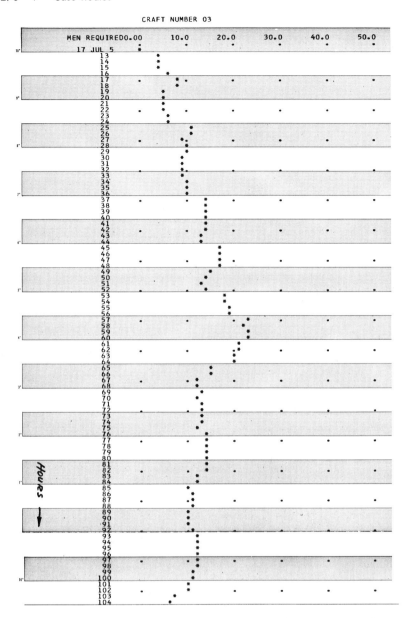

FIGURE 3.   Computer plot of manpower requirements derived from network plan.

The Master Schedule

The final master schedule for the turnaround was prepared as a time-based network, stratified to group activities in work areas in the same horizontal location on the network. The result, shown partially in Figure 4, is an elaborate bar chart indicating interdependencies. This method of displaying the results of the planning effort and computer analyses was very effective in communicating the schedule and the potential problems to all concerned. Responsibilities were assigned by the areas shown on the left side of the chart.

By dividing the plan into work areas to produce several "subnets" the superintendent and the consultant could identify the critical paths for each area. The most critical path was in the reactor area, with a secondary critical path in the valve and piping section because a large vapor line had to be removed and reinstalled. A third was in the kiln area: the difficult work involved in inspecting removed furnace tubes, and replacing defective ones could easily have caused delays which would have made this path the most critical.

Detailed Planning Sketches

The company and the contractor also found it advisable to develop additional detailed sketches which would assist management in communicating requirements to the supervisors and craftsmen. There was a particular need to replace approximately 72 feet of lift pipe in the center of the structure, a process requiring rather intricate rigging operations. To stay within the manpower requirements and assure that the work would be finished in the time alloted for this operation, the managers had a detailed sketch made for each phase of the change out, so there would be no question regarding procedure. The sketches detailed each rigging operation necessary to remove the present lift pipe and install the new one, to make it easy to explain the operation to the craftsmen assigned to carry it out.

Summary and Conclusions

Elmo F. Pace, Vice President, Pierose Maintenance Company, who directed this project, said in summarizing the results: "Network planning enabled the contractor to plan and schedule work and personnel to make each phase of the operations as efficient as possible, while staying within the manpower and time limits for the work. It was estimated that the total operation would require 14,000 man-hours in twelve days of two shifts per day, or 240 working hours, with an average crew of about 58 men. No attempt was made to network the preliminary work of moving

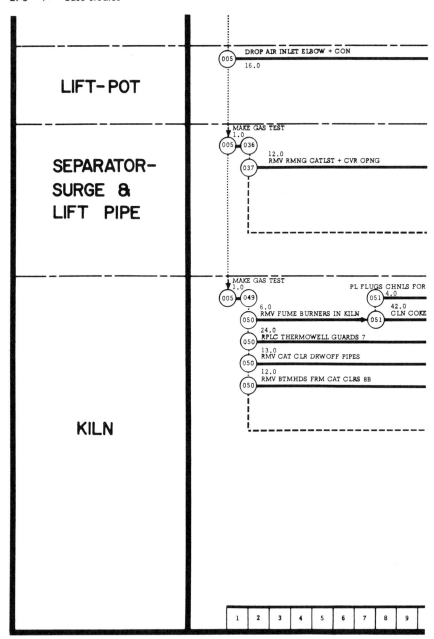

FIGURE 4. Portion of master schedule in time-based network form.

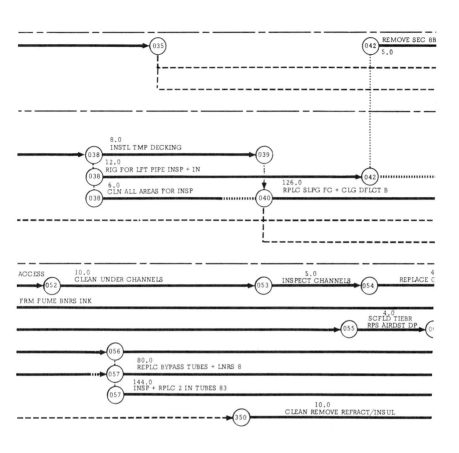

| 10 | 11 | 12 | 13 | 14 | 15 | 16 | 17 | 18 | 19 | **20** | 21 | 22 | 23 | 24 | 25 | 26 | **27** | 28 | 29 |

in, constructing the office and other temporary facilities, or setting up the network; these items were simply planned to take place in a two-week period prior to the shutdown. Similarly, no attempt was made to include the final cleanup and removal of staging, etc., after the job was completed and the TCC unit started back up. These items, and the ordering and delivery of major items of material, could have been planned in network form with some benefit.

"When using CPM for planning and scheduling turnaround work, the manager must realize that CPM is a tool and, as such, only as good as the information fed to the computer. For the owner company and the contractor to reap the maximum rewards from the system, the network must be designed to communicate the maximum amount of information to supervisors and management, while requiring them to make the least-detailed study possible. Progress cards indicating completed activities were processed through the computer after every shift, to update the program and allow the engineer to record progress on the time-based network; by doing so, he was able to communicate to management and all concerned the exact status of the job after each shift or at the end of each day. These progress cards tell what time each job was completed; the manager can analyze the network in the standard fashion, using these completion times as the base line. Before network planning and scheduling were used for turnarounds, the superintendent had to process his plans for such work by hand. Although it has always been necessary to develop a plan for turnaround work, network planning provides a more definite method of varying plans to use manpower most efficiently. A hand-processed plan for a turnaround comparable to one developed by using a network analyzed by computer would cost at least twice as much as the computer plan, and would not lend itself to updating so efficiently. Companies performing this type of work need this tool to enable them to use manpower most efficiently and economically, and thereby to cut costs. Thorough analysis and proper scheduling using the Critical Path Method will shorten shutdown time significantly on any major unit in a refinery or chemical plant."

MACHINE-TOOL REBUILDING OPERATION[2]

The Hughes Tool Company operates one of the largest machine shops in the Southwest for the manufacture and fabrication of oil tools, as well as running a major, commercial, machine-tool rebuilding service; in doing

---

[2]This case study was prepared by J. A. Campise, formerly with Hughes Tool Company, presently with Computer Sciences Corporation, Houston, Texas.

so, it has faced virtually all the possible problems in machine-tool overhaul and rebuilding. A modern maintenance building at Hughes houses all of the facilities for stripping almost any type of production machine to the base, repairing or manufacturing new replacement parts, and rebuilding and repainting the machine to make it "like-new."

In reaching the current scope of operations in this area, characterized by rapid response at optimized cost to demands for service, Hughes overcame many scheduling problems. Facilities and manpower had to be scheduled to provide maximum service to outside customers, without curtailing the normal maintenance and overhaul program for over four hundred machine tools in the production plant. The company had to plan and schedule the warehousing of customer machines to minimize both the wait times and the number of moves for any single machine within the promised lead time. Furthermore, the company had to be capable of manufacturing replacement parts for many types of machines for which parts were not available from the original manufacturer; skilled men in over a dozen different crafts had to be assigned, so that their scheduled times insured their availability during the precise working shift in which their individual skills were required.

Over the years manual planning methods, such as the familiar Gantt chart and milestone techniques, had proved adequate for control. The ground rules were always subject to change, however. In seeking to increase productivity, the Hughes management recognized the need to provide faster and more accurate estimates, tighter scheduling, better communications, and continuous, timely progress reports in which potential problems could be quickly identified; they knew that all those increased capabilities had to increase the probability of meeting deadlines. The company needed the ability to simulate the effects of alternate plans to study their effect upon deadlines, prior to and during the course of the actual work.

The Network-Based System

When Hughes began to improve its systems in this area, its scientific and management staffs evolved a modified version of PERT which considered project consequences from the standpoints both of time involvement, and of manpower requirements and job cost. This HUGHES-PERT Tool System was developed jointly by the Oil Tool Division Plant Engineering Staff personnel (who contributed knowledge of on-the-floor operations in the Machine Tool Rebuilding Center), and by systems-planning and programming personnel from a staff of engineers and scientists designated as the HUGHES DYNAMICS division. Numerous tests produced a modified version of PERT; plans were formulated to expand

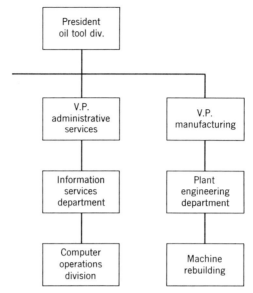

FIGURE 5.   Organizational relationship machine rebuilding and computer operations.

its use to include planning, scheduling, and review of all jobs performed in the Machine Tool Rebuilding Center in Houston.

The critical path technique becomes more advantageous as the size and complexity of the project increases. It is important to note that no change in management organization or policy is required to implement the system, because the system merely makes it possible to organize existing data into a more meaningful form for immediate management use. Since jobs occur almost randomly, PERT must be used in an organization whose structure allows computer time to be freed to process the data for each job. Fortunately, the data processing function in the Hughes Oil Tool Division can give this kind of service. (See Figure 5.) The critical path computations are performed on an IBM 1410 System, with magnetic tapes, a disk file, a printer, a card reader, and a card punch on hand. In its present form the system does not require the use of either tape or disk files. (The procedure described below has been operating successfully for over two years.)

Typical Operation

When an order to rebuild a machine tool is received, or when a scheduled overhaul and rebuild occurs in the maintenance program, the

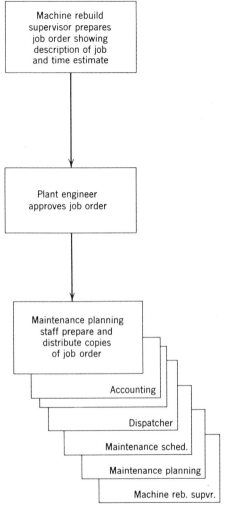

FIGURE 6.   Issue job order.

machine rebuild supervisor prepares a job order showing the description and time estimates for the job. The plant engineer approves the job order and forwards it to the maintenance-planning staff, who prepare copies and distribute them as shown in Figure 6. Phase I of the procedure begins at this time, usually within one day of receipt of the order, as depicted in Figure 7. The maintenance-planning staff prepares a critical path scheduling worksheet and, from this, the arrow diagram (Figure 8). Tasks

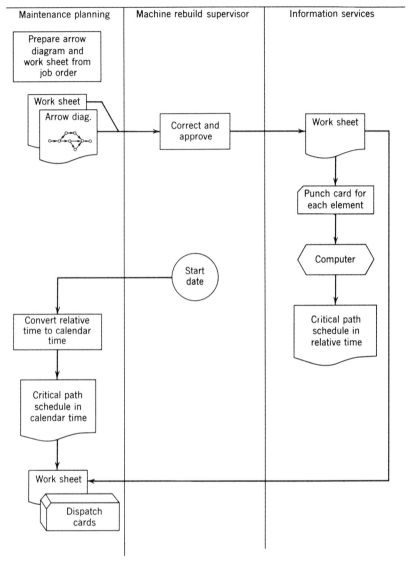

FIGURE 7. Develop critical path schedule and dispatch cards.

are arranged on the work sheet in the general sequence of performance required to complete the job. Nine major groups of tasks are performed in rebuilding most machines. These are grouped by operation numbers as follows:

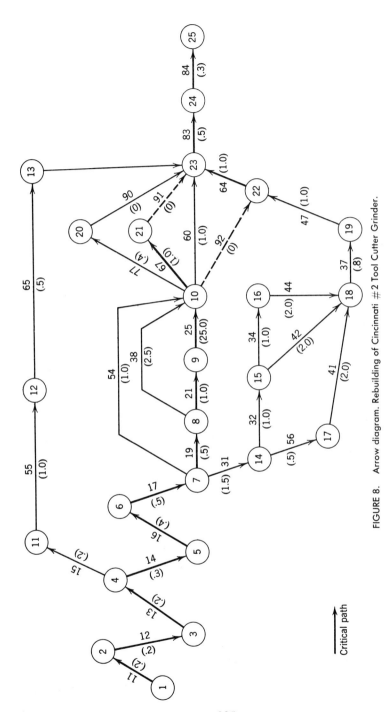

FIGURE 8. Arrow diagram. Rebuilding of Cincinnati #2 Tool Cutter Grinder.

Critical path

	Operation numbers
Preliminary	10–12
Preparatory	13–19
Secure Parts	20–27
Machining	30–39
Scraping	40–48
Painting	50–57
Assemble Components	60–66
Re-Assemble Machine	67–77
Final Test & Load Out	80–83

Each operation listed shows the craft required, and has space to enter the number of men, an estimate of the number of man-hours required, an estimate in days of the time required, and the event numbers of each operation as taken from the head (j) and tail (i) of the corresponding arrows on the arrow diagram. The arrow diagram[3] is constructed by asking three simple questions about each operation listed on the work sheet: first, which operations must precede this one? second, which operations may be concurrent with this one? and, third, which operations must be preceded *by* this one? There are several other straightforward rules to observe in preparing the arrow diagram. Two of these are that there must be no "loops" in the diagram, i.e., each arrow must be uniquely identified by its head and tail event numbers (no two arrows may originate and terminate at the same events), and that there may be only one beginning and one end event for the entire diagram. To conform to these rules it is frequently necessary to insert "dummy" arrows (constraints) into the diagram and to assign zero times to them.[4] Preparation of the arrow diagram is easily mastered with as little as one or two days of training and practice. When the arrow diagram and the critical path work sheet have been completed, they are sent to the machine-rebuild supervisor for approval. Three cards are then punched from the work sheet: card one simply identifies the problem; card two is a program control card, used to enable the computer to allocate memory space and generate certain constants during the computational run; and card three shows the format of each of the data cards. A data card must be prepared for each arrow on the diagram (each operation required for the job) including the dummy arrows. The program is currently operating on an

---

[3]On Figure 8, the numbers in the circles are the event numbers, those above the arrows are the operation numbers and those in parentheses are time estimates in days.

[4]Operation number 92 (on Figure 8) from 10 to 22 shows a dependency of operation 64 upon the completion of operations 25, 38, and 54; and the arrow from 20 to 23 (operation 92) is necessary to insure that operations 60 and 77 have different ending event numbers.

IBM 1410 system operating in 1401 mode, which means that multiple problems may be solved by simply stacking the input cards for each problem in the card reader.

The Computer Program

The program computes the earliest and latest start and completion times, the free float time, and the total float time for each operation to be performed. The following notation is used to state the manner of computation:

Earliest starting time $= T_i$
Latest starting time $= $ L.S.
Latest completion time $= T_j$
Earliest completion time $= $ E.C.
Duration (elapsed time) for an operation $= Y_{ij}$

The latest starting time for an operation is the latest completion time minus the duration of that operation:

$$\text{L.S.} = T_j - Y_{ij}$$

The earliest completion time is the earliest starting time plus the duration:

$$\text{E.C.} = T_i + Y_{ij}$$

The total float is the latest completion time minus the sum of the earliest start time and the duration:

$$\text{T.F.} = T_j - (T_i + Y_{ij})$$

This represents the amount of time that the start of a job may be delayed without affecting any other jobs on the critical path. Those jobs with zero total float then define the critical path. The free float is the amount of time that the start of an operation can be shifted without affecting any other operation:

$$\text{F.F.} = T(j = i) - (T_i + Y_{ij})$$

(Note: $T(j = i)$ is the T_i of the i event corresponding to the j event of the operation in question.)

This critical path scheduling program required about one week to complete, of which only one day was spent in actually writing the program; it required 85 Fortran statements. At present, it requires approximately one to three seconds per event to perform the calculation. Total elapsed computer time from loading to completion of output report is two to three minutes for a single problem of 18 operations. Because running time is not a linear function of the number of operations, the execu-

CRITICAL PATH SCHEDULE

JOB NO. 50787

DATE ISSUED 7/17/62

OPERATION CODE		DAYS REQ D	EARLIEST START	EARLIEST FINISH	LATEST START	LATEST FINISH	DAYS FLOAT	FREE FLOAT
11	1 2	.2	.0	.2	.0	.2		**
12	1 3	.2	.2	.4	.2	.4		**
13	2 4	.2	.4	.6	.4	.6		**
14	3 5	.3	.6	.9	.6	.9		**
15	4 11	.2	.6	.8	27.8	28.0	27.2	.0
16	5 6	.4	.9	1.3	.9	1.3		**
17	6 7	.5	1.3	1.8	1.3	1.8		**
19	7 8	.5	1.8	2.3	1.8	2.3		**
21	8 9	1.0	2.3	3.3	2.3	3.3		**
25	9 10	25.0	3.3	28.3	3.3	28.3		**
31	9 14	1.5	1.8	3.3	21.0	22.5	19.2	.0
32	4 15	1.0	3.3	4.3	22.5	23.5	19.2	.0
34	15 16	1.0	4.3	5.3	23.5	24.5	19.2	.0
37	18 19	.8	7.3	8.1	26.5	27.3	19.2	.0
30	10 18	2.5	2.3	4.8	25.8	28.3	23.5	23.5
41	17 18	2.0	5.3	7.3	24.5	26.5	19.2	23.5
42	15 18	2.0	4.3	6.3	24.5	26.5	20.2	23.5
44	16 18	2.0	5.3	7.3	24.5	26.5	19.2	23.5
47	19 22	1.0	8.1	9.1	27.3	28.3	19.2	19.2
54	7 10	1.0	1.8	2.8	27.3	28.3	25.5	25.5
55	11 12	.5	.8	1.3	28.0	28.5	27.2	.0
56	14 17	2.0	3.3	5.3	22.5	24.5	19.2	.0
60	10 23	1.0	28.3	29.3	28.3	29.3	.0	**
64	22 23	1.0	28.3	29.3	28.3	29.3	.0	**
65	12 13	.5	1.3	1.8	28.5	29.0	27.2	.0
67	10 21	1.0	28.3	29.3	28.3	29.3	.0	**
75	13 23	.3	1.8	2.1	29.0	29.3	27.2	27.2
77	10 20	.4	28.3	28.7	28.9	29.3	.6	.0
83	23 24	.5	29.3	29.8	29.3	29.8	.0	**
84	24 25	.3	28.4	28.7	29.0	29.3	.6	.6
90	20 23	.0	29.3	29.3	29.3	29.3	.0	**
91	21 23	.0	29.3	29.3	29.3	29.3	.0	**
92	10 22	.0	28.3	28.3	28.3	28.3	.0	**

** Critical Operations

FIGURE 9. Sample output of network analysis program.

tion time for an operation increases as the number of operations increases. The output format is shown in Figure 9. The times are all relative to the starting time of the first operation, taken as time zero. The machine-rebuild supervisor furnishes a calendar date for the start, and the maintenance-planning group prepares a schedule converting relative time to calendar time. (The system was later revised to permit the calendar start date to be specified to the computer so that the computer can convert relative time to calendar time, taking weekends and holidays into consideration.) Figure 10 shows a portion of the schedule with this conversion completed.

The Critical Operations

The critical path is made up of the group of operations with no float time; they constitute the longest path through the maze and are termed "critical" because failure to comply with the schedule for any one of them will affect all future operations on the path. Figure 11 shows the arrow diagram depicted in Figure 8 with the critical path indicated by a heavy line.

Operating Procedures

The maintenance planning group then prepares dispatch cards by duplicating the scheduled date and shift for the operation, the order number, the operation description, the craft number, and the earliest and latest start and completion times (see Figure 12). When the operation is critical, this condition is flagged on the card; when the operation is not critical, the total float time is printed on the card. In the new procedure all of this is computer output. Furthermore, the availability of the various crafts is indicated in the input, so that manpower limitations will be considered in the computations during the manpower scheduling phase. Rescheduling because of manpower limitations will be automatic; the output will show that the critical path has changed and will indicate which operations will be affected by the changes; thus, conflicting needs for certain skills may be averted during the planning phases, rather than requiring satisfaction after work on any job has begun.

Since it is frequently necessary to rebuild machine tools of the same type, Phase II establishes an improvement file, which is simply a file of case histories of completed jobs (see Figure 13). With such a file for reference, it is frequently possible to use previous worksheets to improve upon time estimates for the various operations (hence the name improvement file). It is sometimes possible to bypass the computer operation entirely; however, under the new system the jobs which have been previously computed will be run back through the computer for manpower

CRITICAL PATH SCHEDULE   START DATE July 9   ORDER: 50787   DATE: 7-17-62

EQUIP. DESCRIP: Grinder, Cinn #2 Tool & Cutter   SERIAL: 1D2T 1Z 442
EQUIP. NO: AREA CC B & J   PROD. CC

No	OPERATION	JOB	MEN	HRS.	ES	EF	LS	LF	CP	T	I	J
10	DISCONNECT	211										
		181										
11	UNLOAD OR MOVE	175	1	.5			7/9	7/9	*	0.2	1	2
		176	2	.5			7/9	7/9	*			
12	CONNECT POWER & PRE-TEST	181	1	1.0			7/9	7/9	*	0.2	2	3
		185	1	1.0			7/9	7/9	*			
		140	1	2.0			7/9	7/9	*			
13	REMOVE ELECTRICAL UNITS	181	1	1.0			7/9	7/9	*	0.2	3	4
		140	1	.5			7/9	7/9	*			
14	CLEAN MACHINE	179	1	4.0			7/9	7/9	*	0.3	4	5
15	INSPECT & LIST ELECT. PARTS	181	1	1.0			8/14	8/15	27	0.2	4	11
16	REMOVE & DISS MECH. UNITS	140	1	6.0			7/9	7/10	*	0.4	5	6
		147	1	6.0			7/9	7/10	*			
		211										
16-17	CLEAN MACHINE PARTS	150	1	8.0			7/9	7/10	*	0.5	6	7
18	LIST SPNDL.REPLACEMT PARTS	140										
19	LIST MECHANICAL PARTS	140	1	8.0			7/9	7/10	*	0.5	7	8
20	ORDER SPNDL.REPLACEMT PARTS	MP										
21	ORDER MACH. MFR'S PARTS	MP	1				7/10	7/11	*	1.0	8	9
22	RECEIVE MFR'S NO-STOCK NOTICE	MP										
23	DRAFT TOOL ROOM PARTS	FE										
24	ORDER TOOL ROOM PARTS	MP										
25	RECEIVE MACH. MFR'S PARTS	MP	1				7/11	8/15	*	25.0	9	10
26	RECEIVE SPNDL.REPLACEMT PARTS	MP										
27	RECEIVE TOOL ROOM PARTS	MP										
28												
30	MACHINE BED	143										
31	MACHINE SADDLE	143	1	24.0			8/6	8/7	19	1.5	7	14
32	MACHINE SLIDES	143	1	16.0			8/7	8/8	19	1.0	14	15
33	MACHINE COLUMN	143										
34	MACHINE TABLE	143	1	16.0			8/8	8/9	19	1.0	15	16
35	MACHINE RAM	143										
36	MACHINE HEAD	143										
37	MACHINE GIBS	143	1	12.0			8/13	8/14	19	0.8	18	19
38	MACHINE PARTS	143	1	40.0			8/10	8/15	23	2.5	8	10
39	MACHINE TURRET OR CARRIER	143										
40	SCRAPE BED	140										
		147										
41	SCRAPE SADDLE(S)	140	1	16.0			8/9	8/13	19	2.0	17	18
		147										
42	SCRAPE SLIDE(S)	140	1	16.0			8/9	8/13	20	2.0	15	18
		147										
43	SCRAPE COLUMN	140										
		147										
44	SCRAPE TABLE	140	1	16.0			8/9	8/13	19	2.0	16	18
		147										
45	SCRAPE RAM	140										
		147										
46	SCRAPE HEAD	140										
		147										
47	SCRAPE GIBS	140	1	8.0			8/14	8/15	19	1.0	19	22
90	Dummy	-147								0.0	20	23
91	Dummy									0.0	21	23
92	Dummy									0.0	10	22

FIGURE 10. Typical schedule converted to calendar scale.

scheduling and conversion to calendar time. The alternate procedures for Phase II are shown in Figures 12 and 13. The dispatcher maintains

No.	OPERATION	JOB	MEN	HRS.	ES	EF	LS	LF	CP	T	I	J
48	SCRAPE TURRET OR CARRIER	140										
		147										
49												
50	PAINT HEAD	207										
51	PAINT GEAR BOX	207										
52	PAINT APRON(S)	207										
53	PAINT FEED UNITS	207										
54	PAINT CROSS SLIDES	207	1	6.0			8/14	8/15	25	1.0	7	10
55	PAINT MOTOR	207	1	2.0			8/15	8/15	27	.5	11	12
56	PAINT MACHINE	207	1	16.0			8/7	8/9	19	2.0	14	17
57	PAINT UNITS	207										
58												
59												
60	ASSEMBLE HEAD	140	1	14.0			8/15	8/16	*	1.0	10	23
		147										
61	ASSEMBLE GEAR BOX	140										
		147										
62	ASSEMBLE APRON	140										
		147										
63	ASSEMBLE FEED UNITS	140										
		147										
64	ASSEMBLE CROSS SLIDES	140	1	14.0			8/15	8/16	*	1.0	22	23
		147										
65	ASSEMBLE MOTOR9S) & UNITS	181	1	6.0			8/15	8/16	27	.5	12	13
66	ASSEMBLE TURRET OR CARRIER	140										
		147										
67	HANDLE PARTS	150	1	8.0			8/15	8/16	*	1.0	10	21
68	ASSEMBLE HYD. & LUB SYSTEM	211										
70	INSTALL HEAD	140										
		147										
		211										
71	INSTALL GEAR BOX	140										
		147										
72	INSTALL APRON	140										
		147										
73	INSTALL FEED UNITS	140										
		147										
74	INSTALL CROSS SLIDE	140										
		147										
75	INSTALL MOTORS & ELECT.PARTS	181	1	2.0			8/16	8/16	27	.3	13	23
76	INSTALL TURRET OR CARRIER	140										
		147										
77	INSTALL SPINDLE	140	1	6.0			8/15	8/16	1	.4	10	20
		147										
78												
80	MAKE AND INSTALL GUARDS	191										
81	BUILD SKID AND ATTACH	201										
82	WELD	260										
83	CONNECT POWER & POST TEST	181	1	1.0			8/16	8/16	*	.5	23	24
		185	1	1.0			8/16	8/16	*			
		140	1	6.0			8/16	8/16	*			
84	TOUCH-UP & LOAD OUT OR MOVE	207	1	2.0			8/16	8/17	*	.3	24	25
		175	1	1.0			8/16	8/17	*			
		176	2	1.0			8/16	8/17	*			
85	CONNECT MACHINE	181										
		185										
		211										

Total Hr._____

FIGURE 10
(Continued)

two active card files for jobs in process. One of these is a file of job cards, with the cards grouped by job. Individual cards within each job are

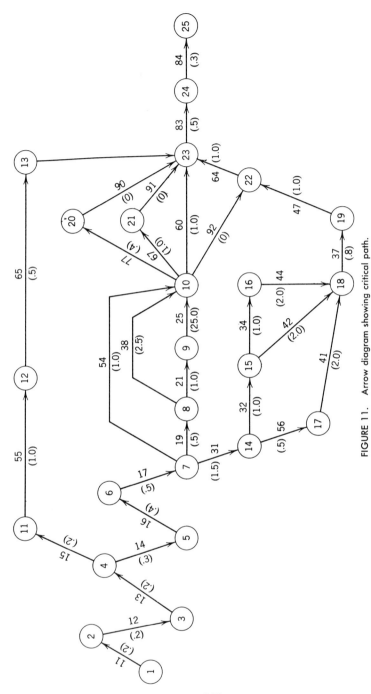

FIGURE 11. Arrow diagram showing critical path.

292

Maintenance planning	Dispatcher	Craftsmen

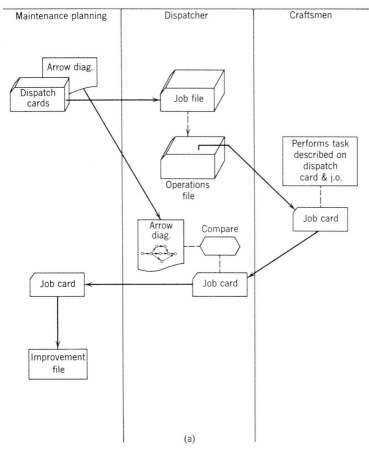

(a)

FIGURE 12. Dispatching.

moved to the operations file in the sequence indicated by the arrow diagram. These are grouped in the operation file by craft and dispatched to the work area each shift. When the craftsman performs the task described he returns the card to the dispatcher, who compares the operation described on the job card to the arrow diagram. This provides a double check on progress and permits the dispatcher to review the status more-or-less continuously. He can easily foresee potential delays and expedite the operations on the critical path when they are potentially behind schedule. He also has before him a record of the amount of slack or scheduling flexibility available in those operations not on the critical path. Thus, he has all the data he needs for on-the-floor revision of the daily schedule.

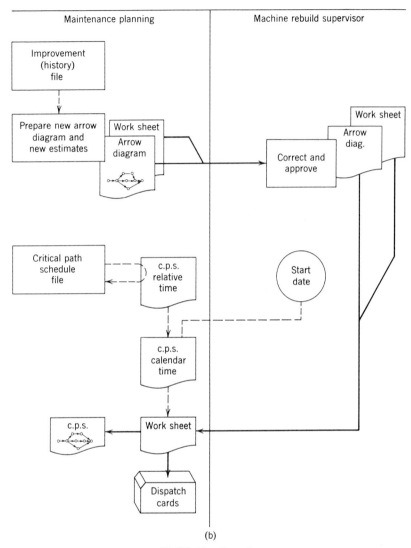

(b)

FIGURE 13.   Phase II.

Future System

Under the new system, all jobs in process are rescheduled periodically, with manpower requirements changed; a list of jobs on which the critical path has been changed is also furnished. These data are used to revise the overall schedule. There is no feedback on jobs which have had no changes on the critical path. Furthermore, the improvement file is stored on magnetic tape along with a statistical history of each operation, so performance may be compared with the plan and with past performance.

The revised system provides for three time estimates from the improvement file: the optimistic time, the most-likely or most frequently occurring time, and the pessimistic time. As the improvement file builds up, it will be necessary to estimate times for entirely new jobs only, or for ones which include the introduction of new resources, changes in personnel, or changes in facilities.

Using these estimates, the computer can produce the mean and variance of each operation performance time, and estimate the probability of compliance with multiple schedules. A gross estimating equation will be used to make these determinations; the mean time will be one-sixth of the sum of the optimistic time, the pessimistic time, and four times the most-likely time; from this value the variance may be readily calculated. Knowing the mean time and variance will permit management to base estimates upon a quantified probability, and to periodically review progress and recycle known data for a new critical path schedule with new probabilities of on-time completion for each operation; thus the technique will serve as a means of monitoring and re-planning as well as of initially planning, thus permitting management to offer the best possible service to all customers.

MAJOR INDUSTRIAL CONSTRUCTION PROJECT[5]

In January, 1964, Brown and Root, Inc., was awarded the contract to engineer and construct major additions to the Pasadena, Texas, plant of Champion Papers, Inc. The project involved the largest single machine installation for the production of coated publication-grade paper then in operation or planned; construction cost was on the order of 27 million dollars. The paper-manufacturing equipment included a 220″ trim Beloit fourdrinier paper machine with on-machine roll coating, and a two-station Rice Barton off-machine blade coater to apply the final coating for production of 325 tons of uncoated paper or 240 tons of coated paper per day. The project included constructing facilities for stock preparation, coating preparation, paper machine, off-machine coater, supercalenders, sheet finishing and storage in a constant humidity-controlled atmosphere. The coating-preparation installation was developed from a Brown & Root design for an automated, program-controlled measuring, mixing, and cooking operation. The contract commitment called for all

---

[5]This case study was prepared by H. S. Coumbe, of Brown and Root, Inc., Houston, Texas.

work to be completed by July 5, 1965, and for a critical path schedule to be maintained; a portion of the computer printout was to be an integral part of the client's progress report to the client.

A project of this size must be divided into smaller segments to facilitate discussion of the work as well as listing for computer input. This job was separated into eight geographical areas along boundaries indicated by the equipment layout. For computer input, these became subnetworks within the network, "Brown-Bilt For Champion Papers, Pasadena." Five of these subnets represented installation of groups of paper-manufacturing equipment within a building, the construction of which was described by the sixth subnet. Construction of the coating manufacturing facility, also contained in the same building, was diagrammed on the seventh subnet, and the eighth represented construction of an adjoining building and the equipment therein. As the logic was developed and interdependencies became apparent, these subnetworks were interfaced at some seventy points. Because the equipment involved in five of the subnet operations was similar, it was practical to duplicate portions of the arrow diagramming with sepias, and part of the computer input listings with Xerox.

A total of 3600 activities and dummies was listed for the initial computer run, describing process design, procurement, and detailed engineering and construction. Special 80-column forms with appropriate column headings were used to list the data by hand for keypunching. Preparing such forms proved worthwhile even had they been used on only one job. The reader should remember that a minimal check of the listings required as much time as writing the original copy.

Arrangement of the In-House Report

Hopefully, CPM reports will be read at all levels and in all departments. Recipients will have widely varying degrees of experience with CPM. They must all be sold the idea that CPM will help them do a better job with what they already know about their own specialty. First, they must be convinced that the report is an unbiased collection of details sorted for their individual convenience; second, that no detailed knowledge of CPM technique will be required of them. An understanding of the arrangement of data in the report is a timesaver to the reader and a list of any special abbreviations and symbols should accompany the reports. The following is quoted from the explanation that was initially provided users of these reports.

The subnets appear in the following order in the computer printout and are so abbreviated:

SHIP   Shipping and Finishing Equipment with Building
BLDG 07   No. 27 Machine Room Building less Floor Slab

NETWORK BRCWN-BILT FOR CHAMPION PAPERS, PASADENA (CHAMPN)   RUN DATE 20JUL64
SUBNET   RICE-BARTON OFF-MACHINE-COATER COL 32-41A (OMC)   DATA DATE 20JUL64

LEVEL DETAIL   CONTRACT NO.   1 WORKING DAYS IN EACH TIME UNIT   BY TOTAL FLOAT

2-121

PRED. EVENT	SUCC. EVENT	C Y	ACTIVITY DESCRIPTION	EST. DUR.	START DATE EARLIEST	FLOAT TOTAL	FREE	COMPLETION DATES EARLIEST	LATEST	SCHED. DATE	DEPT.
28316	32316		DUMMY		31AUG64	-12.0 +	.	31AUG64	D1AUG64		
32316	32321		DFTG ELEC INST TRAY-CBL LAYO	20.	31AUG64	-12.0 +	.	25SEP64	10SEP64		ENGR E
32321	32325		DFTG ELEC INST WIRNG-PR PANL	15.	25SEP64	-12.0 +	.	19OCT64	01OCT64		ENGR E
32325	32329		ORDR ELEC INST MTRL FOR TRAY	15.	19OCT64	-12.0 +	.	26OCT64	08OCT64		PURCH
32329	32333		DELV ELEC INST MTRL FOR TRAY	60.	26OCT64	-12.0 +	.	20JAN65	04JAN65		VENDOR
32333	06926		DUMMY	00.0	20JAN65	-12.0 +	.	20JAN65	04JAN65		
C6926	06927		CONTINUE INSTRUMENTATION	15.0	20JAN65	-12.0 +	.	10FEB65	25JAN65		CONSTR
06927	06965		COMPLETE INSTRUMENTATION	15.0	10FEB65	-12.0	+10.0	03MAR65	15FEB65		CONSTR
16108	10117		OK FLOW SHEET PROCESS	10.	01MAY64	-12.0 +	.	X15MAY64	X29APR64		CHAMPN
10117	12124		1SUE4FLOW SHEET PROCESS	5.	15MAY64	-12.0 +	.	X22MAY64	X06MAY64		ENGR M
25209	29216		APDR MACH AIR HTRS AND FANS	45.0	01MAY64	-12.0 +	.	X03JUL64	X17JUN64		RICE
29216	29224		OKDR MACH AIR HTRS AND FANS	20.	03JUL64	-12.0 +	.	03AUG64	X16JUL64		ENGR M
29224	29233		DELV MACH AIR HTRS AND FANS	85.0	03AUG64	-12.0 +	.	30NOV64	12NOV64		RICE
29233	30233		START OFF MACH-DRYER SYSTEM	25.0	30NOV64	-12.0	-56.0	06JAN65	17DEC64		CONSTR
30233	06949		CONT. OFF MACH-DRYER SYSTEM	25.0	06JAN65	-12.0 +	.	06JAN65	17DEC64		CONSTR
06949	06965		COMP. OFF MACH-DRYER SYSTEM	15.0	10FEB65	-12.0	+10.0	03MAR65	15FEB65		CONSTR
08124	17201		EXPL PIPE OPER	10.	03JUL64	-12.0 +	9.0	X20JUL64	X01JUL64		ENGR M
00000	29233	•		(73)		-12.0 +	.	30NOV64	12NOV64		
52733	31233		DUMMY		16OCT64	- 9.0	-67.0	16OCT64	05OCT64		CONSTR
52733	00000			(71)		- 9.0		16OCT64	05OCT64		
00000	31233	•		(72)		- 9.0	-67.0	16OCT64	05OCT64		
82533	00000			(3)		- 9.0		11NOV64	29OCT64		
41121	41128		QUOT EQMT OMC COLOR SCREENS	5.	01MAY64	- 7.0 +	.	X08MAY64	X29APR64		VENDOR
41128	41201		OKBY EQMT OMC COLOR SCREENS	10.	08MAY64	- 7.0 +	.	X22MAY64	X13MAY64		CHAMPN
41201	41209		ORDR EQMT OMC COLOR SCREENS	10.	22MAY64	- 7.0 +	.	X05JUN64	X27MAY64		PURCH
41209	41216		APDR EQMT OMC COLOR SCREENS	10.	05JUN64	- 7.0 +	.	X19JUN64	X10JUN64		VENDOR

FIGURE 14. Activity time-status report by total float.

297

S. CAL	Supercalendars/Finish Equipment North of Col. 42
OMC	Rice-Barton Off-Machine Coater Col. 32–41A
COAT	Coating Preparation System Adjacent to OMC
REEL	Calendars, Massey Coater, 5th Dryer and Reel
DRYR	1st Thru 4th Dryer Sections Col. Line 11–21
WETN	Stockprep, Fourdrinier, Press Sect. Col. 1A–11

Three distinct reports are produced by computer computation and listed. Under each subnet they are arranged in the following order:

(1) *By Predecessor.* In order of the event numbers used to define the beginning of each activity. This listing is most useful as a dictionary to locate an individual activity.

(2) *By Total Float* (Shown in Figure 14). In order of most critical items first.

(a) A negative value indicates that an adjustment must be made in order to complete the project on schedule.

(b) Items currently on the critical path are so shown with "+ . + ." float (zero understood).

(c) Items shown with a positive amount of total float may be adjusted to end any time within the limits established by earliest and latest completion dates. Values over 99.9 are indicated by +XX.X.

Calculated dates of events are based on estimated durations and certain restraints and sequence placed upon them as logical relationships between interrelated activities. Calendar dates are placed on starting events and the actual completion of activities in order to calculate the current status of the activities remaining to be completed. Five working days per calendar week were used in these computations.

Activities of equal total float are listed together as "families" on the printout and often form a partial chain through the network.

The last item in a chain is sometimes shown to have "free float." This is a float shared by every activity in the partial chain; it is with respect to the next longest path to be found between the end events of the partial chain under consideration. [Ed. note: Large amounts of negative free float shown on some activities in Figure 14 are caused by late dates related to interfact events in other subnets.]

(3) By Earliest Estimated Completion Within Responsible Department (By EEC Within Dept.) (Shown in Figure 15.)

The following responsible departments listed are abbreviated:

BELOIT	Beloit Iron Works
BELO E	Beloit Eastern
CHAMPN	Champion Papers, Inc.
CONSTR	Brown & Root Construction
ENGR C	Civil Engineering by Brown & Root
ENGR E	Electrical Engineering by Brown & Root
ENGR M	Mechanical (Piping) Engineering by Brown & Root
FABR B	Reinforcing Steel Fabricator
FABR S	Structural Steel Fabricator

NETWORK BROWN-BILT FOR CHAMPION PAPERS, PASADENA (CHAMPN)   RUN DATE 20JUL64
SUBNET   COATING PREPARATION SYSTEM ADJACENT TO OMC (COAT)   DATA DATE 20JUL64

LEVEL DETAIL   CONTRACT NO. 2-121   BY EEC WITHIN DEPT

1 WORKING DAYS IN EACH TIME UNIT

PRED. EVENT	SUCC. EVENT	C Y	ACTIVITY DESCRIPTION	EST. DUR.	START DATE EARLIEST	FLOAT TOTAL	FLOAT FREE	COMPLETION DATES EARLIEST	LATEST	SCHED. DATE	DEPT.
42301	42307		SPEC INST PANL ELEC COMPONET	15.0	28MAY64	-36.0 +	.	X18JUN64	X29APR64		ENGR E
43101	43107		SPEC LEVEL INSTR	15.0	28MAY64	+44.0 +	.	X18JUN64	D20AUG64		ENGR E
44101	44107		SPEC CONTROL + RELIEF VALVES	15.0	28MAY64	+29.0 +	.	X18JUN64	D30JUL64		ENGR E
04301	04314		DSGN INST PANL ELEC SKEMATIC	20.0	28MAY64	+ 9.0 +	.	X25JUN64	X09JUL64		ENGR E
04314	20301		DFTG INST PANL ELEC WIRE DIA	10.0	25JUN64	+ 9.0 +	+10.0	X10JUL64	X23JUL64		ENGR E
32316	32321		DFTG ELEC INST TRAY-CBL LAYO	20.0	17JUL64	+44.0 +	.	D14AUG64	15OCT64		ENGR E
35201	86722		SPEC ELEC EQMT AC MTR	02.0	18AUG64	+12.0 +	.	D20AUG64	07SEP64		ENGR E
60701	60708		DSGN ELEC LOADTAB 480VMCC	05.0	20AUG64	+12.0 +	+12.0	D27AUG64	14SEP64		ENGR E
32321	32325		DFTG ELEC INST WIRNG FR PANL	25.0	14AUG64	+44.0 +	.	18SEP64	19NOV64		ENGR E
74708	85514		DFTG ELEC GROUNDING PLAN	05.0	14SEP64	+25.0 +	+25.0	21SEP64	26OCT64		ENGR E
60708	80815		ISUE ELEC SKEMATIC MMC	10.0	14SEP64	+22.0 +	.	28SEP64	28OCT64		ENGR E
60708	60716		DFTG ELEC TRAY + CABLE PLANS	20.0	14SEP64	+ . +	.	12OCT64	12OCT64		ENGR E
60716	67732		DFTG ELEC SKEMATIC	20.0	14SEP64	+18.0 +	.	12OCT64	05NOV64		ENGR E
60716	60723		ISUE ELEC TRAY + CABLE PLANS	05.0	12OCT64	+13.0 +	.	19OCT64	05NOV64		ENGR E
60708	60726		TAB ELEC CABL 9/10 DONE	30.0	14SEP64	+ 8.0 +	.	26OCT64	05NOV64		ENGR E
80815	76822		DFTG ELEC WIRE CUBL 480VMCC	20.0	28SEP64	+22.0 +	5.0	26OCT64	25NOV64		ENGR E
60716	85514		ISUE ELEC RUFIN GREAT + SLAB	10.0	12OCT64	+ 8.0 +	.	26OCT64	26OCT64		ENGR E
60723	59733		DFTG ELEC LIGHTING PLAN	10.0	19OCT64	+73.0 +	+73.0	02NOV64	15FEB65		ENGR E
76814	76822		APDR ELEC EQMT 480VMCC	10.0	19OCT64	+17.0 +	.	02NOV64	25NOV64		ENGR E
60726	60733		DFTG ELEC EQMT PLAN	20.0	26OCT64	+ 8.0 +	.	23NOV64	03DEC64		ENGR E

FIGURE 15. Activity time status report by early completion within dept.

299

GE	General Electric
INSTRU	Instrument Engineering by Brown & Root
PURCH	Brown & Root Purchasing
RELIAN	Reliance Electric
RICE	Rice-Barton
VENDOR	Equipment Vendor
WEC	Westinghouse Electric

Each department is listed on a separate page to facilitate distribution to that department.

The earliest start date is listed first. Estimated duration is added to it to produce an estimated earliest completion date. The latest completion date is the point at which any further delay in completing is reflected as a delay in the completion of the project. When the earliest completion coincides with latest completion, total float is zero and by definition the item is on the critical path. Note that the estimated earliest completion date may print out as a date later than the latest completion. In this case float is less than zero or negative. These items are to be reviewed in an effort to arrive at a revised, but still likely duration or a different sequence that will permit them to be completed by the latest completion date.

Special flags are placed on items due immediate attention. At left of either column of completion dates "D" signifies the item falls due with the next thirty working days; "X" signifies this date has passed in relation to the "RUN DATE" at the top of the page.

Additional CPM work package time reports (see Figure 16) provide the client easily verified "milestones" with which to assess job progress against the job schedule. Accompanying generalized arrow diagrams show the interrelations of activities listed on the work package time report.

(1) *CPM Milestones*

The point in time at the completion of an activity or chain of activities is designated a "milestone." The associated chain of small jobs that are combined to form one generalized activity has been selected to provide a point that is easily discernible in the mainstream of work in progress. In some cases a milestone is best described by the last detailed activity in the chain under consideration.

(2) *Computer Computations*

After detailed CPM computations have been made and calendar dates established for each item in the detailed CPM network a special computer program[6] generates the work package time report based on those calendar dates.

(3) *CPM Work Package Time Report*

Items on the report are defined in the order they appear:

A. *Pred. and Succ. Event* are listed for computer input only. They associate the mileposts on the generalized diagram with a particular event in the detailed CPM diagram.

---

[6][Ed. Note: The program used on this project was a modification of the IBM 7094 PERT Cost Computer Program.]

NETWORK BROWN-BILT FOR CHAMPION PAPERS, PASADENA (CHA/PN)   RUN DATE— 01APR65 DATA DATE 19FEB65

LEVEL 3   CONTRACT NO. 2-121   1 WORKING DAYS IN EACH TIME UNIT   BY DEPARTMENT

PRED. EVENT	SUCC. EVENT	CHARGE NUMBER	ACTIVITY DESCRIPTION	ELAP TIME	EARLST START	***COMPLETION DATES**** ACTUAL	EARLST	LATEST	DEPT	TOTAL FLOAT
BLDG07										
3 00908	00916	.......CLIENT RPT	DRIVE PILING FORTH FR COL 36	10.0	01MAY64	15MAY64		12MAY64	CONSTR	-3.0
3 05948	00971	.......CLIENT RPT		86.0	03AUG64		01DEC64	04NOV64	CONSTR	-19.0
3 00977	00989	.......CLIENT RPT	CONSTRUCT MACHINEWAY COL3019	13.0	01DEC64	18DEC64		08JAN65	CONSTR	+13.0
3 01953	01915	.......CLIENT RPT	ERECT+ROOF COATPREP LEAN-TO	98.0	18SEP64	05FEB65		19FEB65	CONSTR	+10.0
3 04900	01918	.......CLIENT RPT	CONSTRUCT MACHINEWAY COL18A5	.	12JAN65	12JAN65		08JAN65	CONSTR	-2.0
3 01909	01957	.......CLIENT RPT	SEAL SOUTH END OF BUILDING	71.0	30OCT64	10FEB65		05MAR65	CONSTR	+17.0
3 05945	01979	.......CLIENT RPT	ERECT STRU STEEL COL 18A-1A	.	20OCT64			08JAN65	CONSTR	+55.0
3 57533	04930	.......CLIENT RPT	ERECT THREE LARGE CRANES	99.0	20AUG64			08JAN65	CONSTR	+
3 04977	04956	.......CLIENT RPT	FINISH-OFF NORTHEND MACHBLDG	XX.X	20OCT64		16APR65	16APR65	CONSTR	+
3 04984	04977	.......CLIENT RPT	ERECT STRU STEEL COL. 42-50	43.0	20AUG64	20OCT64		20OCT64	CONSTR	+
3 65637	04978	.......CLIENT RFT	ERECT STRU STEEL COL. 51-60	76.0	20OCT64	05FEB65		05FEB65	CONSTR	+
3 05936	05902	.......CLIENT RPT	COMPLETE PILING COL 42 TO 50	10.0	18MAY64	30MAY64		19MAY64	CONSTR	-9.0
3 05902	05921	.......CLIENT RPT	CONSTRUCT FDNS FOR SUPERCALS	XX.X	30MAY64	20OCT64		07OCT64	CONSTR	-9.0
3 00908	05943	.......CLIENT RPT	COMPLETE UNGRD+FDNS TO CCL19	78.0	01MAY64	20AUG64		18AUG64	CONSTR	-2.0
3 05942	05944	.......CLIENT RPT	ERECT STRU STEEL COL. 41A-33	.	20AUG64	20AUG64		28AUG64	CONSTR	+6.0
3 05944	05945	.......CLIENT RPT	ERECT STRU STEEL COL. 32-19	21.0	20AUG64	18SEP64		08JAN65	CONSTR	+76.0
BLEACH										
3 86638	37233	.......CLIENT RPT	CONSTRUCT OPER FL SLAB18A-1A	10.0	22MAR65		05APR65	29MAR65	CONSTR	-5.0
3 03953	03954	.......CLIENT RPT	ERECT AND LINE BLEACHTOWER	35.0	20JAN65		10MAR65	10MAR65	CONSTR	+
3 03953	03955	.......CLIENT RPT	START PIPING-ELEC-INSTRU	25.0	20JAN65		24FEB65	10MAR65	CONSTR	+10.0
3 03955	03956	.......CLIENT RPT	FINISH PIPING-ELEC-INSTRU	20.0	10MAR65		07APR65	07APR65	CONSTR	+
COAT										
3 03951	03956	.......CLIENT RPT	COMPLETE MSTAL CLO2 GENERATR	15.0	05FEB65		26FEB65	10MAR65	CONSTR	+8.0
3 02928	02600	.......CLIENT RFT	INSTALL COATING PREP SYSTEM	72.0	14DEC64		26MAR65	22MAR65	CONSTR	-4.0
3 01931	06931	.......CLIENT RPT	COMPLETE INSTRU PANEL HOOKUP	20.0	23MAR65		20APR65	22MAR65	CONSTR	-21.0

FIGURE 16. Work package time report by department.

301

B. *Charge Number* states the recipient agency, in this case "Client Report."
C. *Activity Description* defines the work within the generalized activity or the last activity to be accomplished in the chain of detailed activities under consideration *within a given SUBNET*.
D. *Elapsed Time.* This is a summation of durations running through the generalized activity. Durations between generalized activities are neglected.
E. *Computed Calendar Dates.* The earliest start date is listed first. Elapsed time is added to it to produce an estimated earliest completion date. The latest completion date is the point at which any further delay in completing is reflected as a delay in the completion of the project. When the earliest completion coincides with latest completion, total float is zero and by definition the item is on the critical path. Note that the estimated earliest completion date may print out as a date later than the latest completion. In this case float is less than zero or negative. These items are to be reviewed in an effort to arrive at a revised, but still likely duration or a different sequence that will permit them to be completed by the latest completion date.
F. *Department.* This is the agency responsible for completion of the activity.

Putting the Data To Use In A Client Report

Project information was accumulated and listed from January 27, 1964, one week after the contract was awarded, until May 20, 1964, the date of the first full-scale computer run. Inconsistencies in the network prevented a complete printout the first time, and so adjustments were made until July 20, 1964. A total of seventeen computer runs was required to complete this course, with forty-nine percent of the computer cost expended in the process. Planners continued to update completed activities throughout this period. About the same time, the planners were required to prepare an additional report by extracting milestones from the prime network and processing them as a summary; the necessary input was prepared and added to the master input-tape file in time for the July 20 run. At the client's request, the milestones reported by printout were presented on a conventional bar chart and included in the progress-report package.

Complete runs were made again in August, September, and October. A considerable increase in scope caused planners to rework the network logic in October. Runs were continued in November, December, January, and March. By then, equipment deliveries and engineering were substantially complete, and construction had been completed on one-half of the facility which had been scheduled for an early start up. Computer printouts were discontinued and the remaining activities were monitored on a punch list basis. Machine installation was complete by June 30. The first marketable product was a sheet made on July 7, 1965, two days later than the original schedule.

People and CPM

Much of the early effort in this project should have been charged to the learning process. For input purposes, planners must make decisions about the direction in which a project is to move, much earlier than engineers customarily make such decisions. This is not a disadvantage, but a prime virtue of using a formal and well-documented planning technique.

As the pieces of the plan begin to fit together and last-minute details flood in, it is natural for the planners to want to overdo improvements as the need for them becomes apparent. This sudden rush of unexpected manhours of preparation increases the pressure to turn out a schedule before the information becomes history. During the busy time extending from the period just before the initial computer run until after debugging, planners must avoid making R & D-type refinements. The main burden at that time is the clerical work involved in listing tasks for keypunching; if planners attempt to polish the logic diagrams to reflect the latest information at the same time, they add to the problem. The only assistance they can expect is that which can be given by persons borrowed from a pool of junior clerical or drafting employees. To offset their total lack of familiarity with the proceedings, these assistants should have traits of curiosity and thoroughness; the planner should watch for such qualities when selecting these persons.

Lessons Which Will Benefit Future Applications

To recover some of the expense of such an elaborate approach to scheduling, the planners had to keep future usefulness in mind. Recurring items were considered with extra care and rearranged to develop standard logic formats which could be used on later projects.

(A) *The Procurement Process.* As procurement logic was refined into a recurring sequence, it was possible to express it as:

Specify.
Approve specifications.
Inquiry.
Receive quotations.
Tabulate quotations.
Approve selection of vendor.
Requisition.
Purchase.
Deliver.

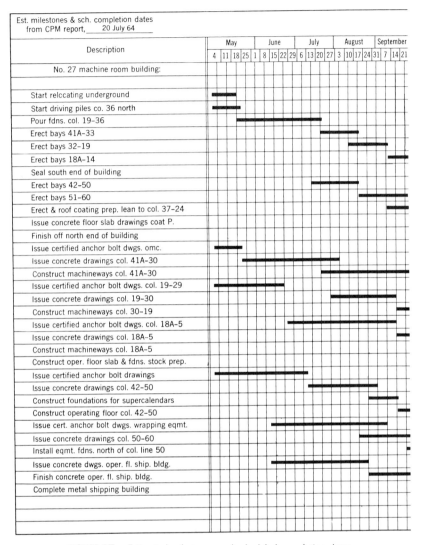

FIGURE 17. Estimated milestones and scheduled completion dates.

When information is arranged to be used again in the subsequent stages of the system it is economically justifiable to prepare an entire equipment list in the critical path schedule, even though it is necessary to include a large number of activities. The details of compiling a thorough list were explored throughout the life of this project. The reader should

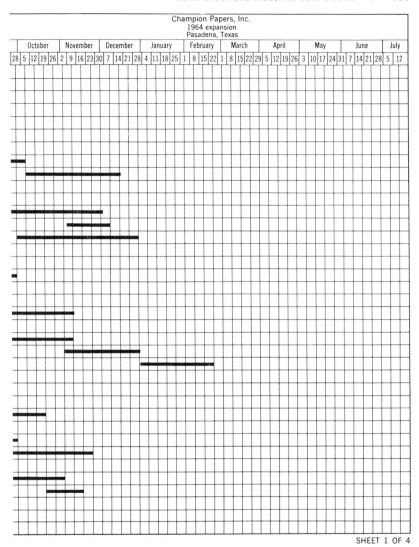

Champion Papers, Inc.
1964 expansion
Pasadena, Texas

SHEET 1 OF 4

notice how the responsibility moves from *engineering* to *client* to *purchasing* to *vendor*, and so on. The total duration of the procurement process from specification to delivery can therefore be monitored as changes in responsibilities are due to occur, because the next responsible agency warns the project manager about a bottleneck, if the work does not arrive in the next area when due.

The information needed to complete the procurement process for the schedule is closely allied to the other standard types of project status reporting. We have previously explored in detail the *client report* which informs the client's planning group of progress. Two other types of status reports are common to the type of project studied in this case: the *equipment list*, which is compiled for procurement control, and the *expediting report*, which helps control delivery. Costs can be divided among these three types of reports, to lower the total cost (when reproduction of many copies is considered) and to provide the user with timely information sorted to his individual needs.

(B) *The Craft Process.* The experience gained in arranging crafts into off-the-shelf logic diagrams can also be used in the future. The engineering crafts considered were mechanical (layout and piping), electrical, structural, and instrumentation. Planners produced good results when they tried to develop an instrumentation package; such results indicate the problem remaining with other crafts is in developing suitable nomenclature. The computer format provided little space for word descriptions of activities. To offset this, the planners provided the users with a system of four-letter words. Giving drawings short titles composed of these "words" led to assigning families of drawings to cost centers; since it is possible to do this before knowing how many drawings will be made or what their complete titles will be, by using this method the planners can divide engineering work into parts describable in network logic. This potentiality could be a step toward providing manpower-loading and closed-loop cost (man-hour) reports for the engineering phase.

(C) *Maintenance of the Computer Data.* Updating can consume the full time of one person from the first full-scale run to the completion of the bulk of the engineering. Updating includes collecting and publishing a progress report in a form which uses at least some information from the CPM reports. In the early stages, updating input was manually listed the same as the original input data. Later practices have shown that the input deck may be duplicated and used to update: cards representing completed activities can be pulled and updated simply by punching in the date completed on the duplicate card. This method guarantees that a completed dummy will not be forgotten or an inconsistency introduced by misnumbering the events on new update cards.

Each computer run output begins with a list of error messages. The list is immediately used to correct input that violates the conditions of the program. A carefully labelled file of these violations must be maintained to provide a backup for computer service billings and a study guide for improving updating procedures.

(D) *Choice of Reports.* The planner who is familiar with any large-capacity computer program will generally have little desire to use the "latest thing" on each new project. The input format, diagnostic output and report output formats cost too much ($500 to $2500) to revise without making a thorough study, especially if he does not intend to use them on other projects. On the project we are describing, it was quite unmanageable to connect subnets manually, and so a computer program was devised to enter the interfaces on the subnets; such a program was conceived and prepared for future use. The variety of computer-generated reports already available was quite large and a choice of any three was offered in the pricing package of the computer service used. However, the planner had to choose reports carefully, because those best serving as a guide to day-to-day direction of the project are not the most desirable to include in a progress report to the client; also, certain sorted lists of all input are only useful for checking and updating. Drawing on their past experience, the planners decided that of the reports already available, the most practical ones are:

Application	Type
Day-to-day guide	Listed by early start within each craft, covering next thirty calendar days.
Client report	Abbreviated milestones for entire project. Also items under client's direct control or responsibility set up as a craft (to cover approvals and procurement of owner-furnished items) covering next thirty calendar days. Listed by earliest start.
Long-range planning	Listed by total float.
Checking	{ Listed by predecessors. { Listed by succesors.
Updating	Listed by early start within each area.

Cost History and Future

The computer service was purchased from the CPM Systems Division of Informatics, Inc., and was priced on the basis of the total number of activities (both completed and uncompleted) in the prime network, with special charges for extra reports, extra copies of reports, runs of individual subnets, editing runs (error messages without reports), keypunching, and consulting time required to debug data for processing.

The direct costs were divided into several categories and subcategories. Labor costs to prepare initial run included:

Scheduling supervisor—based on 5-months' salary	$4,000.00	
Drafting—685 hours @ $2.25 an hour	1,540.00	
Clerical—525 hours @ $2.00 an hour	1,050.00	
Consultation by construction superintendent and engineering supervisors	no charge	
Total labor cost through initial run		$6,590.00
Data processing costs for the initial run, debugging, and one monthly report (over a 3 month period)		4,928.29
Total cost to make initial report run		$11,518.29
Labor costs to update and maintain data for 8 months, and to prepare and publish 6 monthly reports to the client with CPM and 2 reports without included:		
Scheduling supervisor—based on 8-months' salary	6,400.00	
Consultation by construction superintendent and engineering supervisors	no charge	
Total labor cost for last 8 months		6,400.00
Data processing cost for 6 succeeding monthly reports		5,130.65
Total cost for last 8 months		11,530.75
Miscellaneous expenses included:		
Air Freight and long distance telephone calls from Houston to Los Angeles (applies because out-of-town computer service was used)	200.00	
Reproduction expenses above that normally used to make a client report without CPM	700.00	
Total miscellaneous expenses		900.00
TOTAL COST FOR PROJECT		$23,949.05

When using these costs to estimate future projects, the manager should consider the number of updatings and publishings of reports anticipated. The benefits of the experience with this project would be found in the increased timeliness and confidence in the reporting, not particularly in any cost reduction on computer services. By approaching the initial run on the basis of one subnet at a time and starting with off-the-shelf logic diagrams developed on this project, the planners can expect the following results:

Initial run on first subnetwork four weeks after start of project.

Initial run on remaining major subnetworks during the fifth to tenth week after start of project.

Complete debugging, tying together of subnetworks, and updating to make first complete run twelfth week after start of project. At about this age of the project, the quality of information available can be expected to be much improved, and the race can continue, with updating of revised durations and completed activities.

SUMMARY

Three different applications of network planning are described in this chapter, the first using a relatively small computer and an uncomplicated network but scheduling manpower as well as actions; the second using a larger computer and scheduling a manufacturing production process; and the third using a sophisticated computer program and a complex network without directly identifying manpower. These applications illustrate some of the complexities of using the simple logic of network planning in real-life situations.

15

Application to multiple-unit construction[1]

The basic principles of network planning, scheduling, and control are as applicable to a multiple-unit project as they are to a single-unit one. However, networking techniques must be modified in certain ways to apply the principles to a multiple-unit problem in a practical way. When the CPM or PERT technique is applied to such a problem, proven production control techniques are simply added to the basic principles. In this chapter, rather than discussing these techniques theoretically, we will show how the modifications are made, by describing the step-by-step application of the technique to an actual project.

DESCRIPTION OF PROJECT

The project we will consider is the construction of one segment of a condominium, containing 68 dwelling units in 11 buildings. Besides constructing the units and buildings, it was necessary to make certain off-site improvements, such as landscaping and landscape lighting. The condominium, called Loma Riviera, is located in San Diego, and was built by the Deelan Construction Company. There were a number of variations in the plans. The living units contain 3, 4, or 5 bedrooms. The architectural style of the building is Spanish, and there is a mission plaster finish on the exterior; some of the buildings have mission tile roofs, and the others have a concrete boss tile of similar coloring. Some buildings have attached carports and others have related parking structures, although one building has both. There are between 5 and 8 units in a building.

In the original evaluation of the project, the planners determined that it was desirable to start construction of one building every other day.

---

[1]By Richard M. Wrestler, Director of Operations, CPM Systems Division, Informatics, Inc., Los Angeles, Calif.

PLANNING PHASE

As usual, it was necessary to establish what work was to be done. In this case, the planners initially determined that there were three basically different types of products being produced: living units, buildings, and area or site improvements. Therefore, there would be one area network, 11 building networks, and 68 unit networks, 80 in all. Preparing that many networks might seem a horrendous task; however, a typical network plan with variances would suffice for both the building and the unit networks. The typical networks would contain the most complex set of conditions.

The planners decided to separate the internal from the external portion of the building for purely practical reasons. All beginning operations, such as pouring the slab, framing, rough electrical, rough heating, rough plumbing, roofing, and lathing, are performed on the building as a whole. All the finishing operations, such as interior tape and texture, interior finish electric, and installation of appliances may be carried on completely independent of activity on the exterior of the building. Although the application of drywall would normally be considered an interior job, it was included on the building network because of building code requirements: when the drywall had been put in and the lathing had been done, the nailing had to be inspected before any plaster could be applied. After this nailing inspection, the exterior of the building could be worked almost completely independent of any interior unit. Therefore, the unit network began with the taping and texturing of the drywall. One of these, shown in Figure 1, represents the construction of a typical living unit. A similar network was prepared for a typical building, showing the most complex possible arrangement of jobs to be performed on the buildings; it included construction of both carport and parking structures and the application of both mission and boss tiles for the roof (which are laid at different points in time).

The various networks were developed independently in two phases. In the planning phase the sequence of activities and of jobs was determined. At that point we made the first modification of the CPM technique. We did not ask, "how early can something possibly be done?" or, "how late can something be done?" Instead, we prepared the network in the sequence in which the contractor wanted to build. By taking this approach, we will be able to benefit from the contractor's field superintendent's more than 20 years of building experience, and from the experience and technical knowledge of an architect-engineer and of the Loma Riviera project manager; the CPM consultant employed by the contractor had had experience on over 100 different building projects of this nature,

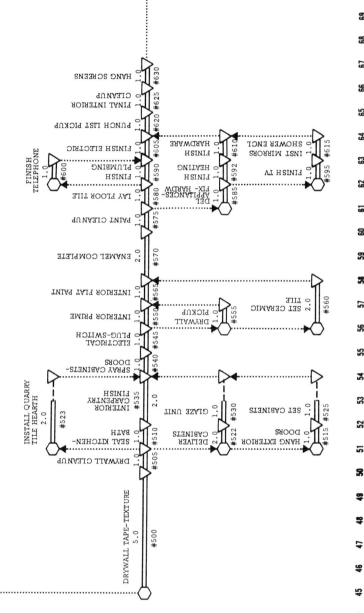

FIGURE 1. Typical network for interior finish operations.

as well. The network we produced showed what our joint experience had indicated was the most convenient, and therefore the most economical, sequence of operations. In the scheduling phase, we went through the network a second time, adding time values. By planning the network in two phases, we had a chance to review the network topography another time without expending any additional man-hours.

At this point, each network was key-punched and processed through a standard CPM computer program to determine the length of the critical path for the various networks. The punched cards representing one network then became the standard deck for that type of network.

Time Estimates

The planners decided that all times should be expressed in even days; although they realized that certain activities would take less than whole days to perform, they knew that they might not be able to predict the specific time of day that things would happen. As an example, they knew that once the inspector is on the job site, it takes him between 15 minutes and one hour to inspect the building thoroughly and to decide whether it would pass. Since the inspection department is part of the local city government, and the particular inspector was overseeing other construction, the project planners had no way of knowing what time of day he would arrive to perform his job. Consequently, they had to schedule activities so that he would begin his inspection on the day following the previous job and so that the next job did not start until the day after his inspection. Most of the subcontractors on this job had been used before and so their performance was well-known. Therefore, the planners predicted that any problems uncovered in the inspections would be minor, and so they decided they would not need a backup day following the inspection.

Cross Relating Networks

In developing the total network plan, the planners went through a third phase of activity: that of relating one network to the other two. For instance, the planners determined that a building must be under construction 45 days before the first living unit could be worked on. They also knew that each living unit within any building must be completed before the last 2 or 3 items of the building network could be completed, and that grading for the landscaping could only be completed the day after the exterior staging from the last building was removed, although it could begin as early as the day after the staging from the first building was removed. With this information and the already-established construction rates, the planners were easily able to calculate when each network should begin so that the networks would tie into one another properly.

314 / Case studies

Lead Times

We should mention an unusual type of activity which was liberally used both in phasing the networks together and in constructing some of the networks: the "lead time" activity, used to show a certain relationship between two other activities. A lead time activity consumes time but no resources, as opposed to a real activity which consumes both time and resources, and a dummy activity which consumes neither. An example could be given in a network showing a concrete curing time or plaster dry time: in the particular area in which this project was built, local codes and good practice dictated that two weeks should elapse between application of the brown coat of plaster and the application of the color coat of plaster; therefore a two-week lead time was inserted between these two activities. The planners decided that some other operation was to occur on the seventh day of the two-week (10 workday) dry time. Consequently, a six-day lead time preceded that activity.

Scheduling Phase

Quite obviously, if we were only going to use one network of each type, we would only have to relate one activity or event to a calendar to create a schedule. However, in the complex situation we encountered, we had more than one of each type and had networks with variations within some of the types.

Let us consider the variations first. It was determined that the difference between a 3- and a 4-bedroom unit consisted mainly of one additional wall; however, one additional wall also created the difference between the 4- and the 5-bedroom unit. After discussing the point with the field superintendent and a few of the subcontractors, the planners determined that for scheduling purposes there was no essential difference between the 3-, 4- and 5-bedroom units, and so we could consider that each of the 68 units followed the same basic network plan. The planners further decided that since we would be constructing buildings at the rate of one every other day, we should have to complete units at the rate of 3 a day (68 ÷ 11 or about 6 units each 2 days). At that point is was necessary to look ahead into production control to determine the most economic means of scheduling such a rate of production.[2]

Speeding Up By Slowing Down

To demonstrate the need to use standard production control techniques, let us consider the paradox of "speeding up by slowing down."

---

[2]Readers who are experienced in the field of construction requirements will probably realize that such a schedule rate will be used on only a few jobs.

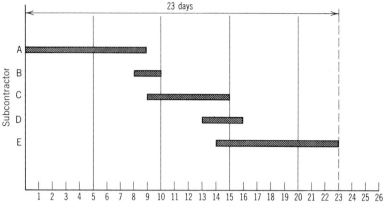

FIGURE 2. Tract schedule with subs working at varying rates.

Let us consider the case of a manager who has 18 items to build, working with a schedule which places five subcontractors on the critical path. The manager goes to subcontractor A to determine how much time A will need to perform his job. After considering the size of the job, the number of people he has available, and the amount of other work he has in process, A tells the manager that he feels he could produce 2 a day. The manager then talks to subcontractor B, who looks over the job and explains to the manager that he is really going to save him money by performing his job very quickly; consequently, he will need only two days to complete the entire project, producing at the rate of 9 per day. The manager also discusses the problem with subcontractors C, D, and E, who tell him that the most desirable rate for them is 3, 6, and 2 per day, respectively. Figure 2 represents the days on which each subcontractor will work. The reader should note that subcontractor B could have worked day 6, 7, 8, or 9, as well as day 10. Day 9 was selected as the first working day because of practical considerations and B's general desires; he wants to move his men and equipment to the site once, perform the job as quickly as possible, and then leave. This job now requires 23 working days, and Contractor B and each of the other contractors have an adequate amount of building "spread" in front of them. The paradox is that this job would be completed sooner if each contractor would slow down to the lowest rate, 2 per day. The result of the slowdown is shown in Figure 3. The schedule in this figure calls for the entire job to be completed in 13 days. Of the other 10 days, seven are the days immediately preceding the day on which subcontractor B begins work, and three are the days preceding subcontractor D's first work day. By slowing down to

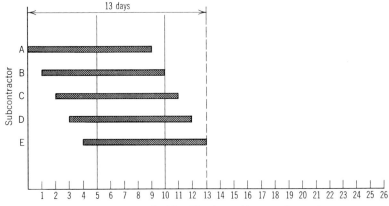

FIGURE 3.   Tract schedule with subs working at uniform rate.

the 2-per-day rate, each subcontractor works on the 2 units completed by the previous subcontractor on the previous day.

On a housing-tract construction job, those days are worth approximately $10 a piece for each house or unit. This $10 breaks down into $4.50 to $5 interest on the construction loan, about $5 interest on the loss of income, and 50¢ to $1 in direct field supervision, overhead, and miscellaneous costs. For a tract of 100 houses, the gain or loss is $1,000 a day. Using these estimates in the case we cited with 18 units, we can see that the cost is $180 per day or $1,800 for the 10 days. It might appear to subcontractor B that he cannot make a profit traveling to the job site several times. In this case the planner can consider several factors:

(a) The additional cost to have subcontractor B come out the additional times is likely to be less than the $180 a day penalty which his preferred schedule imposes.

(b) If subcontractor B is working on other jobs in the immediate area which do not take a full crew-day the extra time can be used on this job.

(c) The planners may be able to replan the job to remove subcontractor B from the critical path, so he has enough days of float or slack to do the job at his rate.

Obviously, the manager will investigate the circumstances before he chooses either to insist that subcontractor B follow his schedule or to resign himself to following B's schedule. Using the experience gained on an actual case, we can estimate that the rate of 3 a day in our example would probably be the least expensive solution; this is not the slowest rate, but is near to it. The rate of 3 a day was the most economical both for the manager and for all the subcontractors in our example. As each subcontractor learned to depend on the schedule, he had less need to keep contingency reserves; since each subcontractor was allocated the same time for each unit, each, in effect, had their work loading performed for them through the schedule.

The Production Schedule

To be of value, a schedule must be printed in a form that is useful to each subcontractor. For the schedule format we turned again to the computer. With the particular computer program we used, we were able to prepare one control card for each network; using it we could prepare a network for each unit on the computer from the standard deck. The control card described the building number, the unit number, and the particular unit floor plan. In addition, one lead time card per unit was inserted between the base date and the start of each unit network. By selecting the proper duration for this lead time, we were able to schedule 3 units per day for every trade in the unit. For those trades requiring two days to complete their task, the schedule simply indicated that the trade was to start 3 units on its first day of work, 3 more units on its second day, complete the first three units on the second day, etc. If it took one crew 3 days to perform an operation, it simply meant that 3 crews were necessary to maintain the rate.

These problems are common to any manufacturing operation. As a matter of fact, one might easily look at the production of houses or apartment units as an operation comparable to a production line. The major difference (if you ignore the appearance of the end product) is that in the production line the product moves past the workers, and in construction, the workers move past and through the product. By varying the time allotted to the starting lead time, it is possible to build up to the 3-a-day rate. For example, on the first day we may require each trade to complete only one unit, on the second day this may increase to 2 units; on the third and following days it may become 3 units, and so forth. In the computer program used, the lead times may be coded so that the lead time activity is not printed, making it unnecessary for a field superintendent or manager to note that the lead time actually did go by.

There are similar problems in setting up buildings on a production schedule. However, there were larger variances between buildings in Loma Riviera than there were between units: some had parking structures, some had carports, and one had both; some buildings had mission tile roofs, and others had boss tile ones. Since the building standard network included activities for all of these variations, it was necessary to omit printing those activities which were not needed for a given building.

Using Dissimilar Networks

Let us consider the sample network shown in Figure 4, which describes a few of the operations required in constructing a 2-story house. The reader should note that the lower rough plumbing, (the process of getting the plumbing soil pipe from the slab to the second floor), the subfloor construction, the layout, and the framing of the second floor are not necessary on a 1-story building. The preceding operation, framing the first floor, and the following operation, stacking the rafters, must be carried out on both 1- and 2-story buildings. If we were to take advantage of the reduced time to build 1-story houses, the case shown in Figure 5 might well be the schedule. The reader should remember that the activity, "framing of the first floor," is, from the schedule point of view, indicative of the schedule of all preceding subcontractors and the activity shown as "stack rafters" is likewise indicative of the schedule of all following activities. The scale along the bottom of Figure 5 represents the schedule of the number of houses that the rafter-stacking subcontractor (and all following subcontractors) must follow. Presented with such a schedule, most subcontractors would not perform at the indicated schedule rate but would rather try to smooth the schedule out so that they too were building at an even rate of one per day; they will do this by delaying work on the tract until they can work on a new house every day. If the rafter-stacking crew does not delay, then one of the 25 or 30 following subcontractors will. Eventually the delays will be large enough so that the timing will be the same as if there were all 2-story houses. If the planners knew such a delay would occur anyway, they could schedule the delay at the point at which it does the most good or least harm; they could further schedule the delay so that only a few contractors are scheduled to produce at an uneven rate. Between 40 and 50 different subcontractors work on a house in a typical housing tract. Using the technique just described (which is shown in Figure 6) the planner can limit the uneven schedule problem to just two subcontractor crews; all other subcontractors who find slight differences between 1- and 2-story houses may maintain an even schedule rate by slightly varying the size of their crews. From the above discussion, the reader can see that if the framing contractor is allowed to frame 1-story houses and the bottoms of 2-story

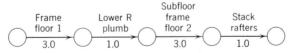

FIGURE 4. Activities related to 2-story house.

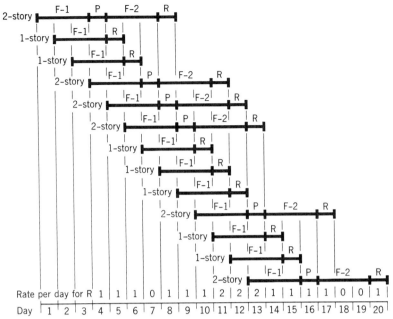

FIGURE 5. Typical schedule for 1- and 2-story houses.

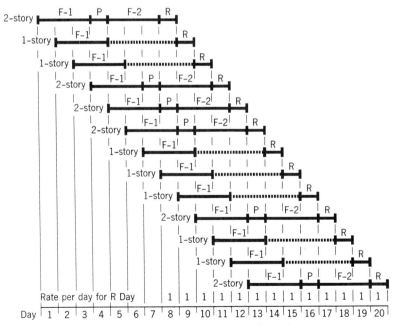

FIGURE 6. Adjusted schedule for 1- and 2-story houses.

319

```
520-530  PAY GROUP 013                                              ACTIVITY 400 PAGE  47
                                        CULVER DEVELOPMENT         PHONE

UNIVERSITY PARK                 ••• WORK SCHEDULE REPORT •••        AS OF 04-14-66

HANG DRYWALL + METAL              3 DAY DURATION    NICOLAS + NICHOLAS    PHONE

  NET      START     COMP.   ACTUAL    UNIT DESCRIPTION        UNIQUE
  NO.      DATE      DATE     DATE                             FEATURES
••••••••••••••••••••••••••••••••••••••••••••••••••••••••••••••••••••••••••••••••••••••••••

 0-193   OK START  05-04-66     LOT 145    PLAN 4-B
 0-194   OK START  05-04-66     LOT 146    PLAN 1-B
 0-195   OK START  05-04-66     LOT 147    PLAN 4-A

 0-196   OK START  05-05-66     LOT 148    PLAN 5-A-R
 0-197   OK START  05-05-66     LOT 149    PLAN 2AA-R
 0-198   OK START  05-05-66     LOT 150    PLAN 4-C-R

 0-199   05-04-66  05-06-66     LOT 151    PLAN 5-C
 0-200   05-04-66  05-06-66     LOT 152    PLAN 3-A-R
 0-201   05-04-66  05-06-66     LOT 153    PLAN 4-A-R

 0-202   05-05-66  05-09-66     LOT 154    PLAN 1-B-R
 0-203   05-05-66  05-09-66     LOT 155    PLAN 3-B-R
 0-204   05-05-66  05-09-66     LOT 156    PLAN 4-B-R

 0-205   05-06-66  05-10-66     LOT 157    PLAN 5-C
 0-206   05-06-66  05-10-66     LOT 158    PLAN 1-B
 0-207   05-06-66  05-10-66     LOT 159    PLAN 2-A

 0-208   05-09-66  05-11-66     LOT 160    PLAN 5-B-R
 0-209   05-09-66  05-11-66     LOT 161    PLAN 3-B
 0-210   05-09-66  05-11-66     LOT 162    PLAN 5-A

 0-211   05-10-66  05-12-66     LOT 163    PLAN 2-A
 0-212   05-10-66  05-12-66     LOT 164    PLAN 4-B-R
 0-213   05-10-66  05-12-66     LOT 165    PLAN 5-A

 0-214   05-11-66  05-13-66     LOT 166    PLAN 1-A-R
 0-215   05-11-66  05-13-66     LOT 167    PLAN 4-B
 0-216   05-11-66  05-13-66     LOT 168    PLAN 3-A

 0-217   05-12-66  05-16-66     LOT 169    PLAN 4-A
ACTIVITY 400               25 JOBS LISTED
```

FIGURE 7. Subcontractor's work schedule.

houses throughout the entire tract, and is then asked to return to do the top of the second stories, the delay is going to be extremely significant.

The Subcontractors' Work Schedule

Once the planner has enough information to establish a computer file and print a schedule, he will have to input only a few additional pieces of information: the designation of what goes on every lot, the designation of each unit, and a start date for the whole project. Figure 7 shows a partial schedule for one operation, "hang drywall and metal." A sub-schedule of this type is available for every real activity on the network and, therefore, for every subcontractor or trade working on the project. The location column specifies the building, unit, and floor plan for each unit listed. In addition to this scheduling by trade, there is also another report which shows all the operations for any particular unit. (Figure 8).

Subcontractor Meeting

We have described how to arrive at an original schedule using the network technique as a planning basis, and production control techniques as a means for meeting a production schedule. When the schedule has been determined, the general contractor calls all the subcontractors together in a single meeting. At this meeting, the contractor carefully explains the system to the subcontractors, describing to them in detail how the schedule was determined and how the network planning technique works. Two important points are made at the meeting. First, the subcontractor is shown how he may use the system to save himself money: the system gives him between two weeks' and 4 months' advance notice of the exact time and place at which he must provide men, materials, and equipment, thereby doing the work loading for him. Payoffs in this area have been quite substantial, mostly in terms of bids received from these same subcontractors on second and third tracts. Many contractors have indicated that on the second and following tracts bids received from certain subcontractors using this system were between 3 and 5 percent lower than they were for the first tract. At least one contractor who has negotiated with each of the subcontractors on a second job has obtained reductions in contracts of $10 and $15 per house from each of the major ones and lesser amounts from the others. This contractor believes that he recoverd the cost of instituting the system five or six times.

RELATING NETWORKS AND DOLLARS

At this time let us see what is required to add job costs to the system. The cost system aims at providing information for the payment of subcontractors, the requesting of loan dollars, and the analysis of cash flow. These functions are not based on time periods, but on physical completion of the project. After the networks have been developed, it is necessary to ask the general contractor's purchasing agent to provide information concerning the payment to be made to each subcontractor upon completion of various items in the network. It is not necessary for the subcontractor to be paid at each step; however, the purchasing agent must provide input to the system telling the payment he wishes to make to the subcontractors and the times (of physical completion) at which he wishes to pay them. For instance, the framer has several operations on 1-story houses: he lays out, raises walls, plumbs and lines the walls, installs joists, stacks rafters, installs the starter and facia, sets the roof, and installs windows, door frames and exterior jambs. Rather than pay him for each operation separately, the purchasing agent may decide to make a certain payment upon completion of layout, another when the building is joisted and a third when framing has been inspected. The important point is that the general contractor must specify the amount and timing of payments. Retentions, overhead, and other miscellaneous items may be added to the network so that the appropriate number of dollars "appear" at the correct time. The particular computer program employed permitted the planners to add dummy activities to the network and to include dollars with these dummies. When the activity preceding this dummy activity is completed, the program automatically completes the dummy. Consequently, when, for instance, the "house complete" activity is finished, the program will automatically complete the dummy activities which follow it and compute the dollars for retentions. Overhead draws and other general costs may also be added to the individual house time networks. (For example, Figure 8 shows the costs for part of a house network for a housing tract.) Once all this information is in the file, it is only necessary to report completions to have costs computed.

Assessing Project Status

In every building project the field superintendent is required to report on the project status to the general contractor's management. The network system can be used to assess this status. The field superintendent walks his tract every day (this is generally required by the general contractor); he is often asked to do so early in the morning, because anything complete at that time must have been finished the previous day; he can note this in his copy of the work schedule (Figure 7) by recording the date opposite the item. Therefore, for each item, the final record will include both a scheduled and a completion date. The system is designed so that if the field superintendent walks the tract from the most-completed house to the least-completed house, the pages in his schedule book will appear in the correct order. While he is walking the tract, the superintendent should note not only the tasks completed but the location to which each subcontractor crew is assigned that morning. If the crews are working on the appropriate structures, it is likely that the work item will be completed on time. If they are not, the superintendent has an early warning that there may be a scheduling problem, allowing him to take corrective action at the earliest possible time. Every other week, the field superintendent's schedule book is submitted to the data processing facility, where the completion information which he has entered is key-punched and processed through the system. At this time, he may change the schedule on the tract by indicating that the whole schedule should be advanced or delayed a certain number of workdays, that the construction sequence of one building related to another shall be changed, or that other aspects of the field operation should be modified. The information is then processed, producing an updated schedule for each trade or activity, listing only the uncompleted items.

PAGE 1

CULVER DEVELOPMENT   PHONE

UNIVERSITY PARK   *** PRODUCTION SCHEDULE ***   AS OF 04-14-66

SEQ NO. 0-006   LOT 039   PLAN 2-B-R   TOTAL SCHEDULE CHANGE 38 DAYS

ACT NO.	H G	DESCRIPTION	UNIQUE FEATURES	SCHED. START	SCHED. COMPL.	ACTUAL COMPL.	CONTRACT DOLLARS	EVENT NUMBERS
005		TRENCH-GRADE-CLEAN		09-24-65	09-24-65	09-29-65		010-030
010		ENG TRENCH CHECK		09-24-65	09-24-65	09-29-65		020-030
015		FORM + REBAR FOOTING		09-27-65	09-27-65	10-04-65		030-040
020		SET SOIL PLUMBING		09-28-65	09-29-65	09-29-65		040-050
025		INSPE... PLUMBING-A		09-30-65	12-03-65	12-26-65		050-07?
030		ROUGH CLEANOUT/SWP-D..		09-30-65				050-440
330		HOT MOP		12-06-65	12-06-65	12-21-65		440-445
340		INSTALL ROCK ROOF		12-07-65	12-07-65	12-21-65		445-480
350		ROUGH ELECTRIC		12-07-65	12-08-65	12-20-65	417.05	450-455
355		PRE...		12-07-65	01-31-66	12-27-65	384.06	450-47?
360		... COMPLETE		12-0?-66	01-31-66	02-03-66		...-710
550		DRYWALL PICKUP		02-01-66	02-01-66	02-03-66		710-720
552		SET PREFINISH CABINETS		02-01-66	02-01-66	01-28-66	435.60	740-780
555		DELIVER WALLPAPER		02-01-66	02-01-66	02-16-66	108.00E	710-730
557		INST SHOWER ENCLOSURE		02-01-66	02-02-66	02-24-66	51.84	710-740
560		INST RETENTION...		02-02-66		03-09-66		720-700
		PAY RETENTION-CLEAN		02-02-66			250.38	720-700
926		PAY RETENTION-GAR DOOR		03-18-66	03-18-66			926-900
931		PAY RETENTION-WETHRSTR		03-18-66	03-18-66		1.68	931-900
960		PAY RETENTION-CARPET		03-18-66	03-18-66		73.89E	960-900

COMPLETIONS THRU 04-14-66   12,429.58
REMAINING TO COMPLETE   2,086.31

TOTAL   14,515.89

FIGURE 8. Partial production schedule for single house.

```
PAY GROUP 010                                                              PAGE  27
                          CULVER DEVELOPMENT              PHONE

UNIVERSITY PARK          *** COMPLETION AND PAYMENT REPORT ***        AS OF 04-14-66

               AIR-CON INC                        PHONE
 C
 H                        UNIQUE      SEQ    ACTUAL    SCHED.    DIFFERENCE   CONTRACT
 G   UNIT DESCRIPTION    FEATURES     NO.    COMPL.    COMPL.    PLUS-MINUS   DOLLARS
 ..............................................................................


ROUGH HEAT + A/C            295
    LOT 147   PLAN 4-A               0-195   03-31-66  04-14-66      10        195.48
    LOT 148   PLAN 5-A-R             0-196   04-01-66  04-15-66      10        201.96
    LOT 149   PLAN 2AA-R             0-197   04-01-66  04-15-66      10        148.50
    LOT 150   PLAN 4-C-R             0-198   04-05-66  04-15-66       8        195.48
    LOT 151   PLAN 5-C               0-199   04-06-66  04-18-66       8        201.96
    LOT 152   PLAN 3-A-R             0-200   04-07-66  04-18-66       7        130.68
    LOT 153   PLAN 4-A-R             0-201   04-07-66  04-18-66       7        195.48
    LOT 154   PLAN 1-B-R             0-202   04-12-66  04-19-66       5        132.30
    LOT 155   PLAN 3-B-R             0-203   04-12-66  04-19-66       5        130.68
    LOT 156   PLAN 4-B-R             0-204   04-12-66  04-19-66       5        195.48
    LOT 157   PLAN 5-C               0-205   04-13-66  04-20-66       5        201.96
    LOT 158   PLAN 1-B               0-206   04-13-66  04-20-66       5        132.30
    LOT 159   PLAN 2-A               0-207   04-13-66  04-20-66       5        148.50


             13 ITEMS LISTED ABOVE            THIS ACTIVITY TOTAL          2,210.76


FINISH HEATING + A/C        600
    LOT 203   PLAN 2-B               0-103   04-05-66  04-01-66       2         99.00


              1 ITEMS LISTED ABOVE            THIS ACTIVITY TOTAL             99.00




PAY GROUP 010 TOTALS **  PAYEE    AIR-CON INC           CHECK NO-----      2,309.76
                     2ND PAYEE
```

FIGURE 9.   Completion and payment report.

Completion and Payment Report

The completion and payment report in Figure 9, shows by activity and subcontractor the items completed in the past period and the amount of money due the subcontractor for these operations. A production schedule for the individual buildings is printed; Figure 8 shows the status of that building on the reporting date. Five copies of this information are provided to the general contractor.

Paying Subcontractors

A copy of the completion and payment report is sent to the project's chief financial officer. Under this system, the general contractor *need not accept any bills or invoices from any subcontractor.* The financial officer can simply write a check for the amount shown, attach it to the page of the report for the subcontractor, and exchange this package for the releases from the subcontractor; the report tells the subcontractor what he is being paid for. The savings to the subcontractor and the general contractor are quite obvious. The subcontractor need not make out and mail an invoice; the general contractor need not receive, verify (generally at 2 or 3 levels), or file it. If there is a difference of opinion between the subcontractor and the general contractor, the subcontractor will bring it specifically to the general contractor's attention. The general contractor has available the field superintendent's report, indicating what the field superintendent believed had been done and, therefore, what was due to be paid. The amounts for each pay item are printed and can be verified against the contract. Disagreement between the subcontractor and general contractor, therefore, should be minimal and easily resolved.

Loan "Draws"

A variation on the payment and completion report can be used to deal with the loan company holding the construction loan paper. At a given point or points in the network the amount of "draw" must be added to input, and so, if the field superintendent reports completion of an item, then a completion and payment report for loan dollars would appear, listing the item and amount in the same manner as is done for subcontractor payments. The general contractor need only prepare a cover letter noting the total number of dollars shown, attach a report copy showing the detail of the request, and present this to the loan company. Experience with lending agencies has indicated that, properly handled, the system allows the lending agency to have the control over the project that it normally has with a voucher system, at a cost equivalent to that of an open line of credit. One company, Wm. W. Main & Associates of Seattle, Wash., has negotiated with their lending agency, J. E. Cooper & Co. of Spokane, so that the direct cost contracts plus the overhead cost are the basis of the draw. Consequently, a copy of the regular payment and completion report is sent to the agency, which uses it to prepare checks for the subcontractor; these checks are returned to the general contractor, who exchanges them with the subcontractors for signed labor and material releases. The project referred to happens to be an FHA project so the lending agency must keep books by individual lot; since the report already provides this information, however, the bookkeeping cost for the lending agency is considerably reduced.

Summary of Project Status—the Line of Balance

A "line of balance" (LOB) chart was also provided on this condominium project, to give the managers a summary picture of the status. This type of chart is shown in Figure 10,[1] and consists of three parts. The lower

---

[1]Figure 10 illustrates a sample project containing only one type of structure. Similar charts for apartment projects show the status of both the buildings and the interior apartments.

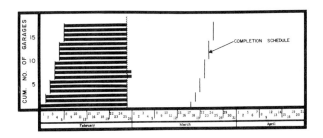

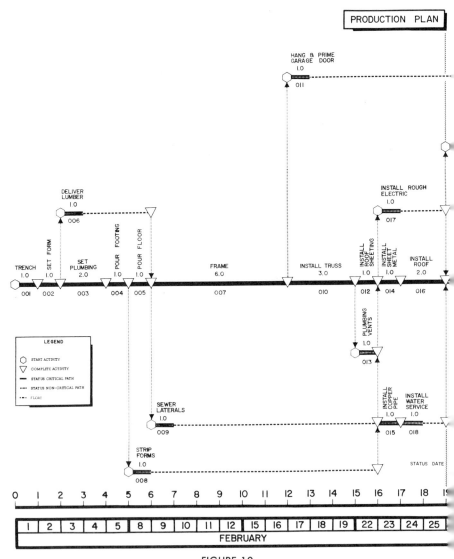

PRODUCTION PLAN

HANG & PRIME
GARAGE DOOR
1.0
011

DELIVER
LUMBER
1.0
006

INSTALL ROUGH
ELECTRIC
1.0
017

TRENCH
1.0
001

SET FORM
1.0
002

SET
PLUMBING
2.0
003

POUR FOOTING
1.0
004

POUR FLOOR
1.0
005

FRAME
6.0
007

INSTALL TRUSS
3.0
010

INSTALL
ROOF
SHEETING
1.0
012

INSTALL
SHEET
METAL
1.0
014

INSTALL
ROOF
2.0
016

PLUMBING
VENTS
1.0
013

LEGEND

◯ START ACTIVITY
▽ COMPLETE ACTIVITY
━ STATUS CRITICAL PATH
•••• STATUS NON-CRITICAL PATH
•-• FLOAT

SEWER
LATERALS
1.0
009

INSTALL
COPPER
PIPE
1.0
015

INSTALL
WATER
SERVICE
1.0
018

STRIP
FORMS
1.0
008

STATUS DATE

0 1 2 3 4 5 6 7 8 9 10 11 12 13 14 15 16 17 18 19

| 1 | 2 | 3 | 4 | 5 | 8 | 9 | 10 | 11 | 12 | 15 | 16 | 17 | 18 | 19 | 22 | 23 | 24 | 25 |

FEBRUARY

FIGURE 10.

326

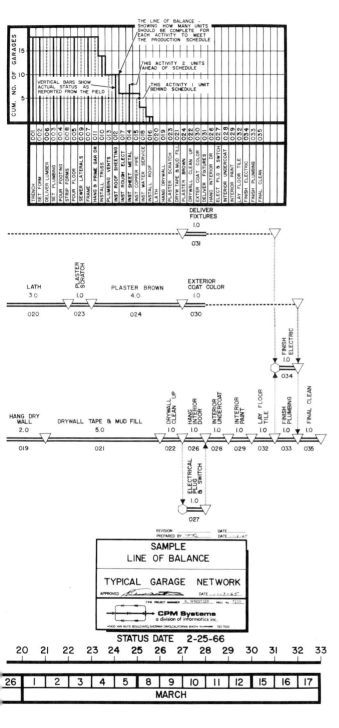

THE LINE OF BALANCE - SHOWING HOW MANY UNITS SHOULD BE COMPLETE FOR EACH ACTIVITY TO MEET THE PRODUCTION SCHEDULE

THIS ACTIVITY 2 UNITS AHEAD OF SCHEDULE

VERTICAL BARS SHOW ACTUAL STATUS AS REPORTED FROM THE FIELD

THIS ACTIVITY I UNIT BEHIND SCHEDULE

CUM. NO. OF GARAGES

15

10

5

001	TRENCH
002	SET FORM
006	DELIVER LUMBER
003	SET PLUMBING
004	POUR FOOTING
008	STRIP FORMS
005	POUR FLOOR
009	SEWER LATERALS
007	FRAME
010	HANG & PRIME GAR DR
011	INSTALL TRISS
013	PLUMBING VENTS
012	INST ROOF SHEETING
017	INST ROUGH ELECT
014	INST SHEET METAL
015	INST COPPER PIPE
018	INST WATER SERVICE
016	INSTALL ROOF
019	LATH
020	HANG DRYWALL
023	PLASTER SCRATCH
021	DRYW TAPE & MUD FILL
024	PLASTER BROWN
022	DRYWALL CLEAN UP
030	EXTER COAT COLOR
031	DELIVER FIXTURES
026	HANG INTERIOR DR
027	ELECT PLG & SWTCH
028	INTERIOR UNDERCOAT
029	INTERIOR PAINT
032	LAY FLOOR TILE
034	FINISH ELECTRIC
033	FINISH PLUMBING
035	FINAL CLEAN

DELIVER FIXTURES
1.0
031

LATH
3.0
020

PLASTER SCRATCH
1.0
023

PLASTER BROWN
4.0
024

EXTERIOR COAT COLOR
1.0
030

FINISH ELECTRIC
1.0
034

HANG DRY WALL
2.0
019

DRYWALL TAPE & MUD FILL
5.0
021

DRYWALL CLEAN UP
1.0
022

HANG INTERIOR DOOR
1.0
026

INTERIOR UNDERCOAT
1.0
028

INTERIOR PAINT
1.0
029

LAY FLOOR TILE
1.0
032

FINISH PLUMBING
1.0
033

FINAL CLEAN
1.0
035

ELECTRICAL PLUG & SWITCH
1.0
027

REVISION _____ DATE _____
PREPARED BY _____ DATE __-__-65

SAMPLE
LINE OF BALANCE

TYPICAL GARAGE NETWORK

APPROVED _____ DATE __-__-65
CPM PROJECT MANAGER R. WRESTLER   PROJ. NO. 7500

CPM Systems
a division of informatics inc.
4530 VAN NUYS BOULEVARD, SHERMAN OAKS, CALIFORNIA 91403 • TELEPHONE: 783-7500

STATUS DATE 2-25-66

20	21	22	23	24	25	26	27	28	29	30	31	32	33

26	1	2	3	4	5	8	9	10	11	12	15	16	17
						MARCH							

327

section is a simplified network of the construction of the structure. The upper left-hand section shows the overall schedule of starts and completions—in effect the critical-path status of each structure. The upper right-hand portion of the chart shows the line of balance, which is the stairstep line on the chart. The reported completions are plotted vertically against the cumulative scale, and are graphically compared to the required number of completions for each activity, shown by the line of balance. The line of balance shows how many structures should have been completed through "finish plumbing," for instance, in order to be on schedule. When the reported actual completions fall below the line of balance for a given activity, that activity is behind schedule, and when the actual completions go above the line of balance, the activity is ahead of schedule.[2] The LOB chart is a useful tool for the project manager who is not in daily contact with the project.

Successful Use of CPM Is Not Automatic

The discussion in this chapter demonstrates that network planning is a tool which management can use to carry on business more efficiently and more economically than is possible without it. CPM is a tool in the same sense that a shovel is a tool. If someone buys a new shovel for $19.95, takes it to his place of business, stands it in the corner, and continues to dig holes by hand, the only thing that has been done is that $19.95 has been wasted. However, if the businessman uses the shovel for the purpose for which it was designed in the manner intended, he can dig holes at less expense than he can by hand. So it is with CPM, a logical tool; the test of time will dictate certain changes and improvements in the method, but the results of its use depend solely on the way and extent to which it is used.

SUMMARY

This case study illustrates how the basic concept of network planning can be used with other management disciplines such as production control methods to produce more effective management systems in a repetitive operation project. It also clearly demonstrates that in certain environments, the linking of dollar costs to activities in the networks is practical and desirable. Using the system described, builders commonly reduce the overall time to construct a housing tract by 3 to 4 weeks, or 13 to 17 percent of the usual 5 to 6 months required. When the planners

---

[2]For a full description of this production control technique, see *Line of Balance Technology*, Navy Department, Office of Naval Materiel (NAVEXOS P1851 Rev 4–62).

set up a master schedule and a constant construction rate and impose the discipline of the system, everyone falls in step. Other benefits enjoyed by both the contractor and the subcontractor result from improved communication, reduced paper work, and increased attention to quality. Delivering the houses to the buyers on schedule is, of course, also an important result. This system is widely used in California, and it appears that its acceptance and use will continue to spread.

16

Decision trees for decision-making

We have considered network planning primarily as a tool to be used on projects. Becoming equally important is the use of network planning techniques to evaluate existing administrative systems and procedures, and to introduce structure and logic into the decision-making process. In the following discussion of the use of networks as aids in making long range choices and gains, and determining risks and goals for capital investment, John Magee has provided an imaginative picture of the broader implications of the system as a management tool. Progressive implementation of network techniques with their inherent capabilities, linked with expanded use of the computer, inevitably leads us closer to realization of the "total systems" management ideal. In this chapter, quoted from the *Harvard Business Review*[1] John F. Magee presents a typical management problem facing a hypothetical company which he calls Stygian Chemical Industries, Ltd., to begin his illustration of the use of decision trees or network theory for decision-making.

The management of . . . Stygian Chemical Industries, Ltd., must decide whether to build a small plant or a large one to manufacture a new product with an expected market life of ten years. The decision hinges on what size the market for the product will be.

Possibly demand will be high during the initial two years but, if many initial users find the product unsatisfactory, will fall to a low level thereafter. Or high initial demand might indicate the possibility of a sustained high-volume market. If demand is high and the company does not expand within the first two years, competitive products will surely be introduced.

If the company builds a big plant, it must live with it whatever the size of market demand. If it builds a small plant, management has the option of expanding the plant in two years in the event that demand is high during the introductory period;

---

[1]"Decision Trees for Decision Making," July-August, 1964, pp. 126–38, and "How to Use Decision Trees in Capital Investment," September-October, 1964, pp. 79–96.

while in the event that demand is low during the introductory period, the company will maintain operations in the small plant and make a tidy profit on the low volume.

Management is uncertain what to do. The company grew rapidly during the 1950's; it kept pace with the chemical industry generally. The new product, if the market turns out to be large, offers the present management a chance to push the company into a new period of profitable growth. The development department, particularly the development project engineer, is pushing to build the large-scale plant to exploit the first major product development the department has produced in some years.

The chairman, a principal stockholder, is wary of the possibility of large unneeded plant capacity. He favors a smaller plant commitment, but recognizes that later expansion to meet high-volume demand would require more investment and be less efficient to operate. The chairman also recognizes that unless the company moves promptly to fill the demand which develops, competitors will be tempted to move in with equivalent products.

The Stygian Chemical problem, oversimplified as it is, illustrates the uncertainties and issues that business management must resolve in making investment decisions. ([The term] "investment" [is used] in a broad sense, referring to outlays not only for new plants and equipment but also for large, risky orders, special marketing facilities, research programs, and other purposes.) These decisions are growing more important at the same time that they are increasing in complexity. Countless executives want to make them better—but how?

. . . [As indicated above, this chapter will] present [the] recently developed concept [of network planning, which in this chapter, will be referred to as] the "decision tree," which has tremendous potential as a decision-making tool. The decision tree can clarify for management, as can [few] other analytical tools, the choices, risks, objectives, monetary gains, and information needs involved in an investment problem. We shall be hearing a great deal about [network planning, or] decision trees in the years ahead. Although a novelty to most businessmen today, they will surely be in common management parlance before many more years have passed.

Later in this [chapter] we shall return to the problem facing Stygian Chemical and see how management can proceed to solve it by using decision trees. First, however, a simpler example will illustrate some characteristics of the decision-tree approach.

DISPLAYING ALTERNATIVES

Let us suppose it is a rather overcast Saturday morning, and you have 75 people coming for cocktails in the afternoon. You have a pleasant garden and your house is not too large; so if the weather permits, you would like to set up the refreshments in the garden and have the party there. It would be more pleasant, and your guests would be more comfortable. On the other hand, if you set up the party for the garden and after all the guests are assembled it begins to rain, the refreshments will be ruined, your guests will get damp, and you will heartily wish you had decided to have the party in the house. (We could complicate this problem by considering the possibility of a partial commitment to one course or another and opportunities to adjust estimates of the weather as the day goes on, but the simple problem is all we need.)

This particular decision can be represented in the form of a "payoff" table:

	Events and Results	
Choices	Rain	No Rain
Outdoors	Disaster	Real comfort
Indoors	Mild discomfort, but happy	Mild discomfort, but regrets

Much more complex decision questions can be portrayed in payoff table form. However, particularly for complex investment decisions, a different representation of the information pertinent to the problem—the decision tree—is useful to show the routes by which the various possible outcomes are achieved. Pierre Massé, Commissioner General of the National Agency for Productivity and Equipment Planning in France, notes:

> The decision problem is not posed in terms of an isolated decision (because today's decision depends on the one we shall make tomorrow) nor yet in terms of a sequence of decisions (because under uncertainty, decisions taken in the future will be influenced by what we have learned in the meanwhile). The problem is posed in terms of a tree of decisions.[2]

[Figure 1] illustrates a decision tree for the cocktail party problem. This tree is a different way of displaying the same information shown in the payoff table. However, as later examples will show, in complex decisions the decision tree is frequently a much more lucid means of presenting the relevant information than is a payoff table.

The tree is made up of a series of nodes and branches. At the first node on the left, the host has the choice of having the party inside or outside. Each branch represents an alternative course of action or decision. At the end of each branch or alternative course is another node representing a chance event—whether or not it will rain. Each subsequent alternative course to the right represents an alternative outcome of this chance event. Associated with each complete alternative course through the tree is a payoff, shown at the end of the rightmost or terminal branch of the course.

. . . [In drawing these illustrative decision trees, we have indicated] the action or decision forks with square nodes and the chance-event forks with round ones. Other symbols may be used instead, such as single-line and double-line branches, special letters, or colors. It does not matter so much which method of distinguishing [is used] so long as . . . one or another [is employed.] A decision tree of any size will

---

[2]*Optimal Investment Decisions: Rules for Action and Criteria for Choice*, Englewood Cliffs, New Jersey, Prentice-Hall, Inc., 1962, p. 250.

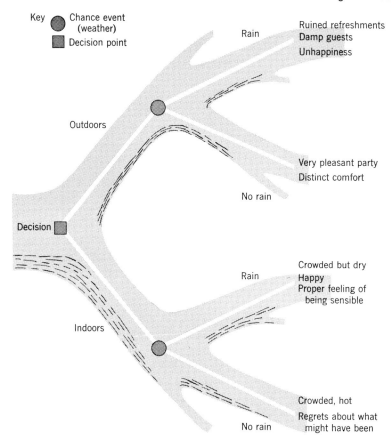

Key

● Chance event (weather)

■ Decision point

Outdoors

Rain

Ruined refreshments
Damp guests
Unhappiness

No rain

Very pleasant party
Distinct comfort

Decision ■

Indoors

Rain

Crowded but dry
Happy
Proper feeling of being sensible

No rain

Crowded, hot
Regrets about what might have been

FIGURE 1.   Decision tree for cocktail party.

always combine (a) *action* choices with (b) different possible *events* or *results* of action which are partially affected by chance or other uncontrollable circumstances.

Decision-Event Chains

The previous example, though involving only a single stage of decision, illustrates the elementary principles on which larger, more complex decision trees are built. Let us take a slightly more complicated situation: You are trying to decide whether to approve a development budget for an improved product. You are urged to do so on the grounds that the development, if successful, will give you a competitive edge, but if you do not develop the product, your competitor may—and may seriously damage your market share. You sketch out a decision tree that looks something like the one in [Figure 2].

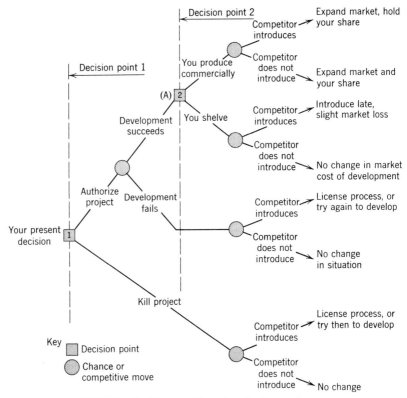

FIGURE 2. Decision tree with chains of actions and events.

Your initial decision is shown at the left. Following a decision to proceed with the project, if development is successful, is a second stage of decision at Point A. Assuming no important change in the situation between now and the time of Point A, you decide now what alternatives will be important to you at that time. At the right of the tree are the outcomes of different sequences of decisions and events. These outcomes, too, are based on your present information. In effect you say, "If what I know now is true then, this is what will happen." Of course, you do not try to identify all the events that can happen or all the decisions you will have to make on a subject under analysis. In the decision tree you lay out only those decisions and events or results that are important to you and have consequences you wish to compare. . . .

ADDING FINANCIAL DATA

Now we can return to the problems faced by the Stygian Chemical management. A decision tree characterizing the investment problem as outlined in the introduc-

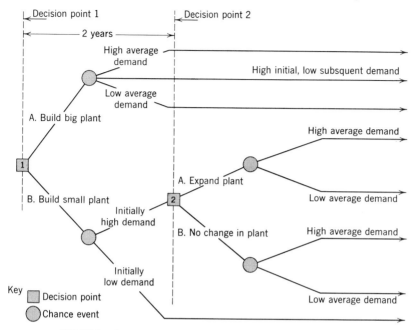

FIGURE 3. Decisions and events for Stygian Chemical Industries, Ltd.

tion is shown in [Figure 3]. At Decision 1 the company must decide between a large and a small plant. This is all that must be decided *now*. But if the company chooses to build a small plant and then finds demand high during the initial period, it can in two years—at Decision 2—choose to expand its plant.

But let us go beyond a bare outline of alternatives. In making decisions, executives must take account of the probabilities, costs, and returns which appear likely. On the basis of the data now available to them, and assuming no important change in the company's situation, they reason as follows:

Marketing estimates indicate a 60% chance of a large market in the long run and a 40% chance of a low demand, developing initially as follows:

Initially high demand, sustained high:	60%	
Initially high demand, long-term low:	10%	Low = 40%
Initially low and continuing low:	30%	
Initially low and subsequently high:	0%	

Therefore, the chance that demand initially will be high is 70% (60 + 10). If demand is high initially, the company estimates that the chance it will continue at a high level is 86% (60 ÷ 70). Comparing 86% to 60%, it is apparent that a high initial level of sales changes the estimated chance of high sales in the subsequent periods.

Similarly, if sales in the initial period are low, the chances are 100% (30 ÷ 30) that sales in the subsequent periods will be low. Thus the level of sales in the initial period is expected to be a rather accurate indicator of the level of sales in the subsequent periods.

Estimates of annual income are made under the assumption of each alternative outcome:

(1) A large plant with high volume would yield $1,000,000 annually in cash flow.

(2) A large plant with low volume would yield only $100,000 because of high fixed costs and inefficiencies.

(3) A small plant with low demand would be economical and would yield annual cash income of $400,000.

(4) A small plant, during an initial period of high demand, would yield $450,000 per year, but this would drop to $300,000 yearly in the long run because of competition. (The market would be larger than under Alternative 3, but would be divided up among more competitors.)

(5) If the small plant were expanded to meet sustained high demand, it would yield $700,000 cash flow annually, and so would be less efficient than a large plant built initially.

(6) If the small plant were expanded but high demand were not sustained, estimated annual cash flow would be $50,000.

It is estimated further that a large plant would cost $3 million to put into operation, a small plant would cost $1.3 million, and the expansion of the small plant would cost an additional $2.2 million.

When the foregoing data are incorporated, we have the decision tree shown in [Figure 4]. Bear in mind that nothing is shown here which Stygian Chemical's executives did not know before; no numbers have been pulled out of hats. However, we are beginning to see dramatic evidence of the value of decision trees in *laying out* what management knows in a way that enables more systematic analysis and leads to better decisions. To sum up the requirements of making a decision tree, management must:

(1) Identify the points of decision and alternatives available at each point.

(2) Identify the points of uncertainty and the type or range of alternative outcomes at each point.

(3) Estimate the values needed to make the analysis, especially the probabilities of different events or results of action and the costs and gains of various events and actions.

(4) Analyze the alternative values to choose a course.

CHOOSING COURSE OF ACTION

We are now ready for the next step in the analysis—to compare the consequences of different courses of action. A decision tree does not give management the answer

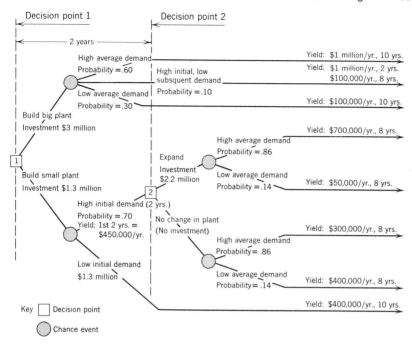

FIGURE 4. Decision tree with financial data. Key: square denotes decision point, circle denotes chance event.

to an investment problem; rather, it helps management determine which alternative at any particular choice point will yield the greatest expected monetary gain, given the information and alternatives pertinent to the decision.

Of course, the gains must be viewed with the risks. At Stygian Chemical, as at many corporations, managers have different points of view toward risk; hence they will draw different conclusions in the circumstances described by the decision tree shown in [Figure 4]. The many people participating in a decision—those supplying capital, ideas, data, or decisions, and having different values at risk—will see the uncertainty surrounding the decision in different ways. Unless these differences are recognized and dealt with, those who must make the decision, pay for it, supply data and analyses to it, and live with it will judge the issue, relevance of data, need for analysis, and criterion of success in different and conflicting ways.

For example, company stockholders may treat a particular investment as one of a series of possibilities, some of which will work out, others of which will fail. A major investment may pose risks to a middle manager—to his job and career—no matter what decision is made. Another participant may have a lot to gain from success, but little to lose from failure of the project. The nature of the risk—as each individual sees it—will affect not only the assumptions he is willing to make but also the strategy he will follow in dealing with the risk.

The existence of multiple, unstated, and conflicting objectives will certainly contribute to the "politics" of Stygian Chemical's decision, and one can be certain that the political element exists whenever the lives and ambitions of people are affected. Here, as in similar cases, it is not a bad exercise to think through who the parties to an investment decision are and to try to make these assessments:

What is at risk? Is it profit or equity value, survival of the business, maintenance of a job, opportunity for a major career?

Who is bearing the risk? The stockholder is usually bearing risk in one form. Management, employees, the community—all may be bearing different risks.

What is the character of the risk that each person bears? Is it, *in his terms*, unique, once-in-a-lifetime, sequential, insurable? Does it affect the economy, the industry, the company, or a portion of the company?

Considerations such as the foregoing will surely enter into top management's thinking, and the decision tree in [Figure 4] will not eliminate them. But the tree will show management what decision today will contribute most to its long-term goals. The tool for this next step in the analysis is the concept of "rollback."

"Rollback" Concept

Here is how rollback works in the situation described. At the time of making Decision 1 (see [Figure 4]) management does not have to make Decision 2 and does not even know if it will have the occasion to do so. But if it *were* to have the option at Decision 2, the company would expand the plant, in view of its current knowledge. The analysis is shown in [Figure 5]. ([We] shall ignore for the moment the question of discounting future profits; that is introduced later.) We see that the total expected value of the expansion alternative is $160,000 greater than the no-expansion alternative, over the eight-year life remaining. Hence that is the alternative

Choice	Chance event	Proba-bility (1)	Total yield, 8 years (thousands of dollars) (2)	Expected value (thousands of dollars) (1) × (2)
Expansion	High average demand	.86	$5,600	$4,816
	Low average demand	.14	400	56
			Total	$4,872
			Less investment	2,200
			Net	$2,672
No expansion	High average demand	.86	$2,400	$2,064
	Low average demand	.14	3,200	448
			Total	$2,512
			Less investment	0
			Net	$2,512

FIGURE 5. Analysis of possible decision 2 (using maximum expected total cash flow as criterion).

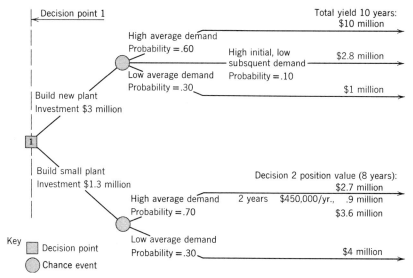

FIGURE 6. Cash flow analysis for decision 1.

management would choose if faced with Decision 2 with its existing information (and thinking only of monetary gain as a standard of choice).

Readers may wonder why we started with Decision 2 when today's problem is Decision 1. The reason is the following: We need to be able to put a monetary value on Decision 2 in order to "roll back" to Decision 1 and compare the gain from taking the lower branch ("Build Small Plant") with the gain from taking the upper branch ("Build Big Plant"). Let us call that monetary value for Decision 2 its *position value*. The position value of a decision is the expected value of the preferred branch (in this case, the plant-expansion fork). The expected value is simply a kind of average of the results you would expect if you were to repeat the situation over and over—getting a $5,600 thousand yield 86% of the time and a $400 thousand yield 14% of the time.

Stated in another way, it is worth $2,672 thousand to Stygian Chemical to get to the position where it can make Decision 2. The question is: Given this value and the other data shown in [Figure 4], what now appears to be the best action at Decision 1?

Turn now to [Figure 6]. At the right of the branches in the top half we see the yields for various events if a big plant is built (these are simply the figures in [Figure 4] multiplied out). In the bottom half we see the small plant figures, including Decision 2 position value plus the yield for the two years prior to Decision 2. If we reduce all these yields by their probabilities, we get the following comparison:

Build big plant: ($10 × .60) + ($2.8 × .10) + ($1 × .30) − $3 = $3,600 thousand

Build small plant: ($3.6 × .70) + ($4 × .30) − $1.3 = $2,400 thousand

The choice which maximizes expected total cash yield at Decision 1, therefore, is to build the big plant initially.

ACCOUNTING FOR TIME

What about taking differences in the *time* of future earnings into account? The time between successive decision stages on a decision tree may be substantial. At any stage, we may have to weigh differences in immediate cost or revenue against differences in value at the next stage. Whatever standard of choice is applied, we can put the two alternatives on a comparable basis if we discount the value assigned to the next stage by an appropriate percentage. The discount percentage is, in effect, an allowance for the cost of capital and is similar to the use of a discount rate in the present value or discounted cash flow techniques already well known to businessmen.

When decision trees are used, the discounting procedure can be applied one stage at a time. Both cash flows and position values are discounted.

A. Present values of cash flows

Choice—outcome	Yield	Present value (in thousands)
Expand—high demand	$700,000/Year, 8 years	$4,100
Expand—low demand	50,000/Year, 8 years	300
No change—high demand	300,000/Year, 8 years	1,800
No change—low demand	400,000/Year, 8 years	2,300

B. Obtaining discounted expected values

Choice	Chance event	Proba- bility (1)	Present value yield (in thousands) (2)	Discounted expected value (in thousands) (1) × (2)
Expansion	High average demand	.86	$4,100	$3,526
	Low average demand	.14	300	42
			Total	$3,568
			Less investment	2,200
			Net	$1,368
No expansion	High average demand	.86	$1,800	$1,548
	Low average demand	.14	2,300	322
			Total	$1,870
			Less investment	0
			Net	$1,870

NOTE: For simplicity, the first year cash flow is not discounted, the second year cash flow is discounted one year, and so on.

FIGURE 7. Analysis of Decision 2 with discounting.

Choice	Chance event	Proba-bility (1)	Yield (in thousands)	Discounted value of yield (in thousands) (2)	Discounted expected yield (in thousands) (1) × (2)
Build big plant	High average demand	.60	$1,000/year, 10 years	$6,700	$4,020
	High initial, low average demand	.10	1,000/year, 2 years 100/year, 8 years	2,400	240
	Low average demand	.30	100/year, 10 years	700	210
				Total	$4,470
				Less investment	3,000
				Net	$1,470
Build small plant	High initial demand	.70	$ 450/year, 2 years	$ 860	$ 600
			Decision 2 value, $1,870 at end of 2 years	1,530	1,070
	Low initial demand	.30	$ 400/year, 10 years	2,690	810
				Total	$2,480
				Less investment	1,300
				Net	$1,180

FIGURE 8. Analysis of decision.

For simplicity, let us assume that a discount rate of 10% per year for all stages is decided on by Stygian Chemical's management. Applying the rollback principle, we again begin with Decision 2. Taking the same figures used in previous exhibits and discounting the cash flows at 10%, we get the data shown in Part A of [Figure 7]. Note particularly that these are the present values *as of the time Decision 2 is made.*

Now we want to go through the same procedure used in [Figure 5] when we obtained expected values, only this time using the discounted yield figures and obtaining a discounted expected value. The results are shown in Part B of [Figure 7]. Since the discounted expected value of the no-expansion alternative is higher, *that* figure becomes the position value of Decision 2 this time.

Having done this, we go back to work through Decision 1 again, repeating the same analytical procedure as before only with discounting. The calculations are shown in [Figure 8]. Note that the Decision 2 position value is treated at the time of Decision 1 as if it were a lump sum received at the end of the two years.

The large-plant alternative is again the preferred one on the basis of discounted expected cash flow. But the margin of difference over the small-plant alternative ($290 thousand) is smaller than it was without discounting.

UNCERTAINTY ALTERNATIVES

In illustrating the decision-tree concept, uncertainty alternatives [have been treated] as if they were discrete, well-defined possibilities. [The] examples . . . have made use of uncertain situations depending basically on a single variable, such as the level of demand or the success or failure of a development project. [We] have sought to avoid unnecessary complication while putting emphasis on the key interrelationships among the present decision, future choices, and the intervening uncertainties.

In many cases, the uncertain elements do take the form of discrete, single-variable alternatives. In others, however, the possibilities for cash flow during a stage may range through a whole spectrum and may depend on a number of independent or partially related variables subject to chance influences—cost, demand, yield, economic climate, and so forth. In these cases, we have found that the range of variability or the likelihood of the cash flow falling in a given range during a stage can be calculated readily from knowledge of the key variables and the uncertainties surrounding them. Then the range of cash-flow possibilities during the stage can be broken down into two, three, or more "subsets," which can be used as discrete chance alternatives.

Peter F. Drucker has succinctly expressed the relation between present planning and future events: "Long-range planning does not deal with future decisions. It deals with the futurity of present decisions."[3] Today's decision should be made in light of the anticipated effect it and the outcome of uncertain events will have on future values and decisions. Since today's decision sets the stage for tomorrow's decision, today's decision must balance economy with flexibility; it must balance the need to capitalize on profit opportunities that may exist with the capacity to react to future circumstances and needs.

The unique feature of the decision tree is that it allows management to combine analytical techniques such as discounted cash flow and present value methods with a clear portrayal of the impact of future decision alternatives and events. Using the decision tree, management can consider various courses of action with greater ease and clarity. The interactions between present decision alternatives, uncertain events, and future choices and their results become more visible.

Of course, there are many practical aspects of decision trees in addition to those that . . . [have been] covered [here]. . . . Subsequent [discussion will reveal] the [further] range of possible gains for management. . . .

Surely the decision-tree concept does not offer final answers to managements making investment decisions in the face of uncertainty. We have not reached that

---

[3] "Long-Range Planning," *Management Science*, April 1959, p. 239.

stage, and perhaps we never will. Nevertheless, the concept is valuable for illustrating the structure of investment decisions, and it can likewise provide excellent help in the evaluation of capital investment *opportunities*.

New Facility

The choice of alternatives in building a plant depends upon market forecasts. The alternative chosen will, in turn, affect the market outcome. For example, the military products division of a diversified firm, after some period of low profits due to intense competition, has won a contract to produce a new type of military engine suitable for Army transport vehicles. The division has a contract to build productive capacity and to produce at a specified contract level over a period of three years.

[Figure 9] illustrates the situation. The dotted line shows the contract rate. The solid line shows the proposed buildup of production for the military. Some other possibilities are portrayed by dashed lines. The company is not sure whether the contract will be continued at a relatively high rate after the third year, as shown by Line A, or whether the military will turn to another newer development, as indicated by Line B. The company has no guarantee of compensation after the third year. There is also the possibility, indicated by Line C, of a large additional commercial market for the product, this possibility being somewhat dependent on the cost at which the product can be made and sold.

If this commercial market could be tapped, it would represent a major new busi-

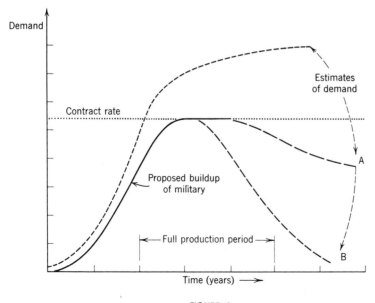

FIGURE 9.

ness for the company and a substantial improvement in the profitability of the division and its importance to the company.

Management wants to explore three ways of producing the product as follows:

(1) It might subcontract all fabrication and set up a simple assembly with limited need for investment in plant and equipment; the costs would tend to be relatively high and the company's investment and profit opportunity would be limited, but the company assets which are at risk would also be limited.

(2) It might undertake the major part of the fabrication itself but use general-purpose machine tools in a plant of general-purpose construction. The division would have a chance to retain more of the most profitable operations itself, exploiting some technical developments it has made (on the basis of which it got the contract). While the cost of production would still be relatively high, the nature of the investment in plant and equipment would be such that it could probably be turned to other uses or liquidated if the business disappeared.

(3) The company could build a highly mechanized plant with specialized fabrication and assembly equipment, entailing the largest investment but yielding a substantially lower unit manufacturing cost if manufacturing volume were adequate. Following this plan would improve the chances for a continuation of the military contract and penetration into the commercial market and would improve the profitability of whatever business might be obtained in these markets. Failure to sustain either the military or the commercial market, however, would cause substantial financial loss.

Either of the first two alternatives would be better adapted to low-volume production than would the third.

Some major uncertainties are: the cost-volume relationships under the alternative manufacturing methods; the size and structure of the future market—this depends in part on cost, but the degree and extent of dependence are unknown; and the possibilities of competitive developments which would render the product competitively or technologically obsolete.

How would this situation be shown in decision-tree form? (Before going further you might want to draw a tree for the problem yourself.) [Figure 10] shows [a] version of a tree. Note that in this case the chance alternatives are somewhat influenced by the decision made. A decision, for example, to build a more efficient plant will open possibilities for an expanded market.

Plant Modernization

A company management is faced with a decision on a proposal by its engineering staff which, after three years of study, wants to install a computer-based control system in the company's major plant. The expected cost of the control system is some $30 million. The claimed advantages of the system will be a reduction in labor cost and an improved product yield. These benefits depend on the level of product throughout, which is likely to rise over the next decade. It is thought that the installation program will take about two years and will cost a substantial amount over and above the cost of equipment. The engineers calculate that the automation project will yield a 20% return on investment, after taxes; the projection is based on a ten-year forecast of product demand by the market research department, and an assumption of an eight-year life for the process control system.

What would this investment yield? Will actual product sales be higher or lower than forecast? Will the process work? Will it achieve the economies expected? Will competitors follow if the company is successful? Are they going to mechanize anyway? Will new products or processes make the basic plant obsolete before the investment can be recovered? Will the controls last eight years? Will something better come along sooner?

The initial decision alternatives are (a) to install the proposed control system, (b) postpone action until trends in the market and/or competition become clearer, or (c) initiate more investigation or an independent evaluation. Each alternative will be followed by resolution of some uncertain aspect, in part dependent on the action taken. This resolution will lead in turn to a new decision. The dotted lines at the right of Figure 11 indicate that the decision tree continues indefinitely, though the decision

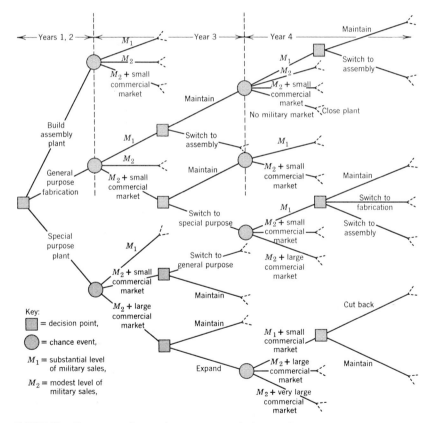

FIGURE 10. Key: square denotes decision point, circle denotes chance event, M_1 = substantial level of military sales, M_2 = modest level of military sales.

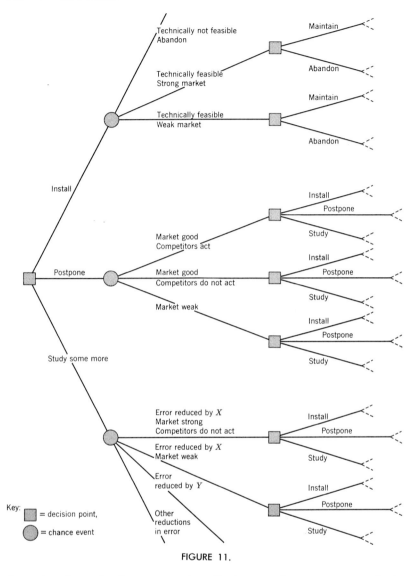

FIGURE 11.

alternatives do tend to become repetitive. In the case of postponement or further study, the decisions are to install, postpone, or restudy; in the case of installation, the decisions are to continue operation or abandon.

An immediate decision is often one of a sequence. It may be one of a number of sequences. The impact of the present decision in narrowing down future alternatives and the effect of future alternatives in affecting the value of the present choice must both be considered.

[To this point] the decision-tree approach [has been] described . . . as a way of displaying the anatomy of a business investment decision and of showing the interplay among a present decision, chance events, competitors' moves, and possible future decisions and their consequences. We [have seen] the values of decision trees in exploring a variety of problems—new-product introduction, plant modernization, research and development strategy, outlays for new facilities, and others.

. . . [The remainder of this chapter will] discuss how to go about using the decision-tree concept—the skills needed, organization of the analysis, and data required. A case example of a business investment question will serve as a vehicle for discussing aspects of the practical use of decision trees. As this example will illustrate, the new approach leads to different and improved answers to investment-analysis questions. It enables management to take more direct account of—

the impact of possible future decisions;

the impact of uncertainty;

the relative values of present and future profits;

the comparative advantages of varying levels of expected profit and corporate flexibility.

PLAN OF ATTACK

Our case example involves a company situation which, though fictitious, incorporates a number of elements of cases I have actually worked with.

The management of Prism Paints, Inc., must decide what to do with one of its manufacturing plants which, though adequate in size, is technically obsolete and unable to supply the quality of products required in the current market. Since the plant in question is rather small by modern standards, and it is argued that a plant of its size is at an economic disadvantage, there is considerable managerial controversy over the proper course of action—whether to modernize the operation by construction of better facilities at that location or to scrap the existing plant and supply the area involved from the company's facilities elsewhere, e.g., plants in adjacent states.

The possible alternatives involve significantly different capital expenditures and appear to lead to substantial differences in operating economy as well. Underlying the controversy over what to do is a concern for future product demand in the area served by the plant.

In this and other cases the key steps in building and using a decision tree for investment-project analysis are:

(1) *Identification of the problem and alternatives*—The information about the investment opportunity may come from a number of sources, including market analysis, operations research and engineering analysis, research and development work, trend studies, competitive intelligence, and executive imagination. Experience indicates that the more intimate the executive is with his analytical support, the more alternatives can be seen. An analysis team combining some of the key disciplines needed for the problem is very useful if the decision to be made is a major one.

(2) *Layout of the decision tree*—This is the formulation of the structure of alternatives underlying the investment decision.

(3) *Obtaining the data needed*—Analyses of various sorts are usually required to estimate cash flows and probabilities of uncertain events. This is another function of the analysis team, if such a team exists.

(4) *Evaluating alternative courses*—Evaluations frequently lead to a restatement of the problem and a reformulation of the decision tree. A good evaluation will test which alternatives appear desirable in light of the standards used. It will show whether apparent conclusions are sensitive to changes in doubtful or controversial estimates. It will examine the effect of choosing alternate standards, which in turn may lead to revision of standards, further analysis, or reformulation.

Let us examine each of these steps. . . . We can then compare the decision-tree approach with other approaches, review various problems and questions that may arise, and see how the Bayesian method of probabilities analysis can be combined with decision trees to analyze the worth of further research or market analysis for a project.

IDENTIFYING ALTERNATIVES

Experience has demonstrated the importance in investment studies of identifying what alternatives, what freedom of action, and what uncertainties exist now and in the future; of estimating the costs, demand volumes, prices, and competitive action anticipated under alternatives; and of narrowing down estimates of the probabilities or likelihood of uncertain events. All future possibilities cannot be identified, of course, but experience shows a reasonable job can be done. The better the job of identifying and estimating, the more intelligently can the investment decision be made.

Marketing analysis, operations research, engineering analysis, and financial analysis have vital roles in investment analysis. To be most effective, however, they should be integrated; that is, management should use them *in combination* to explore alternatives and to open up new possibilities.

Those engaged in the analysis should be encouraged to express doubts and uncertainties and to express estimates of costs, technical feasibility, or forecasts of market conditions in terms of *ranges* or *probabilities*. Much as management might wish the uncertainties would go away, undue criticism of the analyst for being imprecise or insufficiently firm in his estimates will only force the uncertainties and the risks underground. The purpose of the investment analysis is to help management identify alternatives and bring out the facts about them.

Innumerable examples could be given of how staff analyses have helped in the identification of alternatives in various companies. . . . [The] mention [of] just a few [will] indicate the *range* of possibilities:

One company needed a ten-year forecast of the size and character of the market for new labor-saving equipment. The forecast was made by combining the results obtained—first, by an industry sector analysis to estimate how individual industries might be expected to grow in response to primary consumer demand and secondary industrial investment activity; second, by a productivity analysis to estimate productivity increases and labor substitutions needed if demand estimates were to be achieved; and, third, by technical analysis of industry operations to see where labor substitutions of the greatest magnitude were most likely to occur. And out of this came a picture with both a range and a depth that otherwise would have been lacking.

A chemical manufacturer faced the need to expand capacity to meet growing near-term demand. Additions to capacity were extremely expensive, and the manufacturer was uncertain about the longer term load on the plant. Analysis showed that the manufacturer had other alternatives—e.g., revised scheduling procedures combined with a moderate investment in warehouse space would yield an effective increase in production capacity at modest fixed investment, but at some increase in operating cost. Again, the result of the investigation was a broadening in the range of choice for management.

In a study of U.S. Mint facilities investment, the analysis team investigated a series of alternative combinations of plant sizes and multiple-shift operations, to show the tradeoff between operating cost and investment required; that is, the savings in operational expenses that could be achieved by increases in outlays for new facilities. In addition, the costs of excess capacity, overtime, and occasional extra shifts were analyzed to determine the justifiable amount of excess capacity to handle short-term fluctuations in coinage demand.[5]

A proposed field order- and sales-recording system being considered by the management of one company was only marginally attractive. The major part of the investment was in field recording equipment, of which only about half was to be in use more than half the time because of peak-load problems. Investment for handling the peak could be cut, but only with higher operating costs and some congestion. A team of analysts incorporating several types of skill was able to suggest an equipment redesign which permitted more intensive use of expensive electronic gear at peak periods, cut total investment by 25%, and made the investment attractive. The original design engineers were unfamiliar with the size of the peak load and its financial impact, so they had not tried to design around it. The analysis team, in turn, might have missed the redesign alternative if it had not included men with engineering experience.

Analysts trained in marketing, operations research, engineering, and finance have a real contribution to make. However, these disciplines need to be integrated in the project analysis if management is to get a clear picture of the available alternatives and their consequences. [Figure 12] is a schematic portrayal of the right and wrong ways to think of this process.

Returning to the problem of Prism Paints, let us assume that management assigns the plant-investment question to a group made up of an operations research man, a marketing researcher, and an engineer with practical experience and imagination in process-plant design and costing. This team works closely with the company treasurer to investigate alternatives. It also meets regularly with other members of top management to review progress and policy problems.

---

[5] Hearing before a subcommittee of the Committee on Banking and Currency, U.S. Senate, on S.874, March 26, 1963, concerning additional Mint facilities.

Too often the analysis and decision process on a new investment opportunity seems to look like this (the detailed form, of course, depending on the specifics of the project):

The type of interaction management should ask for instead looks more like this:

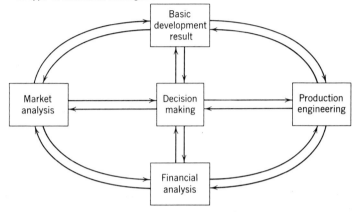

FIGURE 12.   Use of functional specialists in investment analysis.

There may be some interaction between market analysis and production engineering or between production engineering and financial analysis, but not a great deal.

LAYOUT OF THE TREE

Any significant problem can be examined at several levels of detail. The problem in laying out the decision tree is to strike the right level, the one which permits executives to consider major future alternatives without becoming so concerned with detail and refinement that the critical issues are clouded. The illustrations of trees in this and preceding examples indicate the general level of detail that seems reasonable in many typical circumstances.

What time span should be covered for the tree as a whole? The layout of alternatives might go on interminably and be carried out over an indefinite future time. Roughly, the practical time span or "horizon" to consider should extend at least to the point where the distinguishing effect of the initial alternative with the longest life is liquidated, or where, as a practical matter, the differences between the initial alternatives can be assumed to remain fixed.

The time span over which the analysis must extend will vary for *particular decisions*. For some decisions it will be a year or two; for others, ten years or longer. The launching of a new product is an example of the short-horizon type; the net effect of introducing the product, if successful, would last a good deal longer, but it can be characterized as a stream of income continuing for some period at some consistent level, and hence does not need to be portrayed on the tree. There are no visible significant decisions to be considered beyond the first one or two years. On the other hand, plant investment questions, such as Prism Paints' problem of rehabilitating an old plant or building a new one, often have a time span of from five to fifteen years.

Outlining Major Choices

With these generalizations in mind, let us return to our case. Let us suppose the analysis team begins by getting the forecast of demand shown in [Figure 13]. The central curve shows the most likely demand estimate, starting at $9 million in the early years of operation, rising to $12 million in five years, and reaching $18 million at the end of a decade. The lower curve, starting at $6 million and growing to $10 million is the minimum, pessimistic, estimate. The upper curve represents the most optimistic estimate of the rate of market growth.

The team's examination of the demand estimates indicates that there are three basic patterns of operation offering promise:

Program A: To modernize the plant in question and expand elsewhere.
Program B: To close the plant in question and expand elsewhere.
Program C: To modernize and expand the plant in question.

The management, in discussions, points out that one or two major changes in the manufacturing and distribution pattern over the course of the next ten years will be all the company wishes to make, if that many.

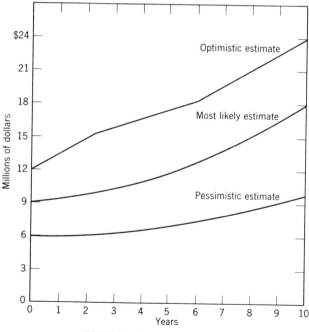

FIGURE 13.   Demand forecasts.

As a practical matter, experience also indicates that breaking up a question into two to four decision stages (with the associated uncertainty alternatives) is reasonable. Thinking out the decision at least through a second decision stage enriches the analysis considerably over a conventional single-stage consideration; but, after about four or possibly five stages, the analysis will bog down in complexity and lack of data. This means that the two, three, or four levels or stages of decision points must be the major, significant points of choice as now seen. It means that the time between stages will vary in a given problem and, of course, will vary from one problem to another, depending on the over-all span of time under consideration.

Costs and Feasibility

As a next step the analysis team refines the alternatives and makes cost analyses of operations under these systems at different demand levels. Costs are worked out by calculating the best choice of warehouse locations, assignment of warehouses to plants for service, choice of item-stocking policy, transportation method, and so forth, under the three alternate patterns of plant operation and volume of demand. Analysis shows that:

Program A is less expensive when demand is less than $10 million annually.

Program B is less expensive when demand is between $10 million and $14 million annually.

Program C is less expensive when demand is over $14 million annually.

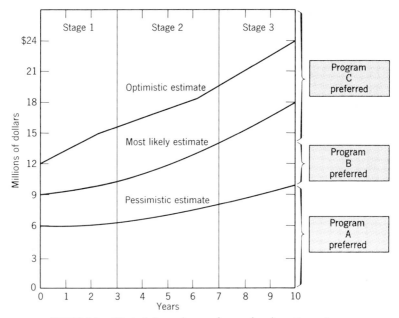

FIGURE 14.   Effect of demand on preference for alternative patterns.

It is decided, therefore, to break up the total time span under consideration into three periods—the first three years, the middle four years, and the period from the eighth year onward—as illustrated in [Figure 14]. During the first stage, demand is likely to be at a level where Program A would be preferred; during Stage 2, demand is likely to be in a range where Program B would be preferred; thereafter, demand is likely to be high enough so that Program C would be preferred.

In theory there are 27 plans over time that might be considered, although in practice some are out of the question. For example, Prism Paints' management may conclude that if Program B were adopted, it would be impractical to go to A or C, and to reopen the plant in question.

The possibilities can be reduced for practical purposes to 13 feasible program sequences, each with its own pattern of investment and operating cost. For example, it is possible to begin to follow Program A, and at a later time—say, eight years later—convert to Program C at some cost of sunk investment, rearrangement, and excess operating expense. The 13 feasible sequences are shown in [Figure 15].

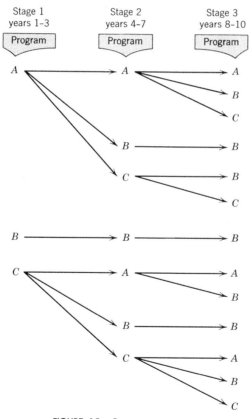

FIGURE 15. Program sequences.

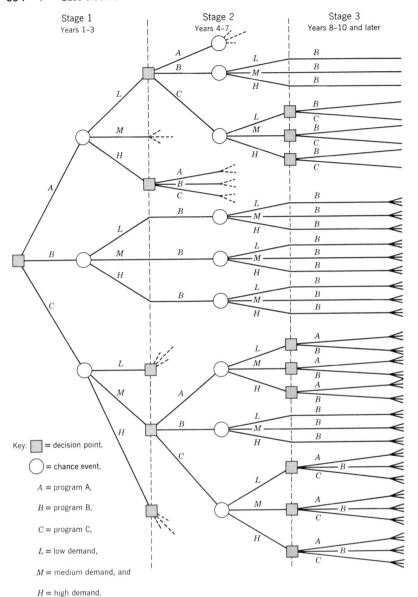

FIGURE 16.   Key: square denotes decision point, circle denotes chance event, A = program A, B = program B, C = program C, L = low demand, M = medium demand, and H = high demand.

Preliminary Drawing

If the company were sure of the sales volumes and operating costs five and ten years hence, it could choose the particular program for plant construction that would make profits largest, after giving weight in some fashion (for example, by discounting future cash flows) to the relative value of present and future profits. Perhaps if sales do develop as projected, Program C is the least expensive over the next decade. But what if sales do not develop as well as, or grow even faster than, forecast? The task of the team is to allow for the relative flexibility of various programs as a factor in meeting uncertain product demand. The choice made now will decide what later alternatives are still feasible. For example, as mentioned earlier, if Program B is chosen, the company will have to live with it for some time. What immediate choice is the best one in view of the interrelationships among the immediate choice, demand levels, and future alternatives?

[Figure 16] gives a preliminary sketch of the decision tree which shows the interaction between Programs A, B, and C concerning plant facilities and the levels of demand in the company's market area. Some of the alternatives are not carried out to complete detail but are indicated by the dash lines. The portion of the tree shown represents 10 of the 13 alternative patterns, many of which would be repeated in the portions of the tree that are not shown.

OBTAINING DATA NEEDED

In my view the appropriate standard for corporate investment decisions is the maximization of expected wealth, or of discounted expected cash profit where future cash profit is discounted to present value at a discount rate equivalent to the market "cost of capital" of enterprises of similarly uncertain future. I recognize, of course, that not all businessmen agree with this standard. Those who do not agree can use the principles to be described, but they will want to adjust them to suit their own criteria. Just bear in mind that in *this* illustration which I am presenting the yield figures to be shown on the decision tree are not accounting profits but cash flows.

The numbers or values needed at each stage under the standard of maximum expected cash profit are:

(1) The probabilities of each of the alternative uncertain outcomes—in [Figure 16], for instance, the probabilities that demand will be low, medium, or high during years one through three.

(2) The cash flow associated with each combination of decision alternative and chance outcome.

(3) Where the time span of the stage is significant—of the order of a few months or more—an estimate of the discount rate to be applied to deferred cash flows in the stage or to the future "position value" of the outcome. . . .

Probability Estimates

The probabilities of the uncertain alternatives may sometimes be estimated objectively from research. For example, the likelihood of surges or slumps in demand

Level of demand	Stage 1	Stage 2: If stage 1 is:			Stage 3: If stage 2 is:		
		Low	Medium	High	Low	Medium	High
Low (under $10 million)	.50	.35	.15	—	.20	.05	—
Medium ($10–$14 million)	.43	.50	.45	.40	.60	.35	.20
High (over $14 million)	.07	.15	.40	.60	.20	.60	.80
Total	1.00	1.00	1.00	1.00	1.00	1.00	1.00

FIGURE 17. Probabilities that demand will be low, medium, or high.

may be estimated objectively from a statistical analysis of demand variations, or the possibility that the market for a proposed product will be one size or another may be estimated from survey data. At other times the probabilities may have to be estimated subjectively from the intuition of an experienced operator or staff man. For example, the likelihood that a development project will succeed may be estimated from the intuitive judgments of two or three skilled engineers or research managers.

What is gained by trying to make subjective estimates of probabilities where limited objective information exists? Why not make the decision on feel, hunch, or intuition in the first place? For one thing, the estimation of elemental probabilities (and other values) permits various parties to the decision to see the basis for each other's conclusions. For another, it permits the executive to make use of the intuitions and skills of subordinate operators as staff members without abdicating his position as the decision-maker. Further, it permits analysis of the impact of variations in the estimate—what is called *sensitivity analysis*—in the conclusion. Finally, as noted later, it provides a means for measuring the value of further work to sharpen up the estimates. This step is called *Bayesian analysis*.

Let us assume that the need for these estimates is felt by the managers in our Prism Paints case. Accordingly, the team analyzes evidence on market demand and the data underlying the demand forecasts (see [Figure 13]) to estimate how likely it is that demand will fall in the low, medium, or high range in each of the three stages of the analysis. The team concludes that these estimates must be made in relation to demand in the preceding stage. That is, if demand is high in Stage 1, that will certainly influence the likelihood that demand will be high in Stage 2. These estimates, for each stage, are summarized in [Figure 17].

Cash Flows

The cash flows are characteristically estimated from the types of marketing, operations, engineering, and financial analyses mentioned earlier. Where the stage covers a considerable time span—one or more years—the schedule should show the cash flow at least on an annual basis.

In our Prism Paints case, distribution analyses show that the relative operating profitability of the alternatives facing management can be expressed by net annual cash flow as follows (in millions of dollars):

	Demand level		
	Low	Medium	High
Program A	$1.5	$1.8	$1.5
Program B	1.0	2.0	2.8
Program C	0.5	1.5	3.3

Initial investment to implement Program A is estimated to be $12 million, compared with $14.75 million for Program B or $14 million for Program C. Program A could thereafter be converted to B at a cost of $6 million, or to C for $8 million. Program C could be converted to A for $4 million, or to B for $6 million.

The management of Prism Paints establishes a desired return on investment or cost of capital of 14% per year. This is the rate to be used to reduce future cash flows to a present value for comparative purposes.

EVALUATING ALTERNATIVES

We can now construct the complete decision tree for Prism Paints' plant-investment problem. The tree is shown in [Figure 18] (ignore for a moment the numbers directly outside the boxes and circles).

Part A shows how the three alternatives at the initial decision, when combined with the possible market conditions in the period one to three years ahead, can lead to six decision situations at the beginning of the second stage in addition to the close-down situation (Program B). The remaining parts (B through D) show how three of these decision situations may develop over the course of the subsequent years.

(The trees for the four situations not illustrated in [Figure 18] would be quite similar in design.)

To evaluate the data in the decision tree, we follow three steps:

Step 1: Evaluate each of the alternatives at the final-stage decision points. Select the alternative with the largest net present value. Assign this value to the position.

For instance, in Stage 3 of Decision 1 (see [Figure 18, part B]), the expected value of each alternative for the uppermost decision can be calculated as shown in [Figure 19].

The alternative with the largest net present value ($12.8 million) is Program A. Therefore, Program A would be chosen *if* this decision at Stage 3 *were* to be made *now.*

Step 2: Evaluate each decision alternative at the next preceding stage.

This is accomplished by calculating the expected discounted present value of the cash flow during the stage, including the value of the following stage as if it were a lump sum received at the end of the period. For instance:

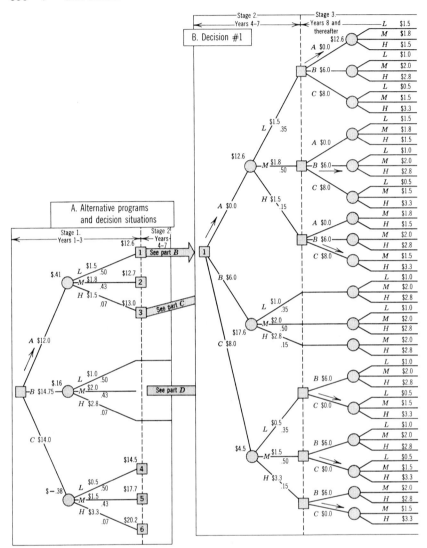

FIGURE 18. Complete decision trees for Prism Paints, Inc. Key: square denotes decision point, circle denotes chance event, A = program A, B = program B, C = program C, L = low demand, M = medium demand, H = high demand. Note: The choices at each decision point are indicated by the letters A, B, or C. The figures following these choices represent the investment required. Two sets of numbers go with each market level—a dollar figure representing cash flow and a decimal representing the estimated likelihood that the level will come about. The numbers outside the square decision-point boxes and the chance-event circles represent present value (in millions of dollars). The arrow at a decision box points to the alternative with the greatest present value.

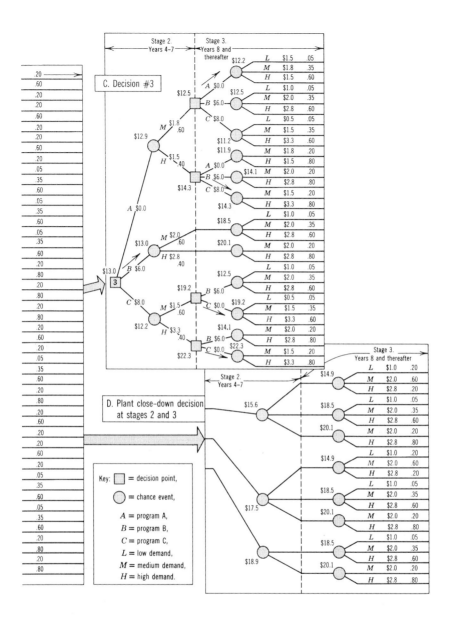

At Stage 2, Decision 1 ([Figure 18, part B]), there are three chance alternatives related to market conditions. When the present value of the cash flow of each alternative—including the following stage values—is calculated and weighted by the likelihood of its occurrence, the expected value of the choice is found. The value of

Alternative	Market size	Likeli-hood (1)	Cash flow per year (2)	Period	Present-value factor[a] (3)		Total present value (1)×(2)×(3)
Program A	L	.20	$1.5	Indefinite	7.6		$ 2.3
	M	.60	1.8	Indefinite	7.6		8.2
	H	.20	1.5	Indefinite	7.6		2.3
						Total	$12.8
					Less investment		0.0
					Net present value		$12.8
Program B	L	.20	$1.0	Indefinite	7.6		$ 1.5
	M	.60	2.0	Indefinite	7.6		9.1
	H	.20	2.8	Indefinite	7.6		4.3
						Total	$14.9
					Less investment		6.0
					Net present value		$ 8.9
Program C	L	.20	$0.5	Indefinite	7.6		$ 0.8
	M	.60	1.5	Indefinite	7.6		6.8
	H	.20	3.3	Indefinite	7.6		5.0
						Total	$12.6
					Less investment		8.0
					Net present value		$ 4.6

[a] The value of $1 per year, received for an indefinite period and discounted at 14% per year on a semiannual basis, is $7.60. Thus, 7.6 times the annual cash flow is the present value of the cash flow discount at 14% per year.

FIGURE 19. Present values of alternatives for a decision at stage 3. (Dollar figures in millions.)

Program A is found to be $12.6 million. Similarly, the value of B is found to be $7.6 million, and the value of C is found to be $4.5 million. The mathematics of this step are the same as described . . . earlier. On this basis, the preferred choice at Decision 1, Stage 2, is therefore Program A, and the position value of Decision 1 is $12.6 million.

Step 3: By repeated application of this process—the process of "rollback" (referred to . . . [earlier in this chapter]—to each stage of decisions, the value of each alternative at the first stage can be found.

In [Figure 18, part A] we find these values for the three programs at Stage 1: A, $410,000; B, $160,000; and C, −$380,000.

These are the amounts by which the expected present-value cash flow exceeds or falls short of the initial investment called for under each alternative. The cash flows under Programs A and B are more than sufficient to earn the required 14% on the investment; the cash flow from Program C is insufficient.

Program A is the preferred choice because it has the greatest value. This does not mean, however, that Prism Paints must live with Program A indefinitely. If the market grows rapidly to high levels, management can shift to Program B (closing the plant, expanding elsewhere) at the intermediate point. Otherwise management can choose depending on the rate of market development during the next seven years or so.

Actually, of course, during the years ahead management will learn more about the market, and conceivably other changes will take place. Thus the *immediate* choice among A, B, and C is all that must be made now. However, the immediate decision is made with a proper eye to its impact on future developments and possible later decisions—with a recognition of the "futurity" of the current decision.

COMPARING OTHER METHODS

Does the use of the decision-tree approach make any difference? Let us compare the result of the Prism Paints analysis with the results of other more conventional discounted cash-flow techniques.

One common method is to evaluate each fixed alternative under the "most likely" cash flow using present-value techniques. This gives us (in millions of dollars):

	Investment	Cash flow per year		
		Years 1–3	Years 4–7	Year 8 and thereafter
Program A	$12.00	$1.5	$1.8	$1.5
Program B	14.75	1.0	2.0	2.8
Program C	14.00	0.5	1.5	3.3

This approach yields net present values as follows (when the cash flows are discounted at 14% per year): Program A, 0; Program B, $1.3 million; and Program C, $2.1 million.

It would appear that Program C is the preferred; this ignores the possibility that the market may not in fact develop as fully as the "most likely" levels.

Another approach is to calculate the net *expected* present value, with the cash flows weighted by the estimated likelihood of their occurrence. This technique, applied to the Prism Paints data, yields the following net present values: Program A, $126,000; Program B, $160,000; and Program C, $380,000.

This approach seems to show Program B as somewhat superior to A. This is due to the fact that the technique does not take into account the opportunity, if A is chosen, to modify the distribution plan later. As the decision-tree analysis shows, the flexibility implicit in Program A makes it significantly superior.

The decision-tree analysis brings out the impact both of uncertainty and of possible future decisions, conditioned on future developments. To emphasize a point made earlier, the analysis *could* be made without actually drawing a tree, i.e., by simply listing the alternatives and stages for computation instead; but in practice the visual form has real value in helping managers understand and discuss an investment problem.

SPECIAL PROBLEMS

In the Prism Paints case example, "discrete" chance events [have been used] for illustrative convenience. A discrete chance event is one where there is some number of clear-cut chance outcomes, each with its probability of happening. For instance, the toss of a coin is a discrete chance event.

In many practical cases, however, the chance events are in fact continuous; a whole range of answers is possible. There may be an estimated probability distribution associated with this range. The daily temperature is an example of a continuous chance event. There is an expected value for any locale and any time of year, though the actual value can be anywhere in a wide range. Using a probability distribution related to temperature, you can speak of the chance the temperature will be between 75° and 80°, below 65°, and so forth.

Continuous-chance variables can be handled readily using the decision-tree technique. There are two ways:

(1) In the Prism Paints case, we arbitrarily sliced the continuous-chance variable (future market levels) into the form of discrete chance levels—low, medium, and high. This approach often is adequate. It oversimplifies, but (in many cases) not so much so as to add serious error.

(2) We can calculate the values of the decision alternatives at any stage as related to the full range of the prior chance outcomes. In this way we can fix the part of the range of possible outcomes for which any alternative is best. In some cases, straight calculation is all that is needed; in other cases, simulation methods are useful (for example, where the relationship between the chance variable and the cash flow following some decision alternative is a complex one).

Another problem is that of many chance variables. In the Prism Paints example, I have used market level as the single-chance variable at each stage, but in some problems there are several uncertain variables at each stage. These might include, for example, total demand, share of market, costs, yield, or results of a development or promotion program. Any one or all of them might be so uncertain that a range of probabilities should be used.

Uncertainty connected with several variables can be handled readily. One useful way is the "Monte Carlo" method, a form of simulation: At each stage—still working back from the last stage—values of the chance variables are picked at random, and the value of each decision alternative is worked out. Through repeated application, the expected values and probability distributions for each decision alternative emerge. The best alternative is chosen, and the "rollback" principle is applied to carry out the "Monte Carlo" analysis, if needed, at the next preceding stage.

Finally, there is the calculation problem. The limiting factor in drawing up a complex decision-tree analysis is not the task of computation but the capacity of the analysts to imagine alternatives and think out the implications of the various possible choices. Even so, computer calculation is often useful, and it has one important advantage: it makes it easier to analyze the sensitivity of the end conclusion to changes in key investment or cash-flow figures or in probability estimates. If the conclusion as to what to do now is not too sensitive to changes in certain estimates, these can be estimated roughly. If the conclusion is very sensitive to changes in certain numbers, these are the ones where further research may be in order.

USE OF BAYESIAN METHOD

One alternative that frequently exists but may not be considered is further research or investigation. This does not mean procrastination but an intensive look by a fresh and objective team of men. Such a look may be especially valuable where the existing basis for decision is a series of semi-independent studies by staff groups committed to a position. It may mean carrying a research or development program one step further, or making a marketing study to try to narrow the range of market uncertainty. The concepts of Bayesian statistics[6] give a means for building in subsequent information to modify estimates of probabilities. The Bayesian method also gives a means for estimating the value of further investigation.

[Earlier in this chapter] . . . we discussed the investment issue facing Stygian Chemical Industries, Ltd.—whether to build a small plant or a large one, in view of the uncertainty concerning the size of the market. The key issue facing management was whether to gamble on the size of the market and build a large plant immediately or to build a small plant with the possibility of expanding it later. That example lends itself particularly well to the present question. . . .

Perfect Foresight?

Using discounted expected cash profit as the criterion and discounting future cash flows at a 10% rate per year, we found in the previous article that the appropriate decision for Stygian appeared to be the large-plant alternative. The discounted expected value of that alternative was $1.47 million, compared with a $1.18 million value for building the small plant.

Suppose the company considers now the possibility of doing some further research —for example, making an independent technical-economic or marketing study to estimate more carefully what the long-term market penetration possibilities are. The marketing study properly and thoroughly done will be expensive; it will cost $100,000. Should the company contract for the research?

First, what would perfect foresight be worth? If management *knew* the market would be large, it would build the large plant. If it *knew* the market would be small in the long run, it would build the small plant. The value of knowing for sure whether the market is large or not can be calculated as follows:

(1) Assuming one of the anticipated market conditions occurs and the preferred strategy for it is followed, calculate the present value of the cash flow which will come to the company (subtracting the cost of plant investment).

(2) The discount rate used depends on management's judgment of the situation. For our illustration here, let us assume a 10% annual discount rate on future cash flows.

---

[6]See, for example, Robert O. Schlaifer, *Probability and Statistics for Business Decisions*, New York, McGraw-Hill Book Company, Inc., 1959.

Market situation (demand)	Preferred strategy	Present value[a] of preferred strategy (1)	Present estimate of likelihood (2)	Value of perfect foresight (1) × (2)
High average	Big plant	$3.7	.60	$2.22
High initial, low average	Small plant	1.5	.10	0.15
Low average	Small plant	1.4	.30	0.42
Total				$2.79

[a]The present value of the cash flow, discounted at 10% annually, including plant investment, and assuming the specified market condition occurred and preferred strategy was followed.

FIGURE 20. Calculations for determining value of perfect foreknowledge. (Dollar figures in millions.)

(3) Multiply the present value thus obtained by management's present estimate of the probability that the market condition will actually occur.

(4) Repeat the procedure for the other market conditions anticipated.

These calculations are carried out in [Figure 20]. They give us, in effect, a kind of average of the discounted cash flows weighted by the likelihood of their occurrence, if management knew in advance how big the market would be. This result ($2.79 million) is greater than the discounted expected value of the preferred large-plant alternative ($1.47 million) by $1.32 million. Thus, the value of perfect foresight compared with present uncertainty is $1.32 million.

New Alternatives

This estimate suggests that there is a sizable potential payoff from good research in reducing the likelihood of a mistake, by narrowing down uncertainty in the market estimates, and thus increasing the expected profit from the action taken. Let us suppose management has confidence in the research team, though considering it conservative. Management estimates that:

If the market actually turns out to be large and demand is high from the outset, the chance the research will show a positive result is 70%. (By "positive result" I mean a high estimated demand.)

If demand is high initially but tapers off to a low level, the research effort might be mislead, and there is a 50% chance it will err and report a positive finding.

If, however, demand is low, the chance of a positive result is quite small—about 5%.

From these estimates it is possible to calculate that:

If the research result is positive, the chance of—
. . . a high average demand is .87;
. . . an initially high, low subsequent demand is .10;
. . . a low average demand is .03.

If the research result is negative, the chance of—
. . . a high average demand is .35;
. . . an initially high, low subsequent demand is .10;
. . . a low average demand is .55.

The advance estimate of a positive finding is .485. (The difference between the estimate of .485 that the research will show a large demand and management's estimate of .60 that the demand will be large is a measure of management's estimate of the research organization's degree of conservatism.)

For those interested in technique, the method of calculating the foregoing data is shown in [Figure 21].

Now management has three alternatives: (1) to build a large plant; (2) to build a small plant, with the later possible option of expansion at Decision 2; or (3) to commission the research. If the third alternative is chosen, then, depending on the outcome of the investigation, the decision alternatives are 1' and 2' at Decision 3 (see [Figure 22]) if the research outcome is favorable; or 1" and 2" at Decision 4 if the research outcome is unfavorable. These correspond to the original alternatives: (a) building a large plant; or (b) building a small plant.

While the modified decision tree looks a good deal more complex, careful examination will show that the portion following Decision 3 and the portion following Decision 4 are each duplicates of the original decision tree, with just the probability estimates changed. The probability estimates are changed, as earlier described, to reflect the results of the research and its estimated reliability.

Maximizing Expected Value

The calculation of position values associated with the various decision points can be carried out . . . as shown . . . [earlier]. Using discounted expected value as the criterion and a discount rate of 10% annually, one finds that:

(1) If management were at Decision 3 (research carried out, positive finding),

---

Let HH signify a high average demand; HL signify an initially high, but low subsequent demand; and LL signify a low average demand.

Let F represent a positive result from the research and N a negative result. Then:

$$P(HH/F) = \frac{P(F/HH)P(HH)}{P(F/HH)P(HH) + P(F/HL)P(HL) + P(F/LL)P(LL)}$$
$$= \frac{.7 \times .6}{.7 \times .6 + .5 \times .1 + .05 \times .30} = .87$$

The probabilities $P(HL/F)$, $P(LL/F)$, $P(HH/N)$, $P(HL/N)$, $P(LL/N)$ can be obtained similarly.

The chance of the research showing a positive result is:

$$P(F) = P(F/HH)P(HH) + P(F/HL)P(HL) + P(F/LL)P(LL) = .485$$

FIGURE 21.   Method of obtaining estimates of research reliability.

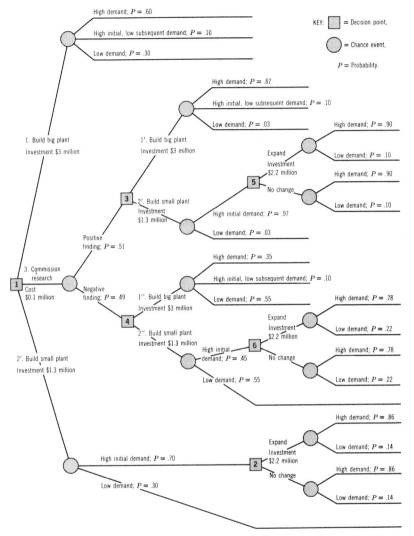

FIGURE 22. Decision tree for Stygian Chemical with research alternative. Key: square denotes decision point, circle denotes chance event, P = probability.

the indicated decision would be to build the big plant, and the discounted expected value—the position value—would be $3.1 million.

(2) If management were at Decision 4 (research carried out, negative finding), the indicated decision would be to build the small plant and hold to it, whether initial demand were high or low. The discounted expected value, or position value, would be $2.6 million.

The expected value of Alternative 3 at Decision 1, therefore, can be calculated as:

$$.49 \times \$3.1 \text{ million} = \$1.5 \text{ million}$$
$$Plus \ .51 \times \$2.6 \text{ million} = \$1.3 \text{ million}$$
$$\overline{\$2.8 \text{ million}}$$
$$Less \text{ cost of research} \quad 0.1 \text{ million}$$
$$\text{Net expected value, Alternative 3} \quad \overline{\$2.7 \text{ million}}$$

The discounted expected profits from Alternatives 1 and 2 are unaffected by the introduction of the third alternative. So Alternative 3, with a value of $2.7 million, is clearly superior to the previously indicated choice, Alternative 1, with an expected value of $1.5 million.

The research increases the expected value of the endeavor by $1.2 million. It accomplishes this by reducing the uncertainty surrounding the demand estimates. It is interesting to note that the expected value of the endeavor will be increased significantly *no matter how the research comes out.*

SUMMARY

Use of the decision-tree concept as a basis for investment analysis, evaluation, and decision is a means for making explicit the process which must be at least intuitively present in good investment decision making. It allows for, indeed encourages, revision of the issue and maximum use of analysis, experience, and judgment. It helps force out into the open those differences in assumptions or standards of value that underlie differences in judgment or choice. It keeps the executive from being trapped in the formalisms of a rigid capital evaluation procedure in which there is little room for feedback, redefinition, or interplay between analysis and decision.

The decision-tree approach will strike some businessmen as complex. Certainly it is more complicated than rule-of-thumb approaches, but of course any realistic method would be. In reality, a decision tree need be only as complex as the decision itself. If the decision is a simple choice among alternatives, then the decision tree reduces to a single-stage analysis, i.e., the use of the present-value technique applied to alternative cash flows. If the situation is more complicated, more stages and alternatives are necessary to reflect that fact.

Explicit use of the decision-tree concept will help force a consideration of alternatives, define problems for investigation, and clarify for the executive the nature of the risks he faces and the estimates he must make. The concept can thereby contribute to the quality of decisions which executives, and only executives, must make.

PART IV

Pitfalls and potentials

17

Problems and pitfalls

Despite remarkable advances in the design and technology of management information systems which seem to make them capable of being most beneficial to their potential users, these systems have too often failed to achieve their objectives. It seems that many of the engineering and manufacturing organizations which could benefit most from such systems find it very difficult to implement the new tools effectively. When the system is implemented correctly, it provides timely and meaningful information to managers which can significantly improve their decision-making and general management.

The introduction of systems such as PERT, CPM, and PERT/COST has often created many problems, despite (or perhaps because of) unprecedented governmental and industrial attempts to develop capabilities in these systems. There is little question that such techniques can be effective, but there is also little question that many attempts to implement the systems have not fulfilled the promise which they seemed to make. Is the fault in the inherent weaknesses of the techniques themselves, or in the manner of implementation and use? Management is increasingly pressed to answer this and related questions, as requirements for improved management methods multiply. In this chapter, we will attempt to discuss some of the problems and pitfalls which make it very hard to implement seemingly sound, well-designed, and rewarding management systems. And, although we will discuss network-based techniques, we will reach some conclusions which can be applied to many other types of management systems.

DEFINITION OF A MANAGEMENT INFORMATION SYSTEM

There has been a great deal written about management information systems in recent years, and we refer the reader to any of the better works on the subject for a fuller treatment. To facilitate this discussion, however, we shall keep our definition of these systems as simple as possible. We may define a management system as a *set of operating procedures* which personnel carry out to acquire needed information from appropriate sources, process the data in accordance with a preprogrammed rationale, and present them to decision-makers in a timely, meaningful form. Most contemporary systems involve manual data collection and input, machine processing, tabular and graphic output production, and human analysis and interpretation. Thus, we can say that the systems collect, synthesize, process, transmit, and display information, which flows from a primary source, through an editing, computation, and selection process, to the manager.

We have used the network-based management information system, specifically the time/cost/manpower application, to illustrate our discussion. The reader must remember, however, that the network-based system is only one of many types of systems which deal with the planning, scheduling, and evaluation of projectized efforts. Other management information systems are applied to vast areas, including production and process control, personnel, financial and legal, marketing, and long range planning. Many contemporary management information systems even include decision-making through the use of judgment or automatic procedures; (when we automate what were previously considered decisions, we are simply acknowledging that judgment may previously have been used in place of adequate information. Indeed, when sufficient information is available in proper form, the rules can, and often will be applied mechanically, so that the "decision" in the old sense is eliminated.) This discussion must limit itself to the narrow definition of a management information system, however, excluding various control systems which apply rules and make "decisions." Within this narrow definition, the management information portion of a process control system, for example, would involve reporting of monitoring and exception information which is not treated by the rules and which calls for the application of judgment by the manager, but would not include the actual exercise of judgment. Hence, we can see that the system discussed in this book may be limited in scope and application compared with information systems generally; however, it is a valid sample of a management system, which should be discussed because it has posed a formidable challenge to those who have sought to use it effectively.

There are many facets involved in implementing a network-based system, and numerous pitfalls lie in the way of success. If we would try to treat them all (if indeed, we could anticipate them all) in this book, we would have to ignore the individual variations in environment and needs. However, we can isolate several fundamental areas of potential difficulty;

if the management recognizes them, they can greatly reduce the probability of failure, but if they overlook them, these problems will almost certainly cause an early breakdown of the system. Among the most difficult problems encountered when implementing network systems and other management systems are:

(a) Making the transition to new systems in the face of resistance to change and the effects of change itself. This problem area includes the difficulties of integrating new management systems with existing systems and procedures.

(b) Technological system deficiencies which lead to misapplication of the new tool.

(c) Integrating new systems with the environments in which they must operate.

Although all these problem areas are interdependent, we will try to discuss each of them separately.

EFFECTS OF CHANGE

Change involves giving up something known and understood for something unknown and either incompletely understood or misunderstood. The transition from an old to a new management system is a delicate operation, something like attempting to rebuild or replace the engine of a moving automobile without reducing its speed; and, in addition to all the other transition problems, we are faced with the powerful and all-pervasive human resistance to change.

In a typical situation involving the application of a new information system, the company may have been in business for some time; procedures and channels of information flow will have been established by design, growth, personalities and circumstance. The evolution of business machines, in particular, has affected this flow remarkably; beginning with the completely manual, quill-pen clerical methods of a century ago, we can trace the introduction and increasing use of desk-top adding machines and calculators, followed by punched-card, electric accounting machines, the early small-business digital computers, and finally the large, powerful computers of today. The early calculators and punched-card equipment mechanized the existing ways of handling information: if a new report was needed, the information was collected and processed, and the report prepared. The computer was first used to improve the speed and accuracy of the many thousands of parallel "batch" processing tasks. These parallel tasks often used much of the same information to produce overlapping reports with only slight differences. Thus, a tremendous amount of duplicate data gathering and reporting often existed. Effectively designed and used management information systems tie together these many parallel information paths into a well-knit system. And, indeed, effective use of today's computers demands the same thing. However, making the transition from myriad bits and pieces to an overall system presents a formidable challenge, even in a small company, because it forces a very penetrating look into the foundations of a company's operations.

Each of the many small subsystems in a typical company is often nurtured and developed into a haven for large numbers of security-minded people. Sections, departments or divisions may have grown up around them, or vice-versa. Facing such a situation, those charged with integrating the system will have to cut across many entrenched organizational lines and to challenge many powerful vested interests. Of course, the existing systems have come into being for reasons which were just as compelling at the time of their introduction, perhaps, as the motivations towards the newer systems. Again, when attempting to illustrate typical problems and pitfalls encountered in the implementation of the network-based system, we do not wish to ignore the significance of other management information systems which are important in industrial operations.

As we have stated, the network-based time/cost/manpower system aims at integrating planning, scheduling, budgeting, and accounting information. We can immediately identify at least five organizations involved:

Business planning or programming will involve those organizations within the company which gather, process, and use information related to the productive effort of the company, concerned with contracts, sales, manpower, facilities, funding, schedules, and so on.

Master scheduling includes the functions concerned with daily and short-term scheduling of men, machines, and facilities to meet contractual or sales commitments.

Project management involves the assignment and bearing of responsibility for a project usually by an individual or individuals which may include tasks performed by several otherwise dissociated operating departments.

Functional or line management is the more traditional line management of a functional department, such as mechanical engineering, receiving, testing, or process operations.

Controller (accounting) operations are those company organizations charged with preparation of financial records indicating the status of various accounts, forecasting financial requirements and conditions, and providing the procedures for budget control.

When we realize that besides these five kinds of company organizations, the systems design and data processing people are involved, we can see the makings of a beautiful donnybrook to determine who will most successfully defend his position and maintain the status quo, or who will emerge triumphantly in control of new systems, ready to build new empires.

People resist change because they fear the unknown, as well as wishing to protect themselves. A man can assume that any change involves a personal threat, because it disturbs the status quo and perhaps reveals the inefficiencies which exist in almost every human (and therefore imperfect) organization. Many middle managers instinctively realize that a good information system can reduce the number of layers of "management"— and thereby eliminate their positions, or at least their large "empire" of human data processors. Thomas L. Whisler, industrial relations professor at the University of Chicago Graduate School of Business, was quoted recently as follows:

> Introducing the computer into the decision process results in fewer men being required for any given output.
>
> In one corporation, substitution of machines and systems for men reduced managerial jobs by 30% in a two-year period.
>
> The organizational stress produced by technological change takes a new twist in the case of information technology. For the first time, those most affected by the change are those responsible for initiating and planning it. [1]

In some cases, some middle managers may promote conflicting information systems and their incompatible results, for the purpose of confusing auditors, the Internal Revenue Service, or government contracting officers or investigators. In short, an effective, integrated information system may produce (or, just as bad, threaten to produce) efficient, clear information, which is not really desired by anyone but a few top executives, the stockholders, and the customers.

An additional problem arising under this type of management information system (at least from the middle management point of view) is that such a system may provide top managemnt with tighter control over project operations than before. Those working on the projects cannot escape recognizing this, yet the successful use of such a system depends on these persons cooperating and giving assistance; in other words, these people are asked to help make successful a system which will put them

---

[1] Thomas L. Whisler, as quoted July 24, 1964 by the Los Angeles *Times*, in an article, "Automation Toll: Now It's Middle Managements' Turn," reporting on a conference on employment problems of automation at Geneva, Switzerland.

under greater control, or at least surveillance. The inherent difficulties here are obvious.

These sources of difficulty can be handled by adequate indoctrination and training.

TECHNOLOGICAL SYSTEM DEFICIENCIES

In addition to the effects of change on the emotions of personnel, another frequent source of difficulty is the system design itself and the manner in which it is developed and operated. Of course, the manner in which the system is introduced and implemented is affected by the problems discussed in the preceding section, but it also is affected by the system characteristics and deficiencies, which are critically spotlighted during the transition phase. The system design must anticipate human resistance and reaction, and must fit the existing organization and environment.

It is easy to retreat to the ivory tower and design the ultimate "total system," proposing use of the most advanced "real-time on-line" concepts and the most powerful computer and related equipment. But, such an ideal system is doomed to exist only in textbooks, without ever being used. In a perceptive article, Philip H. Thurston,[2] Associate Professor of Business Administration, Harvard Business School, discusses the following questions:

Why has control of information-systems work tended to fall into the hands of staff specialists?

What accounts for the failure of specialists to get as satisfactory results with information-systems projects as operating managers can get?

Why should top management assign more responsibility for the control of the systems to operating men?

What unique contributions can staff men make in this area, and how should these specialists be used?

He concludes as follows:

I consider that in the past decade a significant characteristic of information-systems work has been too great a degree of control in the hands of specialists. This situation has developed in part through the failure of top management to place controlling responsibility with operating managers.

This pattern of control has impeded systems work. We are now at a point where the situation can and should be changed. In the next few years we may see a shift to greater control by operating managers, coupled with continued reliance on specialists

---

[2] "Who Should Control Information Systems?" *Harvard Business Review*, Vol. 40, No. 6, November–December, 1962, pp. 135–139.

for their unique contributions. If this shift occurs, as I hope it will, it will contribute greatly to progress in the use of electronic business machines and in attaining improved information systems.

The main point of Thurston's article is that operating managers can often get better results from new information system implementation than specialists can, because the operating managers will accept the environmental realities and produce a system which will be born alive, able to breathe and live in the conditions into which it is introduced.

PERT/COST approaches have often been too sophisticated to solve real-life management information problems. There are many varieties of PERT/COST systems, with varying degrees of sophistication, of course. Imposing one of the varieties on the far end of the scale has often created severe problems because it is an attempt to travel too far in one step; the resulting radical changes often have severe effects. Further, the highly sophisticated approach usually does not fit into an existing environment. On the other hand, relatively unsophisticated network systems, which can still embody all the basic system concepts, have been successful.

Our discussion of technological system deficiencies has been rather general, but in this instance, it would be valuable to explore them in greater detail.

The most important act in developing a successful project management information system is to rigorously and systematically identify and correlate all the elements of the system, both those oriented towards project objectives, and those oriented towards organizational objectives. As discussed in Chapter 2, the information system must be *work-* or *task-* oriented to achieve project objectives. To do so, it must incorporate a most important element: a rigorously defined *Work Breakdown Structure,* correlated with all other planning elements, and providing the framework for all subsequent planning. Although this fundamental step is highly recommended in most applications, it is commonly the step in which the first, fundamental, ultimately disastrous weaknesses appear in many implementation programs, since detailed work plans (networks) and progress measurement points (milestones) are vitally related to this process.

Some Fatal Errors

One serious technological problem encountered in implementing network systems comes when the planners are selecting milestones. Milestone events, the reader will recall, are those specific points in time which management identifies as important reference points in the accomplishment of the project, and against which progress will be monitored. Milestone events include certain dates imposed by customers for the accomplishment of certain tasks, as well as target dates management sets for completion of certain segments of the work; these latter dates are usually imposed to comply with certain incentive clauses in the contracts, which may affect the profits on a project. The occurrence of milestones must be reflected in the plan, but unfortunately, one of the most common mistakes in network planning is to select milestones which have *not* been reflected in the work plans. For example, this occurs when a milestone is represented as a functional accomplishment rather than as a *task* or work-increment accomplishment. Let us look at an actual case:

The project management of a large weapon system project decided that progress could be reported effectively if occurrence of functional milestones at the "system," "subsystem," "group," and "unit" levels of the project Work Breakdown Structure was reported. Each tier was to contain event descriptions with associated identifying numbers, schedule, and cost data. The planners visualized the production of individual units, which would be merged into equipment groups at the next level for estimating and pricing. They planned that at the subsystem level equipment groupings would be integrated into subsystems, and at the "system" level subsystems would be identified to designate the accomplishment of various functions at those levels. For example, a function such as "preliminary design" would be designated at every tier, to show the status of that function for each successively larger aggregation of hardware. The planners apparently thought that the sum of all "preliminary design" milestones at one tier would add up to that function at the level above. In addition, each of the functional milestones at each level had budgets, estimates, and actual costs associated with it, based on the same assumptions. As the program progressed and reports based on this structure were prepared, it became increasingly difficult to determine whether a given milestone had *really* occurred, or whether it was simply reported as occurring because the contributive events at lower levels had all occurred. In addition, planners constantly encountered difficulties when they tried to associate costs with all milestones, or to pinpoint responsibility for the accomplishment of incentive milestones. In spite of elaborate and costly data manipulation and processing which was done in an attempt to overcome these shortcomings of the system, confidence in the validity of the data rapidly deteriorated, individual managers developed their own monitoring techniques or reverted to older ones, and the system degenerated to a costly monthly exercise in the production of reports containing highly questionable data. The company tentatively concluded that the network-based time/cost/manpower was unworkable "in the real world" or that the effort and expense required to make it work would be prohibitive in comparison with results.

Actually, both of these conclusions would be painfully accurate if the implementation process described had been a legitimate example of the network-based system. However, this particular implementation had involved some basic decisions which doomed the system to failure and violated fundamental principles of the network-system concept. A careful evaluation of this example will reveal a number of built-in limitations on the validity of the system; these limitations manifested themselves as conflicting reports gained from various program data and costly and time-consuming mechanical difficulties in the preparation and presentation of reports, both resulting from a basic misunderstanding of the concepts involved and both leading inevitably to a pyramid of inflexibilities and compromises, and finally to abandonment of the system. First, two central flaws in this structure were the identification of functional milestones and the use of an "additive" process of reporting progress against them at higher levels of the Work Breakdown Structure. For example, a milestone designated "complete delivery" (of a subsystem) at the subsystem level, is *not* necessarily the sum of completing unit deliveries; in many instances, integration and testing of the subsystem may be necessary before delivery can actually occur. In the plan described, functions selected as milestones and provided at all levels may or may not have been really meaningful indications of actual calendar progress, or of associated costs. The milestones at various levels did not always entail completion of all the activities involved in true accomplishment of a task which must take place at some working level. The point is, that erroneous conclusions may result from the erroneous premise that the same milestone can appear at each level of the Work Breakdown Structure representing cumulative accomplishments. A milestone event cannot appear at all levels of a program and be checked off as it is passed. Milestones must be real events in the work plans and must be reached, not passed; work is expended to reach a milestone, and all the work is accomplished at a single level, the working level. When planners overlook this basic premise or confuse it with calendar dates as was done in the instance we describe, the resultant reports may appear to present meaningful data, but a closer examination reveals that one cannot really be sure what they mean. Does this milestone *really* occur, or is it simply a summation of contributive events occurring at a lower level? If a *function* at one level disappears, how does one adjust other higher levels without having to "rejigger" the data, and (in the case of government contracts) to extensively justify the change to the customer? Consideration of questions such as these shows how very inflexible the system is made by the misinterpretation described.

The alternative to such a misinterpretation is to understand that to use given milestones to evaluate schedules, the planner must be sure that the milestones represent actual occurrences in the program. If milestones are simply summations, they will not really relate to actual network events; this lack of correlation will constrain the network plans artificially at the working level and enormously complicate the processing of even basic time-only networks.

We must make one more important point in this connection. The difficult and often unworkable interpretation of milestones we have described is based on a lack of understanding of the Work Breakdown Structure as a planning tool. A vertical summation of the Work Breakdown Structure does not represent the total program. Each level is equivalent to each other one: the difference between levels is in the *degree of summarization only*. In other words, the lowest level of a Work Breakdown Structure is exactly equivalent to level 2 or 3, but is simply more detailed. When we understand this, we see that the milestone at tier 2 or 3 *must* appear at the lowest level. Milestones can only be events which appear at the lowest level, but which have particular significance at other levels.

Some Obvious Consequences of Poor Planning

The consequences of such deficiencies in system design as we have described are many and varied; however, we can describe some obvious consequences. *Inordinate amounts of manual manipulation and processing of working level networks are needed to produce the periodic reports.* Because there is no correlation between the working level network plans and the multilevel reporting scheme in this deficient system, routine processing of network data to determine status for reporting milestones is complicated, and sometimes impossible. Personnel in this situation must constantly "adjust" the routine PERT time procedures on the computer or create more-or-less fictional summary networks to back up the reports. Such problems call for unrealistic "crash" efforts to prepare management reports each time they are due, procedures inevitably prone to human error, which may lead personnel to change the type of information they report from period to period, obscuring meaningful trends. Less obvious but perhaps more serious consequences of the deficiency than these essentially mechanical problems are the reductions in management effectiveness. These may be summarized as follows:

(1) *The network system becomes an administrative device used primarily for customer reporting, rather than for management.* When administrative functions (for example, "design review") instead of work-related functions (for example, "unit fabrication") are emphasized, the resultant system may have limited value as a management tool. The work-related function is a true challenge—a goal to accomplish. Selecting administratively oriented functions tends to force project personnel to lump cost and schedule constraints into phases of program development rather than associating them with task increments which add up to accomplishment of project objectives. The system data output is presumably regarded as a management tool for decision-making and direction; we can only conclude that in systems using administratively oriented milestones the

decisions to be made will be administrative rather than technical. Considering the nature and objectives of the network-based system, this may be a severe limitation on the usefulness, validity, and applicability of the system.

(2) *The network system is not a management information system.* Equipped with the unduly cumbersome methodology and the need for manual manipulation which result from a poorly planned system, the resultant product (reports, budget and schedule data, progress assessment) cannot be *timely* or *relevant* enough to stimulate timely and effective management action. In the early stages of a program, this deficiency may not be apparent. However, as the program continues and various areas of responsibility become increasingly separated, milestones will be endangered, potential cost overrun situations will develop and the system will fail to provide timely, relevant information concerning deteriorating situations and trade-offs or alternative courses of action which could favorably affect outcome.

(3) *Ambiguous assignment of responsibility.* Specific individuals cannot usually be held responsible for creating for planning purposes milestones above the working level which do not actually exist in the normal project maturity cycle. Without such responsibility, an overrun or other out-of-phase situation simply leads to evasion of responsibility instead of positive action. This could be a particularly disadvantageous situation in which an incentive-type government contract is involved, because incentive milestones can determine the profitability of the project. If specific milestones are not assigned to specific individuals, incentives for accomplishment may be sidetracked by other, more immediately stimulating but nonproductive, activities. This is a primary reason for missed milestones on any program, and ambiguous responsibility only fosters poor judgment. Also, in such a situation it is likely that management will be unaware of a deteriorating situation until *after* the milestone is missed. This again defeats the goal of providing program management with information on potential problems and allowing them to assess impact *before* crises actually compromise program objectives. Only in this way can management use resourcefulness and ingenuity to devise a strategy to minimize the effect of a potential problem.

(4) *Impact of program redefinition and changes.* Numerous redirections or changes are inevitable in any research and development project, but they can seriously disrupt program control unless the system regularly defines changes and assesses their impact in a timely, adequate, and accurate way. In a system with the limitations that we have discussed, it would be difficult (if not impossible) to determine the nature and effect of changes. Higher-level events in this deficient system represent phases of

program development; therefore, it is most difficult for technical problems concerning work accomplishment at the unit level to be evaluated in terms of other units, because the problems may not come to light at the report level. Also, summarizing costs at higher levels does not enable management to identify the cost of change in technical areas, and so makes it difficult to determine accurately which costs are due to changes and which are not.

INTEGRATING NEW SYSTEMS INTO THE ENVIRONMENT

Finally, there is the problem of integrating new systems into the environment in which they must operate, a problem related to the others we have discussed. Two of the major factors in the environment appear to be most important: realization of need for the new system, and compatibility with existing related systems and existing structures.

Realization of Need

In addition to human reactions to change, there is one environmental factor which pervasively affects the ease with which the new system is implemented: the degree to which personnel realize that a new system is needed, and that the proposed system will produce the desired benefits. Unless personnel realize that the system is needed, they will not exert the effort required to make it work. Operating people will not be receptive to new systems which are developed for the sake of showing what computers can do, or for the purpose of keeping an inadequately used computer or system design staff busy. They will also resist—with sufficient reason—an elaborate system requiring many changes and new procedures, which merely produces the results already obtainable from the old system, unless it results in overall economies, capacity for expansion, or similar advantages. For example, governmental agencies have often encouraged private contractors to use the PERT/COST system, because they have been aware of the need for the system; in the past, however, many contractors have not been aware of it. These contractors have firmly believed that there has been no real need for the new system, and therefore see no reason to expend any real effort to make it work. This kind of environmental factor always produces a real problem.

Compatibility With Environment

We have discussed the need to make the system compatible with the existing organizational structure. Related systems must also be made compatible with the network-based one. The new information system will interlock with other systems, each of which has a role to play in overall operations and each of which frequently handles information which will be fed to other systems. The success of the new system depends largely upon how compatible it is with related, existing systems. Using the PERT/COST System as an illustration of a management scheduling system, we can see that the following systems are related:

PERT/CPM (handling time-only information) or other scheduling systems.
The organization structure.
The resource (labor, materials) estimating, and pricing systems.
The accounting systems.
Other management reporting systems.
The contract administration procedures.
The procurement procedures and regulations.
The work authorization and control procedures.

These are, of course, only a representative sampling of the possibilities.

In attempting to introduce a system such as the network-based one which is related to so many other systems, those in charge may smooth over or ignore many incompatibilities. An integrating approach such as network planning inevitably spotlights any existing deficiencies, and so turns up many sore spots in an organization. The various systems will be likely to be incompatible because of:

Differences in time units, such as hours, shifts, days, working days, weeks, months, accounting months, calendar years, and fiscal years.

Differences in cutoff dates, which can be day of the week, weekend, biweekly, bimonthly, monthly, quarterly, etc.

Differences in cycle time; some systems produce reports within hours of the cutoff date, but others require weeks to produce valid information.

Functional versus product orientation; some information is related only to the type of work being done (function), while other information is related to a specific product; often, there is no cross reference or "torque converter."

Manual versus machine systems; systems vary from the completely manual extreme to the almost completely mechanized extreme. We can debate about which is more easily changed, but attempts to link the extremes result in awkward, inefficient operations.

Identification and coding of information; codes used in one system are often meaningless in another. Different codes are frequently used to identify the same reference within the same company. In today's large companies, merely identifying and correlating these existing codes is often a tremendous task, without even considering the introduction of some order and control over their allocation and meaning.

Diverse methods, procedures, and terminology; various parts of a company often develop diverse methods, procedures, and terminology. It is well-established, for instance, that a company cannot mix governmental contract work with commercial work, not only because of different product requirements, but also because of the different accounting, auditing, purchasing, and other administrative functions required. Terminology can vary considerably within the company, and terms such as "work in process," "commitment," and "budget," often have widely different meanings. All these factors make it difficult to implement a management information system which is intended to have broad impact.

The new system input-output characteristics. Another more tangible source of major difficulty involves the data which go into, and the information which comes out from the system. Generally, the characteristics of these are most important to the operating managers, assuming, of course, that the part of the system between input and output is capably designed to operate economically.

The output reports are the only reason for the existence of the system; however, often these reports are not carefully designed. System output can be displayed in a variety of ways with varying amounts of detail. Poorly designed reports or displays will evoke strong resistance in the managers who are supposed to be using them. Personal preferences are strong, and the report one manager likes may seem useless to the next. Probably the best solution to this problem is to design the system so that new reports can be generated easily and economically while some control and uniformity in report terminology, layout, and frequency is maintained. Perhaps the most fundamental requirement of the system output is that the reports must provide information that is required by the manager in a useful, understandable form. The output must not present data merely because it is possible for the system to do so; there must be a *need* for the information. In addition, the information must be better than the existing reports, or the new system will not be accepted.

The input data characteristics are the next most frequent problem source. Each new input requirement increases the difficulty the system will have in gaining acceptance. If data reporting can be streamlined, and can be based on already existing requirements, and if duplicate reporting of the same data eliminated, the system will be accepted enthusiastically.

METHODS OF MINIMIZING THE PROBLEMS

We have discussed the causes of some of the major problems encountered in implementing a new information system. The approach to system design and implementation discussed in the following paragraphs has prevented or minimized these problems.

Define objectives. The first step is to define the system objectives. What information will it produce? Why is it needed? Who will it serve? How will it operate? Answering these and other related questions will bring out the objectives. If objectives are not established, it will be difficult to design and later to understand the system. It is obvious, for example, that the complex requirements of a large weapons system application of PERT/COST will differ considerably from the network system requirements of a construction project or small business application. The scope of the interface requirements, the processing techniques, and other aspects of the system will vary greatly, as will the accounting procedures, formats, terminology, management reporting requirements, etc. Hence, to decide to implement a network-based system, as well as to design criteria for its implementation, the planner or manager involved must understand the nature and objectives of the particular application; this will prevent him from trying to interpret requirements (as in the case of a weapon system application), or to superimpose a system package to fill a need, without paying sufficient attention to the appropriateness of the approach. He must understand that data acquisition or processing must be capable of being tailored to satisfy real management information needs in diverse situations.

Give operating management responsibility. The alternative to this is to make a staff office responsible for system design and application. Although such an office may employ highly skilled specialists, it rarely contains persons who are intimately familiar with the environment in which the system will operate. This lack of knowledge invariably leads to difficult problems. For instance in a large, decentralized corporation, even a well-conceived staff effort at the corporate level can encounter serious implementation difficulties. For example, a major aerospace corporation recently tried to implement a fairly sophisticated PERT/COST system in the following manner: a highly skilled staff was set up to design and implement a PERT/COST system in the whole corporation; the emerging system was exceptionally well designed, and initial pilot tests produced encouraging results. Subsequent attempts to implement the system failed, however, because the more-or-less autonomous operating units of the corporation were unreceptive to standardized techniques which would directly affect proprietary operating procedures; responses ranged from the rapid development of new and competing systems to satisfy the "requirement" without having to succumb to procedures "not invented here," to outright resistance by perversion of the system which rendered it ineffective. Even cooperation could not overcome certain inherent ivory tower aspects of the system, however; eventually, the perfectly valid and costly effort was largely discredited, and discarded. Hindsight always seems to reveal mistakes clearly, but it does seem apparent that if operating managers had participated more actively in designing and implementing the system, it might have been considerably more successful. Any network-based system should be built upon the various needs of the company and upon the points of correspondence and interaction between the corporation, group, division, and department procedures, and the project-oriented system itself.

Use task force approach. The design team and the implementation team should represent each affected part of the organization. In this way, impractical approaches are avoided, and knowledge of the system is fed back into each organization, causing fewer fears to develop or surprises to be encountered. This team should be a strong core which is realistically concerned with implementing the system as a management tool to achieve project objectives, rather than being, as sometimes happens, a specialized group preoccupied with the esoterics of data processing and technique refinements.

Design for environment. The task force approach provides the detailed knowledge of the environment which is necessary for the successful development and use of the system. All affected and related systems must be identified and their characteristics understood, so that the new system will blend in and become part of the company. Existing sources of input data must be identified and used as the basis for the new system. The new system should be small enough in scope to be acceptable, but large enough to provide significant benefit. Definite benefits to the operating personnel supporting the system must be obtained and clearly identified, and the burden on these people must be reduced, not increased. Replacement of existing reports and procedures should be a definite objective of the system.

Avoid technological pitfalls. The planners should be certain that the Work Breakdown Structure is clearly and unambiguously defined, and that all levels of personnel concerned really understand this important cncept. They should provide a consistent and realistic procedure for identifying events, developing networks, estimating, and recording data, to make it easy to integrate these data for project management and reporting.

Educate all affected persons. Probably the most critically important phase of implementation is the education and indoctrination of all persons affected by the system. To make the new system work, it is absolutely essential for the personnel to be informed about the system; the planner must describe, explain, and demonstrate what the objectives are, how the system will operate, why it is needed, what the effects and results will be, and why it is important both to the individual and to the company that the new system work. Such indoctrination is not done in a one-shot briefing of a few managers; it must be carried on in a continual, detailed, down-to-earth program, which will eventually overcome the human reactions to accepting something new and to changing old ways, and must include on-the-job teaching while the system is in operation. Often, too, a dynamic training program can produce ideas for improvements to the new system or for development of the next generation of system.

SUMMARY

In this discussion we have shown that although well-designed, advanced information systems should provide significant improvements to an organization, it is often difficult, if not impossible, to apply such systems effectively. The effects of change and the deficiencies in technology are the fundamental sources of difficulty. Basic human reactions to change produce many of the problems during the transition to a new system. Technological system deficiencies often result from an ivory tower approach which ignores the real environment. If personnel do not recognize the need for the new system, or if the new system is not made compatible with existing related systems and organizational structures, difficulties are encountered.

To minimize these problems, the planner must first define system objectives adequately. Placing responsibility in the hands of operating management rather than staff specialists also improves the system design and gives it a chance to survive in the existing environment. Using the task force approach brings together representatives of all affected organizations(often for the first time), and insures a realistic design. Designing for the specific environment will result in a system which will fit into existing procedures and systems, minimize incompatibilities, depend on available or reasonable input data, and provide results beneficial to operating managers. Finally, continuing education and training of all affected persons will reduce, if not eliminate, human resistance to change.

If the planners properly understand these sources of problems and pitfalls in implementing management information systems, and properly emphasize the methods of removing these causes, they will be able to attain the very significant benefits offered by advanced information systems.

18

Simulation and Gaming[1]

Most of this book discusses network planning systems as they exist today. In this chapter, however, we will briefly explore the future, in terms of one possible extension of these systems, involving use of existing computer programs as the basis for simulation and gaming models. This potential use of network planning techniques may ultimately be as profitable to management as the revolution in planning itself.

DEFINITIONS AND BACKGROUND

Before going any further, we will give the reader some idea of what is meant by "simulation" and "gaming." Although these two terms have been used in the literature to designate a number of different activities, in this chapter we will use them in a simple, rather narrow sense. To avoid confusion in the subsequent discussion, we have prepared a brief list of essential definitions:

A *"model" is a representation of a system and its operations. Only mathematical models will be discussed in this chapter.*

A *"simulation" is an exercise performed with a model in which all operations and manipulations of the model are performed without any human interaction. (In general this does not mean that a simulation must be computerized, but in this chapter such a meaning is intended.)*

A *"game" is an exercise performed with a model and one or more human beings who dynamically supply data to and receive data from the model.*[2]

The type of model to which we will refer in most of this chapter resembles a company material flow model. In such a model, the movement of all

---

[1]By W. J. Erikson, System Development Corporation.

[2]See Joel M. Kibbee, Clifford J. Craft, and Burt Nanus, *Management Games, A New Technique for Executive Development*, Reinhold Publishing Corp., New York, 1961.

materials (such as raw stock, finished components, subassemblies, and finished products) is represented by equations and tables in a computer. In the simplest form, the tables contain data telling the amount of each type of material in each department of the firm, and the equations are used to change the state of the materials (by machining, assembling, etc.) and to move the materials from department to department. The model is driven by decisions made in response to events. The events are predetermined and contained in a script or scenario written prior to a run of the model.[3] Typical events in material flow models are customers ordering products, suppliers going on strike, (therefore being unable to supply raw materials), and suppliers changing prices. The decision-maker may be a set of equations in the computer, in which case we have a simulation, or a person or persons playing this role, in which case we have a game. In either a simulation or a game, a sequence of events that take place over some set period of time (as written into the scenario) is submitted to the decision-maker according to the scenario time schedule, and the resulting decisions are used to drive the model. A run consists of many of these event-decision-action cycles. We will give a detailed example of how a computer can function as a decision-maker and cycle a model later in the chapter.

Many managers are aware of the potential benefits of simulation and gaming. They realize that such testing of proposed changes in operations, product line, etc., on a computer, then one could make one more aware of the potential benefits or pitfalls involved in such changes. This approach is much more appealing than the alternative of implementing the change within one department or even the whole organization for a short trial period. Most managers would welcome more information before trying programs, because:

Real operations and trials are expensive.

Mistakes in such operations are very expensive.

It takes time to validly assess the effects of changes in real operations. (One part of the Hawthorne experiments consisted of studying variations in working conditions of six workers for five years.)[4]

The effects of personnel reactions cannot be easily assessed in a trial program. (The workers may overproduce to please their superiors, or may underproduce because they dislike the new system, or may react in a multitude of other ways.)

---

[3]The terms "script" and "scenario" (particularly the latter) are often used in simulation and gaming to denote the programmed sequence of events which takes place in the simulation or game.

[4]See F. J. Roethlisberger, and W. J. Dickson, *Management and the Worker*, Harvard University Press, Cambridge, Mass., 1939.

It is difficult to test the new programs in situations that do not occur often, but that can catastrophically affect a company (e.g., suppliers running short, strikes, riots, floods, or war).

The use of simulation does not eliminate the need for trial programs, but it does provide a basis for weeding out grossly infeasible ideas before trial. For example, before deciding to set up a network planning system in an organization, management would like to know if, under ideal conditions (complete employee acceptance and proper use of the new system), the new system would improve operations (lower cost, improve scheduling, etc.). At present the manager must believe either his own judgment about the system, the judgment of those who want to sell the system, or that of other managers who have successfully used such systems in other organizations, even though he may feel that the salesman is heavily biased and really does not understand his manager's problems, and that other managers who successfully used such systems really had different problems. In addition to his skepticism, the manager may not really understand what network planning systems are and how they are used, because of his lack of experience with such systems. The use of simulation or gaming could provide the manager with some of the experience he needs to make a decision.

Another class of examples concerns the multitude of problems faced by a project manager during the course of the project. These problems always seem to require immediate attention, and there is seldom time or funds available for a trial program, so that the solution the manager chooses is used on the project itself, and any losses from the solution are computed for the project instead of for a trial program. What can simulation do to help solve these management dilemmas? Simulation can make it possible for the manager to test his tentative solutions against future conditions typical of his project, showing him how well his organization will function if the decision is implemented, and how the proposed changes will affect the other projects of the company; and will provide a basis for assigning costs to alternate decisions.

However, general models have not yet been developd to fill these simulation needs. A number of problems are involved in model development, including:

The high cost of designing and programming models. (It is not unusual for the development of the type of models we are describing to occupy half a dozen people for two years.)

The difficulty of proving the validity of a model.

The system-specific character of most models.

This chapter is basically an outline or functional design of a model that could be used to overcome these three development problems, to aid managers in assessing the worth of network planning systems, help them gain familiarity with such systems, and provide them with a tool to assess the worth of alternate decisions.

USE OF THE MODEL

This section will discuss the possible uses to which a network planning model could be put by a manager and his staff, including a research simulation, a demonstration simulation, a research game, a demonstration game, and a training game.

First, the reader must understand what is meant by a network planning model. For the rest of this chapter we shall define it as a model that has a network planning program as one of its components. A network planning program is a computer program such as a PERT time/cost/manpower program that reports on the scheduling, costing, and manpower aspects of a project. Most of the examples will be given in terms of PERT, but the concepts of this chapter are general and can be related to any network planning technique.

Research Simulation. In research, a network planning model could be used to study system dynamics.[5] A model of the system under study (say a manufacturing firm) could be run through many years of operation on a computer, and the varying results of using different management information systems would be recorded and compared. In this way, management could select theoretically optimum management information systems for a wide variety of industrial situations. For example, one interesting study would involve determining the optimum reporting period for PERT system outputs. These time periods are usually set up on a basis corresponding to the existing accounting system, or on a very much shorter time span, because the outputs are easy to obtain.

But, are these reporting periods optimum from a management information point of view? No one can really know without detailed study. Anyone can estimate the effects of extreme changes, such as setting one year or one hour as the reporting period, but it is difficult to tell which reports should be daily, weekly, or monthly. It is obviously bad to have all reports coming in on a yearly basis, since most problems require attention much sooner; it is also obviously bad to receive all reports on an hourly basis, since it would be extremely costly, management would do little except

---

[5]See J. W. Forrester, *Industrial Dynamics*, The M.I.T. Press, Cambridge, Mass., 1961, pp. 357–360.

read reports, and the system would be too sensitive either to minor errors in estimates or to minor changes in outputs. However, to determine the optimum reporting periods the network planning model could be used, simulating the operations of the organization over the time span of one complete project. The model could be run many times with different values used for the reporting period of each output. There are a number of ways to evaluate these results, but one of the more likely would be to choose the set of reporting periods that allowed the company to meet the scheduled completion date with the lowest expenditure of funds. Other goals (such as maximum use of in-house capability, minimum use of certain types of manpower, and uniform rate of expenditure) may be more important in some organizations, but determining which goals are the critical ones is a task for the manager, not the model builder. Using a simulation of this type, management could either obtain criteria for setting these critical times, or (on a more applied research—almost a production—basis) use the simulation as a tool to provide the critical time for each report in a new job or project being set up.

Demonstration Simulation. As a demonstration vehicle, a simulation can be used to show management how control of a project takes place with and without a network planning system. Such demonstrations can be either general, or specific, with the system parameters set to simulate the organization and project of the observer. As we will discuss later, the same project could be run through one year of activity in several different modes, each representing a management information system such as:

A PERT time/cost/manpower system.
A PERT cost system.
A PERT time system.
A bar (or Gantt) chart system.
A verbal information system.

The output of the simulation run for each of these systems would be a summary report of factors such as project dollar cost, man-years expended (by worker category), number of overtime hours, schedule dates met or slipped, and current project status. With these outputs a manager could compare, for example, project dollar cost for each of the five management information systems listed above and determine which system is the best.

Research Game. The research that could be carried out with the model being used as a game is basically the same as that discussed under research simulation. However, since the game would have a person operating in the role of decision-maker, it would allow management to study additional problems, such as the effects of varying information loads on performance.

It would also be possible to have people operating as time estimators to provide inputs to the PERT program as part of a game.

Demonstration Game. A game would provide a manager with the same type of demonstration as that discussed under simulation, except that the manager could actually play his usual decision-making role. In this way, the manager would be better able to assess the advantages and disadvantages of each of the information systems being demonstrated.

Training Game. As a training vehicle, the network planning game would be an excellent addition to a course in network planning techniques, or to general business administration courses in which games are used. The trainee could function as a decision-maker who is supplied with information from one of the five types of management information systems, and could learn how to use network planning systems by actually functioning in the positions that exist in the real systems.

THE BASIC MODEL

A basic model is shown in Figure 1. The environment and network planning programs will be used whether the model is used in a game or a simulation, but the decision-making program will be used only for a simulation. The network planning program discussed in this chapter is a PERT time/cost/manpower program capable of producing the usual set of out-

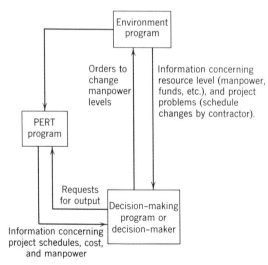

FIGURE 1.　Basic PERT model.

puts, including tentative changes as well as actual changes in such items as manpower loadings, scheduled due dates, and rate of expenditure. The environment program can take many forms, but the simplest is a complete script (or scenario) written prior to the run of the model, that contains a time-phased list of actual events. The network planning program and the decision-making program (or decision-maker) would be driven by and respond to these inputs. The designer can use the scenario for a series of experiments over which he wants to maintain very tight experimental control, or for a series of demonstrations within which he wants to maintain consistency; however, the primary use of a rigid scenario is for training: the trainees can be presented with a fixed set of situations to which they must react. For most purposes a more valuable environment program would use samples from parameterized distributions to produce actual activity completion dates, manpower availability and use, and fund availaibility and use. For example, one could use the three time estimates (optimistic, most-likely, and pessimistic) of activity completion dates to set up a beta distribution. A sample can be drawn from this distribution with the aid of a random number generation routine, and can be used as the actual completion date. A series of runs would be made to insure the statistical significance of the results.

The decision-making program can consist of a set of highly parameterized algorithms set to achieve goals such as minimum time to completion, minimum total project cost, minimum manpower use, or some function of these three. These minimization algorithms could be subject to constraints involving such factors as fixed completion dates, minimum available funds, maximum available manpower, and lowered efficiency of transferred personnel.

For an example of the operation of the basic network planning model, consider the following case of construction project management. A message is sent from the environment program to the decision-maker, stating that a subcontractor who was supposed to begin work today has experienced some difficulties and does not expect to begin work for one week. The decision-maker requests from the network planning program a listing of the present critical path and a listing of the critical path as it would look if the subcontractor in question were delayed by one week. If there is no change, the decision-maker may ignore the input, but if the delay causes slippage that the decision-maker does not wish to incur, he selects from one of three for alternatives: to accept the slippage, to switch some of the present workers to the task that was assigned to the subcontractor, or to hire another subcontractor. The first alternative results in time slippage, but no dollar losses (unless there are schedule slippage penalties). The second alternative may result in no losses (if enough workers are capable of the task and are currently working on tasks that are on slack paths) or may be totally infeasible (if the workers cannot do the tasks or are all on critical paths). The third alternative may also result in no losses (if another subcontractor is immediately available and has a price no higher than the original subcontractor), but in general the second subcontractor will not be immediately available and will cost more than the original. Choosing the alternative is a task for the decision-maker; the network planning program provides him with the information that allows him to make the decision. Once a decision is made, the orders are sent to the environment program and the network plan is updated by a message sent to the network planning program. Messages concerning completed tasks are sent along the link shown between the environment program and the network planning program.

THE FULL MODEL

As we described earlier, the basic model operates by having both the decision-maker[6] and the network planning program receive perfect information from the environment program as soon as the information is available. However, this is an idealized situation, not really representative of the world as it exists for most managers. To better model actual conditions, the full model will include the effects of this problem. The power of the full model comes from its ability to convert the network planning system outputs into other forms, and from the ability of the experiment designer to degrade the quality of the information that is given to the decision-maker. Figure 2 shows a flow-chart of the basic network planning model with the converter and degraders added; it also shows the basic network planning program broken into two parts: the operations program and the query program.

The operations program is exemplified by the basic PERT time/cost/manpower program that keeps track of the project status according to data fed from the environment program. Periodic reports are sent to the decision-maker, and special reports are sent when requested. In addition, changes in the scheduled project due dates, or other changes in the network are submitted by the decision-maker whenever he produces such data.

The query program is used to answer the question, what will happen if A is changed to B in the project? (where A and B may be, for example, the number of men assigned to one activity). The query program obtains a

---

[6]For the remainder of this chapter, the term "decision-maker" will be used to refer either to a person or to the decision-making program.

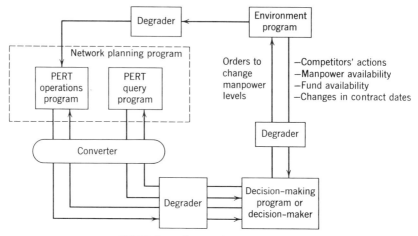

FIGURE 2. Full network planning model.

report on the current state of the project from the operations program whenever the decision-maker asks such a question. Even though such a capability may be an integral part of a PERT time/cost/manpower program, it will be discussed separately in this chapter.

The converter transforms the outputs from the two network planning programs to represent PERT cost outputs, PERT time outputs, bar charts, foreman's opinions, or staff assistant's opinions. The next section will suggest how this may be done.

The degraders are used to vary the quality of the outputs from the converter or the environment program. With this capability, the experimenter can control the quality of all inputs to and outputs from the components of the model and determine the effects of such variations. He will be particularly interested in experiments made to determine the effects of estimation accuracy and of time delays throughout the system. Errors and time delays can be stochastic and can be described by present parameters. For example, one might describe all time delays in terms of a mean and a standard deviation, and during a run of the model this distribution can be sampled to determine actual time delays.

We can consider the example previously discussed in terms of the basic network planning model. When information comes from the environment program concerning a subcontractor who cannot start on schedule, the message may be delayed for as much as a day by subordinates who attempt to get more information about the problem. The manager will probably encounter shorter delays when he wants to get reports from the query program concerning possible changes, but delays caused by computers being unavailable (in the simulated organization of the manager), or by normal processing and handling will still force the manager to wait for outputs. The converter will allow the manager to function as if the operations program and the query program were in one of the five modes mentioned earlier. (Each of these modes will be discussed in more detail below.)

Model Modes

The reader probably has some vague understanding of what this model is supposed to do, but probably finds it hard to understand how the model can be used as we described earlier. Therefore, this section will briefly discuss five different management information systems and show how the network planning model can be adjusted to simulate each. In all of these models the degraders can be set to transmit with a preset degree of error and also with a preset time delay.

PERT Time/Cost/Manpower Mode. This is the basic unconverted operation of the model. The decision-maker will be asked to manage a project using a PERT time/cost/manpower information system. The full set of outputs normally obtained from such a system will be available to him.

PERT Cost Mode. In this mode, the converter would delete all manpower data from the PERT outputs. If desired, these data could be made available to the decision-maker, but could be labeled as the opinion of a staff expert; this staff expert can be made as accurate as desired by appropriate settings of the degrader. The staff expert data can be given directly to a decision-maker as a machine output, or (in game) these data could be fed to another individual who answers the questions of the decision-maker.

PERT Time Mode. Basically the same conditions are set up here as were set up for the PERT/COST mode, except that cost data are also deleted by the converter. Inputs of cost data can also be given to the decision-maker as the opinion of a staff expert.

Bar (or Gantt) Chart Mode. It is also possible to run the game as a bar-chart controlled operation. The basic procedure in this mode is to first present the overall plan in bar-chart form; then, as the model moves through time, each of the completed activities in the PERT plan is translated into a line on the bar chart and this updated material is sent to the decision-maker. The converter also inhibits any communication with the query program. If desired, however, the query program can be used to simulate the opinion of a staff assistant. The degrader can be set to vary the quality of this output independent of the quality of the bar chart outputs.

Verbal Information Mode. In this mode, the decision-maker is furnished with a rough verbal plan of what is to be accomplished and is given a list of activities that have to be completed. It might be interesting to run a subject through several projects of the same type so that he may become "experienced" in this type of project management. As noted earlier, the query program can be used to represent the opinions of a staff expert who provides estimates of completion dates, costs, and manpower needs. It would be possible to provide the decision-maker with several of these sources of data and to have them vary in optimism.

Model Details

In this section, we will see how the job can be done, using each of the model programs discussed previously. All inputs to the network planning programs will be transfers of data (within the computer) in the simulations, and will be console inputs by the subjects in the games.

Operating Program. This program is equivalent to the basic PERT time/cost/manpower program, and the full complement of reports and charts will be available as outputs from it. A typical set of such outputs is[7]:

Financial plan and status report, which is a display of budget requirements for future months as well as of actual expenditures to date.

Manpower loading report, which is a projection of manpower requirements by months.

Operating unit status report, which is a report of the time and cost status of the accounts within each unit manager's area of responsibility.

Project status report, which is an integration of time and cost information summarized for each level of project management.

These outputs will be produced within the network planning program in the same way that they are produced in a real situation, the only difference being in the way in which the outputs are triggered. In a real situation, an individual or group is in charge of getting out the reports; in the model, a parameter is set for each reporting period so that each output is generated automatically during the course of a model run.

Query Program. This program is similar to the PERT operations program; the full complement of reports and charts that are normally available from this program will also be available as output in the model. There is no major difference between the operations of a query program in this model and in an actual situation. In both, a computer program provides an output of a specific table or chart. In both, the input request may contain some changes that should be made in the data used to produce the

---

[7] *DOD and NASA Guide—PERT Cost Systems Design*, Office of the Secretary of Defense, and National Aeronautics and Space Administration, June 1962, p. 83. See Appendix F.

output. The only minor difference is that in a simulation the input may come from another computer program.

Converter. This program converts the network planning program outputs into the mode desired by the experiment designer. Each of the modes discussed previously was obtained by operations of the converter. The converter will operate by testing every output to see what it is; this will be relatively easy if every input to the operations of the query program is stored in the computer and used as a tag for the output. Thus, if a request is made for a manpower loading report and the program is operating in the PERT time mode, the report that is generated will be labeled as the opinion of a staff expert and not as a network planning system output. The one basic difference between the converter outputs sent to a subject decision-maker and those sent to the decision-making program is that the decision-making program is not capable of receiving charts as inputs. This is not a great loss, however, since the decision-making program can be made capable of extrapolating trends from raw data (something that humans find rather difficult without the aid of a chart).

Degrader. This program allows the experimenter to introduce either time delays or actual changes in the output data. Both of these types of degradation can be produced by sampling previously determined distributions. For example, consider a verbal information-reporting system that contains two sources of information: one, X, is very quick to respond but gives answers which vary considerably in accuracy; the other, Y, is slow to respond, but is generally a good, but slightly low, estimator. The distributions that represent the opinions of these two individuals are shown in Figures 3 and 4. Sampling from these distributions will generate responses of the type described. For example, assume that X and Y are two members of the manager's staff who are asked to supply data on manpower loading. The manager makes the request and the network planning program pro-

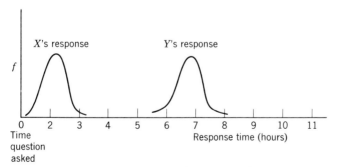

FIGURE 3.   Response time of two sources of data.

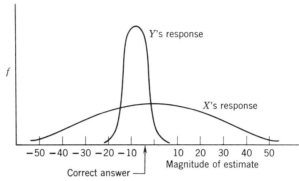

FIGURE 4.   Response accuracy of two sources of data.

duces the report. As discussed earlier, this program output is sent first to the converter, and then to the degrader. The report from X is sent to the manager first (see Figure 3) but the accuracy of the manpower loading estimate may vary considerably from the correct answer (see Figure 4). The report from Y is sent later (Figure 3 again), but his estimate is much closer to the correct answer (Figure 4). What the program does in these two cases is add values to the network planning system outputs. For response time, we can assume that each has a response time equal to his mean response time; thus X turns in his report after two hours and Y turns in his report after about seven hours. For response accuracy, we can assume that each is operating at the plus one standard deviation level; thus Y's prediction of manpower requirements is low by 5 men and X's prediction is high by 20.

Environment Program. As discussed previously, the environment program can take on many forms, the simplest of which is a complete scenario with no stochastic or adaptive elements. When such a method is used to define the environment, the program becomes merely a time- or event-triggered output program; no processing or analysis of data would take place. A second type of environment program is a bit more complicated. In addition to producing actual completion dates with samples from a beta distribution,[8] it uses other distributions to produce the host of other catastrophes that usually plague a project manager. The problems generated could include such occurrences as:

Strikes.
Key personnel quitting.

[8]*PERT Summary Report Phase I, Appendix B*, Special Projects Office, Bureau of Naval Weapons, Department of the Navy, Washington, D.C., July 1950. See also Chapter 5.

Delays in expected funding.
Material shortages.
Cutbacks of portions of the project.
Changes in required completion dates.

These distributions would operate approximately the same way as those shown in Figures 3 and 4 and discussed in the last section. Some of these problems, such as strikes, either occur or do not occur. These would be determined by sampling from a distribution such as the ones shown in Figure 3; the strike would occur only if the sample was above a specified threshold. Thus if the strike threshold is set at the 3-sigma level, a strike will be generated, on the average, about once every 370 samples. In this way the model can be set to sample from this distribution every three days of model time and a strike will be generated, on the average, once every three years. More or less frequent strikes can be set up by changing either the threshold or the sampling frequency.

Decision-Making Program. This program contains a set of highly parameterized algorithms that attempt to achieve goals such as minimizing the total project cost, achieving a uniform rate of expenditure, maximizing manpower utilization (by numbers and skills), minimizing changes in manpower levels, minimizing time to completion, minimizing dollar expenditure, or some function of these. The achievement of stated goals will be subject to constraints involving such factors as fixed completion dates, maximum available funds, maximum available manpower, and lowered efficiency of transferred personnel. Or if the decision-making program is supposed to represent an "intelligent optimizer," the problem structure lends itself to a linear programming solution. However, most real decision-makers do not actually optimize, since this implies a complete knowledge of all choices available and of the exact consequences of each choice.[9] Thus, a more realistic model of a human decision-maker will represent one who attempts to solve the problems of project management by trial and error methods. This is the approach used in the sample decision-making problem described later in this chapter. The decision-making program could also contain features to stop the search for better alternatives after a certain time period has elapsed, to represent the decision-maker getting "fed up" with the problem. It would also be interesting to incorporate features of adaptive behavior into this program. For example, the program could keep data concerning the accuracy of each of the outputs from all sources of information. These data would be used to weight future information coming from the same source. Thus, after some number of cycles, the decision-making program would tend to use

---

[9]H. Simon, *Administrative Behavior*, Macmillan Co., New York, 1960, pp. 61–110.

accurate sources of information more than inaccurate sources. A detailed example of one part of such a program operating on the problem of minimizing time to completion is discussed in the sample decision-making problem later in this chapter.

Input/Output Conversion Program. In the operation of a simulation or a game, it is necessary to be able to change the network planning program parameters easily. Also, if the game is to be used as a training device, it is desirable to have more than one student operate the program at one time. For both these reasons, it seems advantageous to use a console input/output system which operates in a time-shared mode. Such a system enables trainers to change parameters easily, and allows a large number of students to operate their own network planning models.[10]

When operating as a game, an input/output conversion program must be provided to convert a teletype input message into data usable by the computer (and vice versa). This presents no problem, because most network planning programs have been designed to accept card inputs, and presently produce outputs that can be printed out on the subject's console. This is another area in which use of existing network planning programs will save a considerable amount of time when compared to the conventional model development procedure.

SAMPLE DECISION-MAKING PROBLEM

This section is included to show one example of how a computer program can be used to simulate the actions of a decision-maker solving a problem. It will also show some of the interprogram operations of a network planning model. The basic decision situation faced by this decision-maker is given in terms of a network planning program output message that contains the predicted project completion date. If this date is later than the desired completion date, the decision maker must assign additional personnel to activities on the critical path. These personnel can be shifted from other activities within the project, transferred from other projects within the organization, or hired from other organizations. (The reader should note that this is only one way that the given problem can be solved, and that a complete decision-making program would include other approaches, such as changing the network to provide more parallel paths and thereby producing an earlier completion date.)

---

[10]Time sharing is a way of using a computer in which more than one individual can use the same machine at the same time. For a discussion of one such system, see Jules J. Schwartz, Edward G. Coffman, and Clark Weissman, "A General Purpose Time-Sharing System," Proceedings of the 1964 Spring Joint Computer Conference, Spartan Books, Inc., Baltimore, Md., 1964, pp. 397–411.

The subroutine shown in Figure 5 is entered every time the network planning program predicts the project completion date. This date comes in as "START" at the top of the figure. In box (A),[11] this date is compared with a previously stored desired completion date. If the desired completion date is later than the predicted completion date, nothing is done and we exist from the subroutine. However, if it is not later at (B) the decision-maker requests a listing of the activties on the critical path. At (C) the operations program supplies this output. At (D) the decision-making program adds one man to each activity in turn, and determines the effect of this upon the completion date by checking with the query program. At (E) the man is allocated to the activity that results in the greatest improvement in project completion date. At (F) the activity to which the man should be allocated is stored for future reference. At (G) if the predicted completion date is still too late, the program keeps adding men one at a time until this date becomes earlier than the required completion date. At (H), when the predicted completion date is found to be earlier than the desired completion date, the subroutine exits, and the decision-maker searches to see if the manpower needs can be filled by transferring men from slack paths. This search involves a loop similar to (D) and is not shown except (H). The personnel found in this search are allocated in (I) to the activities stored in (F). If more men are needed (J), a loop (K) similar to (H) is used, to find all the men that can be transferred from other projects. These men are assigned to the necessary activities (L), and the program checks to see if any more men are needed (M). If they are, the decision-maker hires them from other organizations (N).

It is obvious from this outline that more detail must be added before anyone would confuse the operations of this decision-making program with those of a real decision-maker. Much more detail would also be required before such a subroutine could be coded. However, it was felt that including these details would only make it harder for the reader to visualize the general logic that can be employed in the design of subroutines that form key elements of models such as the one described in this chapter.

USEFULNESS OF THE MODEL

The development of a network planning model of the type described in this chapter will benefit a wide variety of people: Managers, network planning systems specialists, model developers, and students. Specific uses for each are detailed on p. 406.

---

[11]Each of the capital letters in parentheses in the discussion refers to parts of the program labeled at the top or upper left-hand corner of boxes in Figure 5.

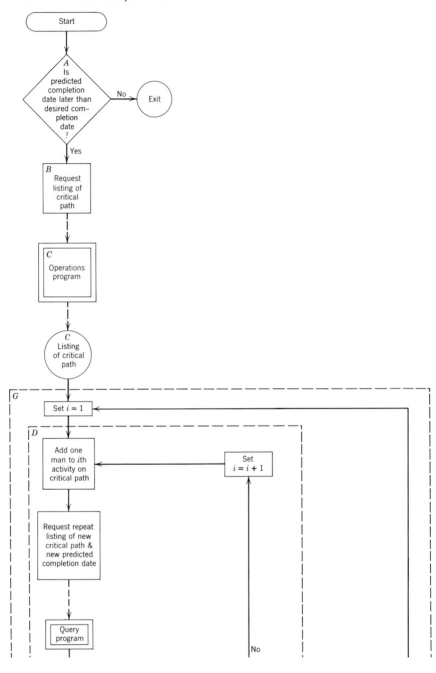

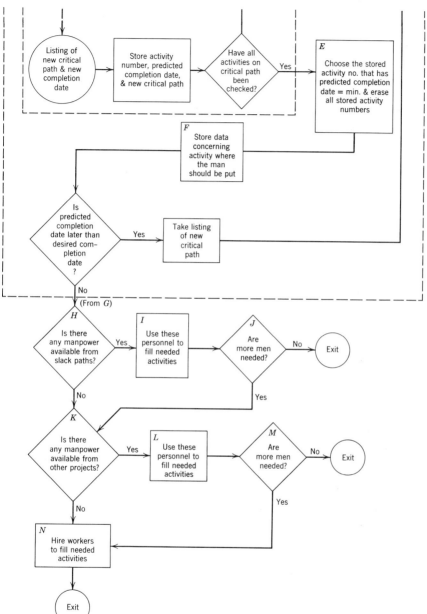

FIGURE 5. Decision-making logic to handle time overruns.

406 / Pitfalls and potentials

Managers. In general, managers will use an operating network planning model that has its parameters set so that the model represents the organization in which the manager is functioning. Such a model could be initially used to assess whether such a capability would be useful to the organization. When management decides to use a network planning system, the network planning model can be used by the organization personnel, management included, to gain familiarity with the use of such a system. Finally, managers can use the network planning model to test alternate ways of running a project before deciding on one plan. For example, consider the decision faced by a manager who must choose between a functional or a product-oriented department structure for a new project. Most such decisions are defended on the basis of such arguments as:

"We always do it this way."

"It costs too much in time, money and confusion to move everyone out of his present department into a new one."

"The new project will be completed in too short a time to make it worthwhile to restructure departments."

The implication is not being made that these arguments are always invalid and should therefore not be considered in the decision, but rather that there should be a way to test their validity. Use of a network planning model will provide a means of performing such tests.

Network Planning System Specialists. One problem faced by systems specialists is that of convincing a manager that PERT or CPM can really work in his organization. Being able to work through a model of a specific organization or project will enable such specialists to show management how these systems will work in their (management's) specific organization. It will also provide the specialists with a tool that can be used to teach the proper use of these systems to personnel at all levels in the organization. The major benefit, however, will probably be as a tool for research, to answer the type of questions that were presented at the beginning of this chapter.

Model Developers. Available operating network planning programs provide the model developer with a model in large part complete, and therefore provide the basis for faster and less costly development of a new model. Once a model has been developed, it is easier to prove that it is a valid representation of the organization under study, because the network planning programs used in the model can be the same as the network planning programs used in the organization. Finally, the use of the converter allows the developers to simulate a wide variety of management information without building a new model.

Students. Students can benefit from such a model by using it in conjunction with the study of management systems, management decision-making, and organization theory. They can use the network planning model to become familiar with the use of network planning systems, and can use the model as a tool for research.

The type of model described in this chapter is not being used because none is presently programmed. However, this type of model is technologically feasible as is demonstrated by the existing models that contain the same type of features as this one. What is required to bring such a model into being is a recognition of need that is great enough to result in funds being allocated to the task.

SUMMARY

Network planning algorithms and computer programs can be used as building blocks to create useful simulation and gaming management tools. This chapter has discussed how this can be accomplished, and has presented an insight into how these tools can be used to improve business operations and increase our knowledge of the way in which the enterprise functions.

19

Relating the time/cost/manpower system to the management function

In this chapter we will see the network-based time/cost/manpower system against the background of the total management function. Our approach will be, first, to relate the system to the elements of the managing process, and second, to see how the system performs within the relevant elements. By so doing, we will be able to understand the role and nature of this system more clearly, and to define its capabilities and areas of application.

THE NETWORK-BASED SYSTEM AND THE ELEMENTS OF THE MANAGING PROCESS

The actual managing process may be described as a simple procedure, consisting of three steps: establishing objectives, directing the attainment of objectives, and measuring the results. Figure 1 represents this as a circular process, which continuously repeats the three steps. When the results obtained from carrying out the originally established objectives are measured, the new objectives become evident. The managing process goes on and on, and is never completed. Within this three-step process, eleven elements of the managing process have been identified, as indicated in Figure 1; these elements are common to the work of every manager, although a top manager spends more time managing and covers a wider area than a department head. The nine elements on the inner circle in Figure 1 are usually incorporated into the management process in the chronological order in which they appear. The other two, "promote innovation" and "develop people," affect the quality and the continuity of management within the manager's jurisdiction. The layout of Figure 1 reflects this concept.

We find that the network-based system predominantly affects two of these elements of the managing process: "plan," and "measure, evaluate, and control." The system has, however, definite relationships with the elements called "gather information," "synthesize information," and "communicate." Before discussing the relationship with the former two, therefore, we will briefly examine the relationship with these three.

Gather Information

When the information-gathering function is related to an action plan, the network diagram is an effective aid. Since the network graphically displays the information in a meaningful way which shows interrelationships between activities, it makes it possible to gather related data quickly and easily.

Synthesizing Information

When a network plan is prepared for a total project, it forces the integration of many diverse activities and actions. This integration is one

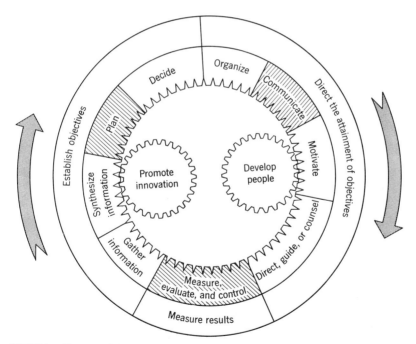

FIGURE 1. The network-based system and the elements of managing. \ \ \ \ \ Denotes primary elements in which the network-based system operates.

means of combining a large number of information elements into a meanigful whole. The network plan is thus a highly effective synthesizing tool when used properly.

Communicate

Because the network plan has a visual impact, it can rapidly communicate the plan to anyone who is interested. If the network is properly prepared, it will present the overall plan and all the intermediate elements, concisely and accurately. It has often been reported that a new manager was able to take over a program in a military weapons project in a fraction of the time previously required, because a network plan was used to show him how the job was to be done.

THE TIME/COST/MANPOWER SYSTEM AS A PLANNING AND EVALUATION TOOL

Let us now look at the two elements of the management process in which the integrated system operates predominantly: "plan," and "measure, evaluate, and control." When we look at the integrated time/cost system as a planning and evaluation tool, we will of course be limiting our discussion to the area of planning and evaluation of actions leading to the accomplishment of an identifiable physical or intellectual achievement with a recognizable beginning and end.

Definition of a Project Plan

To see how the network-based system relates to project planning, we must first define the term, "project plan." For our purposes, we will identify three parts of a complete project plan: the action plan, the resource plan, and the product-characteristics plan.

The Action Plan. This portion of the project plan contains all the actions necessary to complete the project. In the PERT/CPM sense, this means all events and activities in the project network, and so when we use the term "plan" in this context, we mean to include the schedule related to activities and events, as well as the sequential network plan.

The Resource Plan. This second part of the project plan consists of a number of interrelated plans for all the resources required in the project:

Manpower plans, including labor over time, estimates by skills or crafts, manning requirements, and sources of labor.

Dollar expenditure plans, or budgets.

Dollar source plans, such as sales forecasts, progress payment schedules, and borrowing plans.

Material and component part procurement plans.

Facilities plans, such as plans for office, shops, test facilities, equipment, and furniture.

Manufacturing plans, for fabrication, assembly, testing and inspection procedures and instructions.

Additional items can probably be identified for specific situations.

The Product-Characteristics Plan. The third major part of the project plan relates to the end product itself, defining what it is, how it functions, its physical and operating characteristics, and the quantity needed. If the end product is an item of equipment or a building, these questions are answered in the drawings and specifications. If the project is related to applied research or other types of study effort, however, the end product may be a report. Even in this case, however, the characteristics of the end product should be well planned. Planners of the research effort should know what is to be studied, and what answers are being sought. By giving sufficient attention to this part of the project plan at the same time as he is planning the other two parts, the project manager can assure himself that the interrelationships between the action plan, the resource plan, and the product-characteristics plan will not be overlooked, so that, for example, if a schedule is shortened, he will know that the quality or other characteristics of the product may be affected.

THE NETWORK-BASED SYSTEM AS A PLANNING TOOL

If we review the characteristics and capabilities of the network-based system presented in earlier chapters and compare them to our definition of a project plan, we can identify the areas to which the network system contributes directly as a planning tool.

The need for an action plan is fully satisfied by a well-developed set of network plans depicting the events and activities required to accomplish the project. The system enables (and in a sense forces) the planner to establish a logical, thorough action plan, integrating all the action plans related to the project.

Many administrative, indirect, or overhead actions will occur in support of a project, which will probably not be shown on the network plan, although some may be; it is assumed that the ongoing organizational functions are established and operating to provide such support. The resource plan will usually incorporate the support requirements however, even when they are not evident in the network plan. The need for a resource plan is partially satisfied by the network-based, time/cost/manpower system: in the most broadly developed forms of the system, part of the manpower and dollar expenditure plans are included. Although these are major parts of the resource plan, other important parts of that plan are not covered by the network-based system.

The third portion of the project plan, the product-characteristics plan, is not within the scope of the network-based system. The actions depicted on the network and the quantity and quality of the resources used are directly affected by the product-characteristics plan, but these characteristics are not handled directly by the system.

Evaluation of Plans

One important capability of the network system as a planning tool is its capacity to evaluate proposed plans. Laying out a set of plans is one thing, but evaluating its validity is another. We can use the system to plan our project, and then, because of the nature of the analytical results, we can evaluate the plans using various criteria important in a particular situation,   such   as:

Time:

> Total project duration.
> Realism of critical path activities.
> Specific dates of intermediate milestone events.
> Subnet or project element duration.
> Feasibility of conducting simultaneous activities: can the number planned be managed at one time, will there be intereferences which are not shown on the network, etc.

Manpower:

> Labor expenditure, both total and subtotal, by specific skills or crafts and organization element.
> Rate of labor expenditures over the life of the project.
> Head count, both total and by skills and organization element, over the duration of the project.

Cost:

> Dollar expenditures, both total and subtotals for labor, materials, by project, and organizational elements.
> Rate of dollar expenditure over the life of the project, both total and subtotal, as above.

Because the system can link together the elements of time, cost, and manpower, can cross-reference the project breakdown, activities, organization, budgeting, and accounting, and can use electronic data processing properly, it makes effective evaluation of proposed plans possible. Such evaluation aims at development of the best plans by revision and modification based on the criteria being applied.

THE NETWORK-BASED SYSTEM AS A PROGRAM EVALUATION TOOL

To show the network-based, time/cost/manpower system clearly as a progress evaluation tool, we have pictured schematically the operation of a total project control system. The portion of this control system which can be processed by the network-based system is identified on the schematic diagram in Figures 2 through 7. The figures show this system at various stages from start-up through stable operation.

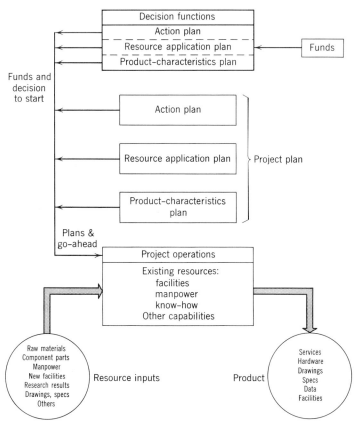

FIGURE 2.   Project control system—start operations.

Start Operations

Figure 2 shows resource inputs flowing through the project operations box to produce the required product. The control elements include the decision function box (through which the special resource of money is controlled) and the threefold project plan consisting of the action plan, the resource application plan and the product-characteristics plan. In this "start operations" phase, management makes funds available and decides to begin the work. The plans and initiating go-ahead are transmitted to the project operations box.

Collect and Transmit Feedback

At the stage shown in Figure 3, operations are proceeding; for control to be exercised, feedback information must be collected and transmitted to the proper portion of the plan for comparison. The figure portrays feedback collection and transmittal in three parts, related to the parts of the project plan; in reality, however, this feedback information is contained in many diverse written and verbal reports, test result summaries, status reports, technical reports, cost accounting reports, and so on. The network-based project management system can perform this function for the action and resource application plans.

Feedback Integration and Progress Evaluation

Information about progress, status, expenditures, etc., must be compared with the plan so that the planners can determine whether the progress is being made according to plan. In Figure 4 we have shown an "exception filter" for each of the three parts of the project plan. These exception filters pass the information which satisfies the plans within the prescribed limits of tolerance, but turn back the information which does not. For example, on the action plan a one-week delay in completion of an activity with six weeks of slack time can be accepted as satisfactory progress; on the other hand, a one-day delay on the critical path may be considered an exception to the plan, and so this item would be screened out by the filter. Computers presently perform this type of screening or filtering. Once the exceptions are identified, they must be transmitted to the decision function for evaluation. The feedback which shows adherence to the plan passes through the filter and is transmitted back to operations to indicate acceptance and to stabilize the operations.

Make and Transmit Decisions

The exceptions to the plan, such as ahead- or behind-schedule completions greater than a certain time, variances from budget of a given

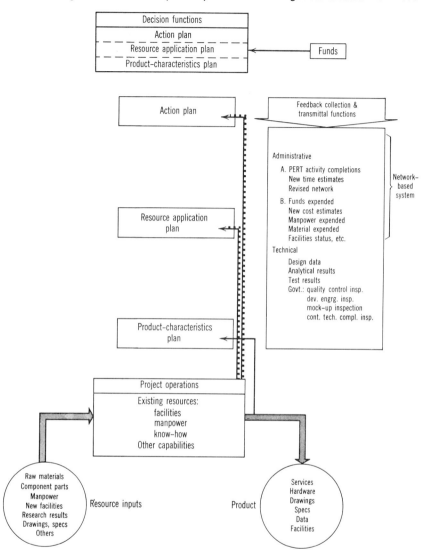

FIGURE 3. Project control system—collect and transmit feedback. Key: dotted and solid line denote network-based project management information system.

percent, or test results not meeting required performance specifications are reported to the managers who are then alerted to the need for decisions (see Figure 5). The manager combines this information about ex-

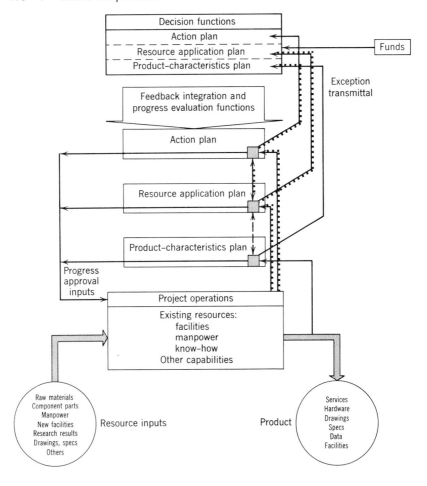

FIGURE 4. Project control system—feedback integration. Key: dotted and solid line denotes network-based system; squares denote exception filters.

ceptions with a multitude of information, and uses his experience and judgment to make a decision. He may even decide to do nothing about the exception, which is a valid decision. Once he makes a decision, it must be transmitted to operations to initiate corrective action. However, it must not be transmitted directly, but must be sent back to the project plan through the exception filters, to determine whether the decision will produce additional exceptions to the plan. If so, these would be re-

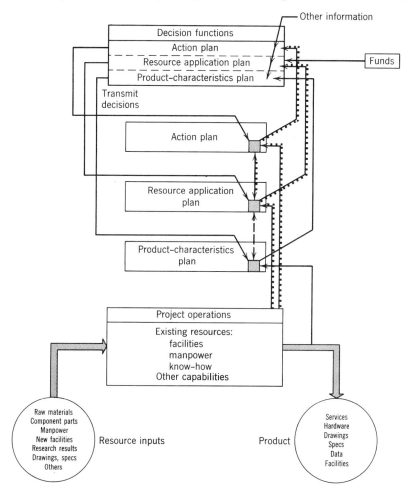

FIGURE 5. Project control system—make and transmit decisions. Key: solid and dotted line denotes network-based system.

ported to management for evaluation and additional decisions; this process continues until acceptable plans are obtained.

Implement Changes

The final step which closes the feedback control loop consists of converting the management decisions into action-directing form related to the project plan, and transmitting these results to operations as control

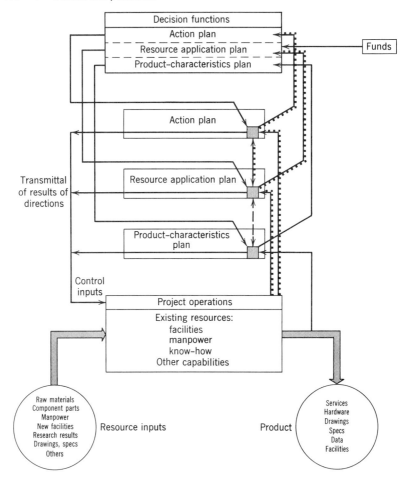

FIGURE 6. Project control system—implement changes. Key: solid and dotted line denotes network-based system.

inputs (see Figure 6). Such inputs take the form of work orders, memos, or other detailed instructions.

Stable Operation

The ongoing control system is shown in Figure 7. Feedback is derived from the operations and from the product; it is compared to the plan and filtered. Acceptable feedback is transmitted to operations as approvals; exceptions are transmitted to the decision-makers, and when the

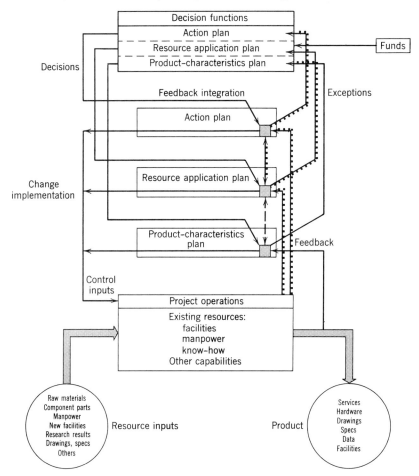

FIGURE 7.   Project control system—stable operation.

decision has been made, it is transmitted back to the plan through the exception filters. Acceptable decisions are translated into operations control inputs to initiate changes.

RELATING THE NETWORK-BASED TIME/COST/MANPOWER SYSTEM TO THE PROJECT CONTROL SYSTEM

The schematic portrayal of the project control system which we have developed enables us to identify fairly clearly the role the network-based system plays in evaluation and control. Earlier in this chapter we saw how the network-based time/cost/manpower system is related to project planning: how it aids in the formulation of the project plan, as well as in the evaluation of such plans. We can now see that the system operates in project control to collect and transmit feedback, and to integrate feedback and evaluate progress. We have discussed each of these previously, but we wish to point out that even a fully developed network-based system performs only a part of the total project control function. It does not make decisions, provide all the information required to make true management decisions, transmit these decisions, or translate them into action-producing directives. Finally, it does not transmit such directives to the operating people. In the future, however, such a total system may be developed, complete with data processing equipment and programs, but with the decision function retained as the manager's key role. Such systems have been successfully developed for production control purposes, but considerable additional development and operation experience is required before such a total control system can be developed for project management. Even in the limited role that the network-based system presently plays, however, it is a great advance over the "separate system" approach. Not only are the actions linked together through the Work Breakdown Structure and the network, but the resource information is also linked to the time element by the same structures. Thus, the effects of changes in one element (actions, time, cost, or manpower) are reflected in the other areas, and the underlying calendar scale is constantly related to all other actions and expenditures. Success or failure in many projects is a matter of timing and timeliness. Rates of resource expenditure are often much more crucial than total expenditures. Therefore, the interlocking of these elements by the network-based system gives the user a powerful method of effectively evaluating the actual progress against both his original and his current plan.

SUMMARY

The network-based, time/cost/manpower system relates to several, but not all of the basic elements of the managing process. It is primarily a planning tool, and secondarily a program measurement and evaluation tool. It does provide distinct benefits in gathering and synthesizing information, and in communication.

In this chapter we have developed the concept of a project plan which consists of three parts: the action plan, the resource plan, and the product-characteristics plan. The network-based, time/cost/manpower system can include all of the action plan and much of the resource plan, but does not directly include or benefit the product-characteristics plan. As a planning tool, the network-based system provides a framework for effective, integrated planning of time and resources. It also provides an excellent means for evaluating plans prior to their implementation, by virtue of the integrated analysis of interdependent actions and resources on a time-scaled base.

In this chapter we have also represented the project control system schematically, to identify the role which the network-based system plays in progress-monitoring and evaluation; this control system is even sufficiently general to be termed an operations control system. We have found that the network-based system does play certain vital parts in overall control, in the collection and transmission of feedback, and in the integration of feedback and evaluation of progress related to the action plan and portions of the resource plan. However, the network-based system is not a complete project management or control system, because it does not make decisions or transmit the results of decisions to control the operations. Instead, it remains for another group of inquiring managers to develop the total project management system. The foundation provided by the network-based, time/cost/manpower system appears to be one on which such a total system could be built, however.

20

Outlook for the network-based system

The previous chapters have discussed and explained the basic network-based planning, scheduling, and monitoring system, the extension of it to include dollars and manpower, and further extensions into optimization, simulation, and gaming. We have also examined the problems involved in implementing such systems, and have described several typical applications. To conclude the book, we will devote this chapter to a discussion of the future of these systems.

The management information system which is the subject of this book is a generalized management tool used for planning actions. It provides a means for recording and communicating the actions necessary to convert ideas, concepts, desires, and objectives into results. The systematic discipline inherent in the network approach forces properly correlated planning, scheduling, and monitoring of progress. Two key factors to be emphasized in the use of the network-based system are the application of data processing to these management functions, and the correlation of time planning with all other aspects of the management process.

THE OUTLOOK

Several dimensions of present network-based systems can be expanded by practicing managers, management scientists, and information system specialists, including:

The areas of application.
The scope of application within an area.
The degree of integration with other management information systems.
The degree of automation.
The development of a general system theory for project and business management.

The following paragraphs discuss each of these dimensions in turn.

AREAS OF APPLICATION

In addition to the applications discussed in Part III, the basic technique is being applied continually to many other new situations. And, it is safe to say that every industry and every government agency carries on some type of effort which could be improved by proper use of the network planning system. While the most obvious uses are in project management, other functional areas within an enterprise could also use network planning effectively.

Table 1 outlines seven functions: four of these are concerned with getting the job done, and the other three largely determine the working climate. While the terms used in Table 1 are applicable to a typical manufacturing company and to a typical military (U.S. Air Force) organization, equivalent functions exist in most production and service organizations: for example, the marketing function in industry is almost equivalent to the operations function in a military organization.

We can envision the use of network planning in every item listed in Table 1. However, this does not mean that network planning would necessarily be the major management tool used for each function. In certain functions, the system could form the base for an entire operation, even a repetitive one; in others, it could be used to plan particular projects within the total function. In most, it is useful as an analytical tool which will enable management to examine procedures in detail, to determine if better ones can be developed. Therefore, we can identify three modes of application: continuous or ongoing applications, one-time projects, and analytical ones. To illustrate these three modes, let us look at some brief examples.

Continuous Applications. In certain phases of financial functions, cycles are repeated over and over on a monthly basis. Books or accounts are closed, reports are made to feed other reports, reports are analyzed, etc. Management can develop a network plan of this monthly operation showing the required actions in proper sequence, and can use it to monitor progress as it is made, to effectively control operations for this phase of the financial function. (This is discussed in more detail in Chapter 13.)

One-Time Project. Almost every function requires the accomplishment of particular projects from time to time. A concerted advertising campaign promoting a new product is an example. The network for such a project should include subnets for engineering, production, marketing, finance, personnel (perhaps), external relations, and legal. Another example is the planning involved in the introduction of a major organizational change; this planning could be done with a network, to minimize disruption and lost motion during transition. The complicated legal and financial actions required to issue new stock could also justifiably be planned with a network. Other examples of the project-type application have been discussed in previous chapters.

Analytical Applications. Any phase of an operation can be laid out in a detailed network plan. The planner controls the degree of detail, and when he wishes to examine the operation microscopically, he merely continues to break down the network activities into smaller pieces. This procedure provides for a better analysis than the typical systems and procedures flow charts which lack the time dimension. For example, the flow of paper and actions in a sales operation can be planned in detail on a network, revealing areas for improvement; usually, these are self-evident once the network is drawn. In one large aerospace firm, the actions performed on an engineering change order were laid out in network fashion for analytical pur-

TABLE 1

Major functions of typical organizations

Typical manufacturing company	Typical military (U.S. Air Force) organization
Functions Concerned with Getting the Work Done	
A. Research and development	A. Research and development
1. Research	1. Research
2. Development	2. Development
3. Product engineering	3. Weapon engineering
B. Production	B. Materiel
1. Plant engineering	1. Civil engineering
2. Industrial engineering	2. Maintenance engineering
3. Purchasing	3. Procurement and production engineering
4. Production planning and control	4. Supply
5. Manufacturing	5. Manufacturing management
6. Quality control	6. Transportation
C. Marketing	C. Operations
1. Marketing research	1. Intelligence
2. Advertising	2. Plans and programs
3. Sales promotion	3. Military systems operations
4. Sales planning	
5. Sales operation	
6. Physical distribution	
D. Finance and control	D. Comptroller
1. Finance	1. Accounting and finance
2. Control	2. Budget and control
	3. Management analysis
	4. Financial data and services

TABLE 1 (Continued)
Major functions of typical organizations

Typical manufacturing company	Typical military (U.S. Air Force) organization

Functions which Determine the Climate in which the Work is Done

E. Personnel administration	E. Personnel
1. Employment	1. Recruiting and employment
2. Wage and salary administration	2. Personnel records
3. Industrial relations	3. Personal relations
4. Organizational planning and development	4. Manpower and organization
5. Employee services	5. Personnel services
F. External relations	F. Public information
1. Communication and information	1. Communication and information
2. Public activities coordination	2. Legislative liaison
	3. Public affairs coordination
G. Secretarial and legal	G. Administrative and legal
1. Secretarial	1. Administrative services
2. Legal	2. Judge advocate general
	3. Inspector general

poses. The effort required ten men and covered four months, and the resulting network stretched completely around a large conference room, strikingly illustrating why the company had so much trouble with change orders. Many drastic revisions in the procedures resulted.

In discussing additional uses, we must also consider the type of organization as well as the type of industry involved. Essentially, all industrial organizations can profit from network planning. Local, state, regional, and federal government organizations have myriad opportunities for improving their action planning, scheduling, and control by using the network system. (Related to these uses is the experience of a politician who used a network planning consultant to plan and manage his reelection campaign: he won the election). In short, wherever management, and more particularly, planning, is needed, the network can be of value.

SCOPE OF APPLICATION

A second dimension in which the network-based system will be increasingly used relates to the scope of the application in a particular situation. When networks are used for project or repetitive operations, they generally cover the hard core of the operation; for example, a developmental engineering project network will center on the engineering aspects, usually excluding related areas, such as procurement, special tooling, and production facility modification. Similarly, a construction project network will usually show the field operations and, perhaps, material purchasing and delivery, but will generally not include the actions of the owner, the architect, or the engineer or of the local government agencies which approve plans.

The future will bring an increase in the scope of application, as managers learn to broaden the network plan to relate *all* actions which contribute to the project into the plan. The benefit of the network-based system increases directly (and perhaps exponentially) as the scope of the network itself increases.

For example, network plans for construction projects usually center on the field work of excavation, forming, pouring, installing, and so on. The scope is increased by including purchasing actions, architect and/or owner approvals and actions, material deliveries, and many other items not visible in the field. One very broad scope network is a 700-activity plan developed by Informatics Inc., for the Donald L. Stone Company, homebuilders in San Jose, California, which covered all important financial, architectural, engineering, construction, and sales actions for the development of a residential housing tract. Unfortunately, such a network is impractical to produce in this book because of its size.

DEGREE OF INTEGRATION WITH OTHER SYSTEMS

A third way in which use of the network-based system will increase is in the degree to which it is integrated with other related information and management systems. In earlier chapters, the network-based system was related to other systems which make up the total management information system. At the present time, these various systems are not really integrated. Data are taken from one of these systems and transcribed in one manner or another to be fed into the related systems. This transcription often is done by taking parts of a finished report and filling in new forms by hand for manual analysis or for key-punching for computer analysis. It is rare for the computer to perform this transcription, as it does, for example, in a well-designed time/cost/manpower system which can read a tape produced by the cost-accounting system and select the information desired.

As we have mentioned before, the network-based system introduces the key elements of time and sequence of actions into all the related systems, revealing *what* actions must be accomplished, *when,* and in what *sequence.* More complete integration of these various systems would greatly improve the validity of the information in other systems, as well as in the network-based system itself. The specific pattern which this integration will follow is not clearly evident, however; perhaps the network-based system will form the end item, action-oriented core around which the other systems will be clustered, or perhaps, the accounting system will be the system at the center of the cluster. In any event, the artificial barriers blocking the free flow of data among these systems must and will be eliminated to achieve a total system. Specific points of attack for this purpose will include:

Master File. A total file of information encompassing all aspects of the organization and its business must be developed. This file would be maintained in sections, and so cross-references between sections would be required. An initial file would include the following major records:

Accounts.
Manpower.
Facilities.
Organization.
Markets.
Contracts.
Inventories.
Products.
Action Plans (Work Breakdown Structure, network plans).

Coding Schemes. Integrated and unambiguous coding schemes must be developed and used throughout the organization in order to use the master file concept effectively.

File Updating. A continuous flow of input data, acquired as automatically and painlessly as possible at the proper sources, must be used to update the master file. Source data collection devices already in use in production control will be used widely in other areas for this purpose.

Report Generation. Flexible, fast-response report generation, which selects the proper data from any segment of the master file and processes them as required, will enable managers to comb the total integrated system to obtain the necessary information for decision-making.

Degree of Automation. Automation will continue to spread in government and industry. The increased use at reduced cost of more powerful computers will enable companies to move rapidly towards the total integrated management information system. Large mass storage devices with direct access will use time-sharing to make apparently real-time response possible and source data collection devices to provide on-line file maintenance; these devices will vastly broaden and increase our knowledge of the way in which a business enterprise actually functions, and the way we should direct, manage, and control it. When network-based systems are used in the future, the completion of activities may be recorded through telephone-like devices (or more elaborate consoles such as the Stromberg Carlson Transactor) the moment the person responsible learns that the activity is complete. Cathode ray tube display of the network or any portion of it, to whatever detail desired, is now a reality; in the future, the manager will be able to call for the display of trouble areas, propose a corrective action, feed it to the computer through a typewriter keyboard or with a "light pencil" on the display scope, and obtain rapid computer analysis of his proposed action. If unforeseen and/or undesirable consequences result, he can keep revising the plan until he works out a solution acceptable to both the project and the entire organization.

DEVELOPMENT OF A GENERAL SYSTEM THEORY FOR PROJECT AND BUSINESS MANAGEMENT

The automated and integrated information system described in the preceding paragraphs will enable management to do significant research into the enterprise as a total system. The insight produced by such research, coupled with advancing operations research, management sciences, and so on (represented, for example, by Jay Forrester's "Industrial Dynamics"[1] concept), should permit the development of a useful general systems theory for project and business management.

Generally, the key impact of the networking disciplines will be on management thinking concerning the planning, processing, and control of work. We can expect increasing use of network techniques to go hand-in-hand with increasing use of the computer in business. In the past, much of the information available to top managers has often been too inaccurate, incomplete, or late to be used in decision-making; the new techniques will offer an increasingly "real-time" picture of operations. In the past, top managers have had to develop great talent for making intuitive judgments, or for assessing by rules of thumb, shrewd guesses and sharp feelings the variables and unknowns in planning and control; in the future, the computer and the network will force them to *think* more explicitly and analytically, to formalize their decision-making processes and spell out their judgments. This new mastery of operations will give top management more time for strategic work such as long range planning, policy making, staff selection, new program and product decisions, capital investments, financing, public relations, and labor relations. In short, top

---

[1]Jay Forrester, *Industrial Dynamics*, The M.I.T. Press, Cambridge, Mass., New York, 1961.

managers will be taking a more detailed interest in operations than they ever have before, without spending an undue amount of time on them.

The development of the general systems theory, linked with computer processing, continuous remote input, and dynamic display of management information will lead to more-or-less total systems which will integrate all data pertaining to the operation and provide almost any analysis, or answer any reasonable question. Thomas Whisler, of the University of Chicago Graduate School of Business, predicts that the trend toward such systems will result in a corporate planning technique which "will . . . become one of creative interrogation," as opposed to the subjective approaches used by today's manager.

Further, as continued efforts are made to describe and analyze operations using techniques such as network planning which have an almost limitless capability to integrate data, we can expect fundamental, irrevocable changes in the structure and functions of organizations, in the types of skills required to operate those organizations, and in the processes employed to get things done. As the present trend towards using network techniques more broadly in more diverse areas continues, new relationships will be found among these areas. A particularly interesting aspect of this will be the expanded use of sophisticated resource allocation techniques, which should have revolutionary effects in economic planning. Although network planning and its related disciplines are relatively new, they have such fundamental effects that we may conservatively expect them to have profound social, political, and economic impacts on a vast spectrum of human endeavor in the years to come.

SUMMARY

The application of network planning to many kinds of human endeavor provides a significant link between the dynamics of time-consuming actions and all other information related to the endeavor. In the future, network planning will be applied to new areas of endeavor, and its scope will increase within all areas of application. Integrated information systems will be made possible through master file definition, proper identification and coding to correlate all files, and automation of file updating and report generation. These developments will lead to the establishment of a general systems theory for business enterprises.

PART V

Appendices

APPENDIX A

Glossary of terms

Network-based systems, particularly PERT and CPM, have been responsible for coining a number of specialized terms and expressions which are encountered frequently. We have used some of these in this book, and so we are including a brief glossary of the most significant and widely used ones, arranged in alphabetical sequence.

Account code structure	The framework of numbers which follows the pattern of the Work Breakdown Structure, and is used for charging (charge numbers) and summarizing (summary numbers) the costs of a project.
Activity	An element of a project represented on a network by an arrow. An activity cannot be started until the event preceding it has occurred. An activity may represent: a process. a task. a procurement cycle. waiting time. In addition, an activity may simply represent a connection or interdependency between two events on the network.
Activity description	Definition of an activity which is identified by the predecessor and successor events.
Activity slack	The difference in time between the earliest completion time or date, (EEC) and the latest completion time or date, (LAC) for a given activity. The activity slack indicates the range of times within which an activity can be scheduled for completion. When the EEC for an activity is later than the LAC, the activity is said to have negative slack; either the current activities or subsequent activities must then be replanned or the project schedule will slip. When the LAC for an activity is later

than the EEC, the activity is said to have positive slack, which indicates that additional time is available for performng the activity without causing the project schedule to slip.

Activity start date   The expected calendar start date of the activity, based upon the expected occurrence (T_E) of the predecessor event.

Activity time   The estimate of the time required to complete an activity. Activity times are represented as:
   Most-likely time estimates.
   Optimistic time estimates (used only with the three-estimate concept).
   Pessimistic time estimates (used only with the three-estimate concept).

Actual man-hours   The actual man-hour expenditures incurred by or assigned to a summary cost category or work package.

Actual completion date The calendar date the activity is actually completed.

Actual cost (work performed to date)   The actual expenditures incurred, plus any prespecified types of unliquidated commitments (unliquidated obligations or accrued liabilities) charged or assigned to the work packages within the summary cost category.

Arrow diagram   The early CPM term for a network.

Backward pass   The reverse (right to left) computation of the latest allowable date (T_L) for each event and/or activity on the network, starting with the end event.

Beta time   Expected (or estimated) time (t_e) for an activity when three time estimates are used, calculated by the following formula:

$$t_e = \frac{a + 4m + b}{6}$$

   a = optimistic time estimate.
   m = most-likely time estimate.
   b = pessimistic time estimate.

Budget   The planned expenditures and commitments by time periods.

Burst point   Single events which have several activities succeeding them.

Charge number   The charge number (shop order number, work order number) used to identify the work package or summary cost category for estimating and accumulating of costs.

Commitment   A transaction which constitutes a firm open order placed for work to be accomplished, goods or materials to be delivered, or services to be performed, and for which the ordering source is thereby obligated to disburse funds.

Condensed network	A network in which selected events taken from a detail network are shown in relationship to one another. It accurately represents all of the characteristics of the detailed network with a reduced number of events. Lines connecting events on a condensed network may not be true activities, but are intended only to portray chronological interdependencies and restraints.
Constraint	A relationship of an event to a succeeding activity in which the activity may not start until the event preceding it has occurred. The term "constraint" is also used to indicate a relationship of an activity to a succeeding event in which the event cannot occur until all activities preceding it have been completed. A zero-time activity.
Cost activity	An activity related to the project that employs resources, the cost of which is normally directly charged to the project.
Cost category	The name and/or number of a functional cost category for which costs are to be summarized. See summary cost category.
Critical path	The sequence of activities and events which takes the greatest amount of time to complete and which has the greatest negative (or least positive) activity slack.
Cutoff date	The accounting cutoff date for the period of actual costs being reported.
Directed date (T_D)	The date on which management has directed that an event will be completed. The T_D may be assigned to any event on the network and may influence the computation of the T_L for the affected event(s).
Dummy activity	This is a network activity which represents a constraint, i.e., the dependency of a successor event on a predecessor event, but which does not have activity time, manpower, budget, or other resources associated with it. A dummy activity is illustrated on the network by a broken line.
Earliest expected time, or date, (T_E) (EEC)	The earliest time or date an event can be expected to occur. The T_E value for a given event or the EEC value for an activity is equal to the sum of the expected times (t_e) for the activities on the longest path from the beginning of the project to the given event or activity.
End items	The hardware, services, equipment, or facilities which the company has commited itself to deliver.
Estimate-at-completion	The estimated man-hours, costs, and time required to complete a work package or summary cost category. The estimate-at-completion amount represents the future effort in addition to that expended to date required to complete the work.

Estimate-to-complete	The cost estimated to be expended from the present to the end of the task or project being estimated, including labor, material, services, and commitments.
Event	A specific definable accomplishment in a project plan, recognizable at a particular instant in time. Events do not consume time or resources, and are normally represented in the network by circles or rectangles.
Event list	A form which lists all the information concerning the event, e.g., the title, contact, time the event occurs, notes, revisions, changes, security classification and issue dates. It is used as a catalog of events in support of networks.
Event slack	The difference between the earliest expected time or date (T_E) and the latest allowable time or date (T_L) for a given event. When the T_E for an event is later than the T_L, the event has negative slack. When the T_L is later than the T_E, the event has positive slack. When the T_E and T_L are equal, there is zero slack.
Expected (estimated) time (t_e)	The expected time for completing an activity. When using the three-estimate method, the t_e is a statistically weighted average time for an activity, incorporating the optimistic (a) most-likely (m), and pessimistic (b) time estimates: $$t_e = \frac{a + 4m + b}{6}$$
Float (CPM) (See also slack)	Total float is the spare time available when all preceding activities occur at the earliest possible times and all succeeding activities occur at the latest possible times. Free float is the spare time available when all preceding activities start at the earliest possible times and all succeeding activities occur at the earliest possible times. Independent float is the spare time available when all preceding activities occur at the latest possible times and all succeeding activities occur at the earliest possible times.
Forward pass	The forward (left to right) computation of the earliest expected time or date for each event (T_E) and/or activity (EEC) on the network, starting with the first event.
Individual cost activity	A cost activity that, by itself, constitutes a work package with identifiable resources.
Integration	A subdivision of the Work Breakdown Structure representing work directly connected with the process of integrating the major subassemblies or subsystems into the next highest assembly. Integration also refers to the consolidation of two

or more subnets representing portions of a total task into major consolidated networks.

Interdependency	The dependency of one event upon another, either within the same network or between two different networks.
Interface data	Those data which flow from one management organization to another.
Interface event	An event in a network which is also functional in one or more associated networks of the master network.
Latest allowable time or date (T_L) (LAC)	The latest time or date on which an event can occur without creating an expected delay in the completion of the project. The T_L value for a given event or the LAC for an activity is calculated by subtracting the sum of the expected times (t_e) for the activities on the longest path from the given event to the end event of the project, from the latest allowable date for completing the project. The T_L for the end event in a project is equal to the directed date (T_D) of the project. If a directed date is not specified, the $T_L = T_E$ for the end event.
Looping	Looping causes a backward or right-to-left flow on the network, which creates an endless circle or loop among events, and is therefore not allowed.
Merge point	An event which is a terminal point for several preceding activities.
Milestone	A key or major event in a project based on planned work accomplishment rather than a fixed time interval. Milestones are used to provide positive reporting points for effective management control. Interfaces are normally considered milestones.
Most critical slack	The worst (least algebraic) slack with respect to the designated program or project end-points for any of the activities on a network.
Most-likely time estimate (m)	The time which it is estimated that the activity will require to complete, assuming that planned resources are available, and all phases of the operation run normally, with only the usual obstacles.
Negative slack	A condition in which the earliest expected time or date (T_E) for an event is later than the latest allowable time or date (T_L) for the event, or in which the earliest completion time or date (EEC) for an activity is later than the latest completion time or date LAC for an activity.
Network	A flow plan of all the activities and events that must be ac-

complished to reach the project objectives, graphically depicting the planned sequences in which they are to be accomplished and their interdependencies and interrelationships.

Node	An event (early CPM term).
Optimistic time Estimate (a)	The time which an activity will require when all phases of the operation run smoothly, and a minimum of setbacks occur. In quantitative terms, this duration will be realized one time out of a hundred.
Other elements	Those summary cost categories on the Work Breakdown Structure, such as reliability, quality control, documentation, and field support for which costs are accumulated at the next highest hardware level because they are not readily identifiable to the specific hardware level at which they appear.
(Over-) under-plan	The planned cost minus the latest revised estimate. When the planned cost exceeds the latest revised estimate, there is a projected underplan condition. When the latest revised estimate exceeds the planned cost, there is a projected overplan condition.
(Overrun), underrun	The value for the work performed to date, minus the actual cost for the same work. When value exceeds actual cost, there is an underrun. When actual cost exceeds value, there is an overrun.
Pessimistic time estimate (b)	That time which an activity will take when phases of the operation encounter unusual difficulties or major setbacks. In quantitative terms, this duration will be realized one time out of a hundred.
Positive slack	The condition when the latest allowable time or date (T_L) for an event is later than the expected time or date (T_E) for the event, or when the latest completion time or date LAC for an activity is later than the earliest completion time or date EEC for the activity.
Primary slack (EEC)	The difference between the earliest expected time or date, (T_E) or EEC, and the latest allowable time or date, (T_L) or LAC for the event or activity, expressed in weeks.
Probability (P_r)	A function of the standard deviation and the difference between the scheduled and expected dates, expressed as the likelihood of meeting a scheduled date.
Reporting organization	The name or identification of the organization responsible for the work.
Resource (skill) code	The organization code for a particular skill.

Run date	The date on which a report was prepared or printed by data processing equipment.
Scheduled completion date (T_S)	The calendar date on which an activity is scheduled for completion. The T_S is established by management as an internal control on the completion of the work. (Where no specific date is assigned, the EEC $= T_S$.)
Secondary slack	The slack computed with respect to certain directed times or dates (T_D) in the body of the network.
Slack	The difference between the time expected and the time allowed for an event or activity. There are three conditions of slack: negative, zero, and positive. If the expected time exceeds the allowed time, there is negative slack. If the allowed time exceeds the expected time, there is positive slack. When the allowed time and the expected time are equal, there is zero slack. The slack along a given path calculated from the terminal point of the path is the primary slack. Within the network, events may have scheduled or directed dates assigned to them. During the backward pass calculation, slack can be computed using these scheduled dates as well as using the terminal event. The slack computed using certain scheduled or directed dates in the body of the network is called secondary slack.
Subnetwork, subnet	A subdivision of a major network. Interrelationships between subnetworks for consolidation or integration into the major network are maintained by common interface events.
Subsystem	The major items of hardware making up the total system. Subsystems may occur at several Work Breakdown Structure levels.
Successor event	The ending event for an activity.
Summary cost category	The functional level appearing below the hardware levels on the Work Breakdown Structure. Typical summary cost categories are design, fabrication, and tests.
Summary network	A summarization of detailed networks. Normally, only milestone and interface events are plotted on summary networks.
Summary number	A number used to identify the work package to be grouped together in summarizing man-hour and cost information for an item appearing on a higher level of the Work Breakdown Structure.
Symbols	$a =$ optimistic time estimate for an activity. $b =$ pessimistic time estimate for an activity. $m =$ most-likely time estimate for an activity. $EEC =$ earliest completion time or date for an activity.

LAC = latest completion time or date for an activity.
Pr = probability.
T_A = actual completion time or date for an activity.
T_D = directed date for the occurrence of an event.
T_E = earliest expected time or date for an event.
t_e = expected (estimated) time for an activity.
T_L = latest allowable time or date for an event.
T_S = scheduled completion date for an activity or event.
t_s = scheduled time for an activity.
σ = (sigma) mathematical symbol for standard deviation.
(σ^2) = An expression of the degree of uncertainty (variance) associated with the derivation of the expected time.

Time now	The current calendar date, used primarily when updating networks.
Top level network	A network prepared at the beginning of a project by the program manager to portray the major events and activities required to accomplish the total task.
Underrun	The amount by which the current approved estimate exceeds the sum of the actual costs and the estimates-to-complete.
Value (work performed to date)	The planned cost for completed work, including that part of work-in-process which has been finished. This value is determined by summing the planned cost for each completed work package. If a work package is in process, the part of its total planned cost which applies to work completed is approximated by applying the ratio of actual cost to the latest revised estimate for that work package.
Work Breakdown Structure (WBS)	The progressive breakdown of the project end item into smaller and smaller increments, to the lowest practical level to which PERT/Cost is to be applied. The Work Breakdown Structure graphically depicts the summary cost categories applicable to each major subassembly or the subsystem.
Work Package (wp)	The division of work resulting in a manageable task from the viewpoint of cost, time, and functional responsibility. The work package begins and ends with real events on the network.
Zero-cost activity	An activity that represents a relationship of precedence or dependence but which does not generate direct project costs. A zero-cost activity is normally represented by a broken line on the network. (See dummy activity.)
Zero slack	A condition when the latest allowable time or date (T_L) for an event is equal to the earliest expected time or date (T_E) for an event or when the latest completion time or date LAC for an activity is equal to the earliest completion time or date EEC.

APPENDIX B

Precedence diagramming—an alternate method of network construction[1]

Several network planning practitioners have simultaneously developed similar methods of preparing network plans which have certain advantages over the event-activity method presented in this book. This is called by various names, including precedence diagramming and activity-on-node diagramming. The basic concept is that the activities (not events) are placed within a circle or square, and the dependencies between activities are shown by lines or arrows. This is illustrated in Figure 1.

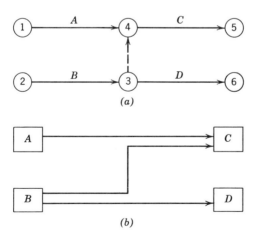

FIGURE 1. Comparison of event-activity network with precedence diagram. (a) Event-activity network portion; (b) the same network portion in precedence diagram form.

---

[1] Much of this material is based on the work of J. David Craig, IBM Corp., Houston, Texas.

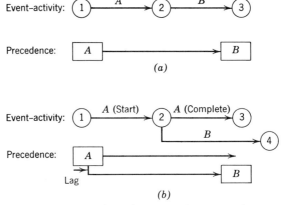

FIGURE 2. (a) An activity must precede another activity; (b) a portion of an activity must be complete before a succeeding activity can start; completion of the preceding activity has no effect on the completion of the succeeding activity.

ADVANTAGES

Precedence diagramming has certain distinct advantages over the event-activity network, as well as certain disadvantages. The advantages include:

1. Elimination of dummies or restraints.
2. Ease and rapidity with which field or operating personnel often grasp the concept.
3. Simplification by the elimination of events.
4. Ability to show lead or lag times, thus eliminating the need for breaking up activities merely for network construction purposes (this is described more fully later on). This usually reduces the number of activities in the network.

DISADVANTAGES

Compared to event-activity networks, precedence diagramming has the following disadvantages:

1. For many applications, events are of great importance. Precedence diagramming eliminates events.
2. Networks cannot be integrated through the use of interface events.
3. Specialized computer programs are necessary, and only one is known to be widely available (for the IBM 1440 Computer).
4. Path tracing is difficult, since the linkage between event numbers is not present.

ELIMINATION OF DUMMY CONSTRAINTS

This is one of the most important advantages of precedence diagramming, and is coupled with the ability to show lead and lag times. Figure 2 illustrates the four basic

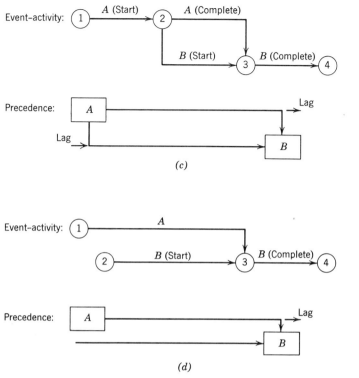

FIGURE 2. (Cont.) (c) A portion of an activity must be complete before a succeeding activity can start, and the preceding activity must be complete before the succeeding activity can complete. (d) The start of a preceding activity has no effect on the start of a succeeding activity but the succeeding activity cannot complete until the preceding activity is complete.

relationships between activities in any project. The possible advantages in simplifying certain types of networks via precedence diagramming are illustrated by Figure 3.

David Craig explains the major advantages of precedence diagramming as follows:

"Theoretically, there is nothing wrong with breaking an activity down into several parts as is required by the arrow diagramming technique. However, this is where confusion arises at the field level. An activity, whether it is trench excavation, pouring concrete, or erection of steel is really a continuous operation performed by one crew operating continuously from the beginning of the activity to its completion. The foreman recognizes it as only one job; yet he may have to show it on the diagram as two activities or more in order to properly show the activity relationships. This causes confusion since the foreman must report 100% complete on an activity when he knows the total activity is only 5% complete.

"Too, if he is trying to associate costs with the arrow diagram, he will either have to

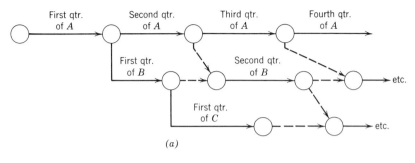

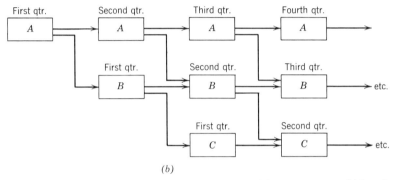

FIGURE 3. (a) Event activity network requiring complex use of dummy constraints. (b) Precedence diagram of same network.

report costs to all of these sub items or use some form of combining activities as is done in the PERT/Cost system.

"Let's consider a typical construction project involving the pouring of a concrete street and see how the different systems would perform. Normally the following activities would be involved:

Fine Grading
Form Setting
Placing Reinforcing Steel
Pouring and Finishing Concrete
Concrete Curing
Form Stripping

"For the project to be performed efficiently many activities must be overlapped. The form setters, for example, could start setting the forms before all of the fine grading is complete. Figure 4 shows how one contractor might plan this project using an arrow diagram. The diagram shows 13 work activities represented by the solid lines and two delays represented by dotted lines (delays are like dummy jobs, since work may not be involved, but unlike dummies in that time is associated with them).

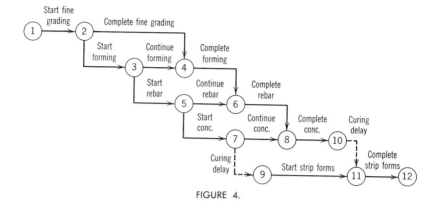

FIGURE 4.

"Figure [5] shows the same work sequence by the precedence system. In the precedence diagram, we only have five activities and two delays. However, the delays are implied in the input to the computer and do not show in the printout of the schedule.

"Precedence diagraming by itself would not allow the foregoing to take place, since it would be restricted by the same scheduling problems as arrow diagramming. However, since the precedence system requires only one number for each activity, and since all preceding jobs for each activity must be listed as input data to the computer, a network can be developed in such a manner that a percentage, a quantity, or a portion of the duration of an activity can precede another activity.

"Table I demonstrates how the street paving problem would be input to the computer. As can be seen in the "preceded by" column, activity 200, Set Forms, is preceded by 10% of activity 100, Fine Grade. Activity 300, Set Rebar, is preceded by 200 linear feet of activity 200, Set Forms. Activity 400, Pour Concrete, is preceded by one day of activity 300, Set Rebar. All of these activities operate under the relationship shown in Figure 2c. This means that both the start and finish of these activities are re-

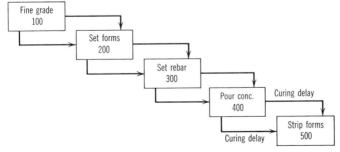

FIGURE 5.

TABLE 1

Computer input based on the relationships shown in Figure 2c

Std. Code	Description	Duration	Preceded by
100	Fine Grade	15	—
200	Set Forms	6	0.10–100
300	Set Rebar	10	200 LF–200
400	Pour Concrete	10	1 Day–300
500	Strip Forms	3	4 Day–400

stricted by the percent, quantity, or time of the preceding activity. For instance, activity 400, Pour Concrete, which is preceded by one day of activity 300, Set Rebar, would start one day after Set Rebar started and would finish one day after Set Rebar completed. In effect, precedence diagramming allows the user to network diagram by the old bar chart method, while maintaining all of the Critical Path Method. It allows this in such a way that the average construction foreman can accurately draw a network that truly represents his thinking with little possibility of error.

"Further, it allows the capturing of costs on a one cost per activity basis by simplifying the reporting required under the arrow diagram procedure and minimizing the number of activities."

CONCLUSIONS

Precedence diagramming has distinct advantages over event-activity networking for applications such as construction projects which are completely activity-oriented. The major benefits are better understanding by field personnel, a reduction in the number of activities in the network, and the elimination of dummy (zero-time) constraints. Limitations include the elimination of events and the difficulty in tracing slack paths using computer reports because of the absence of linking event numbers. In working with precedence diagrams, the authors have found that the total number of activities in a network can be reduced by 20 to 30 percent, primarily due to elimination of dummy constraints.

APPENDIX C

An analytical study
of the PERT assumptions[1]

During the past several years, new techniques based on network models have been developed to aid management in planning and controlling large-scale projects. One such technique, which is discussed in this paper, is PERT (Program Evaluation and Review Technique). The PERT technique has received widespread interest and is currently being used for many types of projects.

Network Models

Network Structures. In general, the type of project for which PERT is often used comprises numerous activities—sometimes thousands—many of which may be interrelated in complex, and often subtle, ways. One of the significant features of PERT and other similar techniques is that the activities as well as the interrelations are depicted in their entirety by a network of directed arcs (arcs with arrows, which denote the sequence of the activities they represent). The nodes, called events, represent instants in time when certain activities have been completed and others can then be started. All inwardly-directed activities at a node must be completed before any outwardly-directed activity of that node can be started. A path is defined as an unbroken chain of activities from the origin node (the beginning of the project) to some other node. An event is said to have occurred when all activities on all paths directed into the node representing that event have been completed.

Time Element in Networks. Each activity takes time to perform. Thus, it will have some duration associated with it. The time at which an event occurs is the maximum of the durations of the inwardly-directed paths to that event, since all of the activities directed into the event must have been completed. The project duration is, then, the maximum of the elapsed times along all paths from the origin to the terminal

[1]This appendix is a reprint of a paper which is a condensed and revised version of a report written while the authors, Kenneth R. MacCrimmon, of the Carnegie Institute of Technology, and Charles A. Ryavec, of the University of Michigan, were with The Rand Corporation. It was sponsored by the U.S. Air Force under Project Rand.

node (the event marking the completion of the project). The path with the longest duration is called the 'critical path,' and the activities on it, 'critical activities.' Any delay in a critical activity will obviously cause a corresponding delay in the entire project.

The duration associated with an activity can be a single number (the deterministic case), or, as in PERT, it can be a random variable with some distribution (the stochastic case). The times used for each activity duration are based on time estimates made by the managers or engineers most directly concerned with the performance of the activity.

The PERT Procedure

The study reported herein will deal with those aspects of the PERT procedure which come after the establishment of the network itself. The following analysis assumes that a unique network representation has been established. However, whether or not a unique network representation can be established a priori is open to question. Nearly all the assumptions made in the PERT mathematical model will be analyzed in this paper. It has been found insightful to consider the assumptions on two separate levels—the level of the individual activities and the level of the whole network.

I. ACTIVITIES

Uncertainty in Activity Duration

Uncertainty in Research and Development Activities. The activities in complex research and development programs are usually unique to a particular program and are seldom of a routine or repetitive nature. Those people most directly involved in the performance of these activities, however, usually have some experience in doing similar jobs. Thus, on the basis of their experience, it is felt that they can estimate how long some new activity will take to complete. On the other hand, the activities often require creative ability—something which is hard to measure in individuals. By the nature of these activities, then, any estimate of their length must be an uncertain one.

A Stochastic Model to Reflect Uncertainty. In order to reflect this uncertainty, a stochastic model may be used; that is, one in which some measure of the possible variation in activity duration is given. This may take the form of a distribution showing the various probabilities that an activity will be completed in its various possible completion times. Alternatively, it may just be some number that represents the standard deviation, range, or some other concept of variation. This latter method would not make any assumption about a distribution form.

PERT Activity Duration—The Beta Distribution and Its Parameters. PERT handles uncertainty by assuming that the probable duration of an activity is beta-distributed. The probability density function of the beta distribution is $f(t) = K \cdot (t - a)^\alpha \cdot (b - t)^\beta$. A few examples are plotted in Figure 1.

In order to determine a unique beta distribution, the endpoints a and b, and the exponents α and β must be specified. PERT uses two time estimates (the optimistic time and the pessimistic time) to specify a and b. The optimistic time is that time

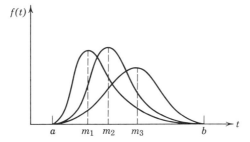

FIGURE 1.   Examples of beta distributions.

earlier than a time in which the activity could not be completed, and the pessimistic time is the longest time the activity could ever take to complete (barring 'acts of God'). A third time estimate m, the most likely time, is also obtained. The value of m is the mode of the distribution, and this value, in conjunction with the PERT assumption that the standard deviation of the distribution is ⅙ of its range, serves to determine the two parameters α and β.

PERT Activiy Duration—Mean and Standard Deviation. It is often convenient to consider only the mean and the standard deviation of a distribution, rather than the entire distribution. Sometimes these two values determine a unique distribution, as in the case of a normal. With other distributions, such as the beta, the mean and standard deviation alone do not determine a unique distribution. Even though PERT deals with a beta distribution, it is convenient to characterize the activity duration in terms of a mean and standard deviation. As noted above, the standard deviation is assumed to be ⅙ $(b - a)$. In general, the determination of the mean involves the solution of a cubic equation. Various values of the mean were calculated from the roots of cubic equations, and in order to simplify the future calculations of activity means, a linear approximation, ⅙ $(a + 4m + b)$, to these values was made. These expressions for the mean and standard deviation are used to represent the activity duration in all future PERT calculations.[2]

Actual Activity Distributions. Although the PERT model makes specific assumptions about the form of the activity distributions, the true distributions are unknown. However, once an activity has been specified precisely, the distribution of that activity's duration has, thereby, been determined (although the distribution may be, and probably is, unknown). To the extent of the authors' knowledge, no empirical study has been made to determine the form of activity distributions. Indeed, there would be many problems connected with such a study, not the least of which would be the

---

[2]Although these two expressions were derived from a beta distribution, the PERT literature is inconsistent about whether the activity durations are now normally or beta-distributed. This inconsistency appears in the original PERT report.[6] Appendix A of that report centers its discussion on a normal distribution (p. A2: "Each activity has a *time*. The time is stochastic and normally distributed . . ."), whereas Appendix B discusses the beta distribution (p. B6: "As a model of the distribution of an activity time, we introduce the beta distribution . . .").

nonrepetitive nature of the activities. The choice of a particular distribution, such as the beta, while seeming rather arbitrary, does have certain features that an actual activity distribution could be expected to possess.

Expected Properties of the Actual Activity Distribution. Three properties that might be postulated for an actual activity distribution are unimodality, continuity, and two nonnegative abscissi intercepts. If the probability that an action will be completed in some small interval around some intermediate value of the activity duration is greater than the probability in a similar interval around some other point, then unimodality is a reasonable assumption. Even if an unknown distribution is discrete, a continuous distribution generally serves as a good approximation. The assumption that the distribution touches the abscissa at two nonnegative points reflects the property that an activity cannot be completed in a negative time.

Possible Activity-Based Errors

Three possible sources of error (caused by the PERT assumptions) in the PERT calculations of activity means and standard deviations will be considered:

1. The true distribution of an activity (and its mean and standard deviation) is probably not known. Given that the distribution is continuous, unimodel, and that it touches the abscissa at two nonnegative points, how much of an error would be introduced into the over-all PERT calculations of an activity mean and standard deviation by the assumption that the activity duration is beta-distributed?

2. If it is assumed that an activity distribution is a beta, and that the expression for this function is known exactly, what errors are introduced into the PERT calculations by the assumption $\sigma_e = \frac{1}{6} (b - a)$ and the estimate $t_e = \frac{1}{6} (a + 4m + b)$, if a, m, and b are known exactly?

3. Finally, if it is assumed that an activity is beta-distributed, with mean and standard deviation given by $\frac{1}{6} (a + 4m + b)$ and $\frac{1}{6} (b - a)$, respectively, what error can be introduced into the PERT calculations if the estimates of a, m, and b are inexact?

Possible Error Introduced by the Assumption of a Beta Distribution. If the actual activity distribution possesses the aforementioned three properties (i.e., unimodality, continuity, and two nonnegative abscissa intercepts), then the beta approximation to this distribution is at least correct with regard to its general shape. Different distributions, which possess these properties, however, could well have very different means and standard deviations; and hence—at least theoretically—an imprecise knowledge of the actual activity distribution could contribute significantly to any over-all error between the PERT-calculated mean and standard deviation of an activity and its actual mean and standard deviation.

Consider, for example, three distributions shown in Figure 2. Each of these distributions possesses the three properties discussed previously; and if the use of a beta distribution, D_1, proceeded from intuitive grounds—from the belief that the actual activity distribution should satisfy the aforementioned three properties—then it can be assumed that D_2 and D_3 can also be possible activity distributions. With this assumption, the extent and direction of any possible errors resulting from the

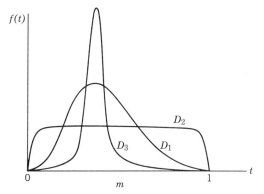

FIGURE 2. Examples of possible activity distributions.

use of a beta distribution can be determined. The three distributions have the range $[0, 1]$[3] and have their modes at m. We further assume $0 \leqq m \leqq \frac{1}{2}$.

D_1 represents a beta distribution with a mean of $\frac{1}{6}(4m + 1)$ and a standard deviation equal to $\frac{1}{6}$ of the range (the standard PERT assumptions). D_2 represents a quasi-uniform distribution. Therefore, its mean and standard deviation will be very close to $\frac{1}{2}$ and $\sqrt{\frac{1}{12}}$, respectively. D_3 is a quasi-delta function with its mean very close to its mode, and its standard deviation very close to zero. Although D_2 and D_3 are extreme examples of possible activity distributions—and hence rather unlikely—they serve to put bounds on possible errors in the calculation of an activity mean and standard deviation caused by the use of an incorrect activity distribution.

Under these conditions, the worst absolute error[4] in the mean is

$$\max \left[|\tfrac{1}{6}(4m + 1) - \tfrac{1}{2}|, |\tfrac{1}{6}(4m + 1) - m| \right] = \tfrac{1}{3}(1 - 2m)$$

The worst absolute error in the standard deviation is

$$\max \left[|\sqrt{\tfrac{1}{12}} - \tfrac{1}{6}|, |0 - \tfrac{1}{6}| \right] = \tfrac{1}{6}$$

It may be noted that the possible error in the mean is a function of the mode. If the mode is near the endpoint of the distribution, the error could be as much as 33 per cent. If the mode is more centralized, say $|\tfrac{1}{2} - m| \leqq \tfrac{1}{6}$, then the error could be around 11 per cent. The worst absolute error in the standard deviation, about 17 per cent, does not depend upon the mode.

The above expressions give the worst *absolute* error, but errors can be both positive and negative; and, thus, it could be expected that some degree of cancellation would occur when the individual activities were combined in series in a network. The extent, and the net result of such cancellation, are dependent on three factors:

---

[3]Zero and one were chosen as endpoints of the range for computational ease. The results, expressed as a per cent or proportion of the range, can be extended to an arbitrary range $[a,b]$.
[4]For simplicity, errors in formulas are given as a proportion, rather than a per cent, of the range.

(1) the number of activities in series, (2) the ranges of the activity durations, and (3) the skewness of the activity distributions. If, in a network, there are a large number of activities in series, their ranges are about equal, and the extent and direction of their skewness is arbitrary, then a relatively high degree of cancellation can be expected.

It is often noted in practice, however, that the skewness of the activities tends to be biased to the right (i.e., the mean is to the right of the mode), and that the range of activity durations can differ by an order of magnitude. Moreover, many networks have a large number of activities in parallel, thus offering no chance for error cancellation. For these reasons there may be little cancellation.

Possible Error Caused by the Standard Deviation Assumption and the Approximation of the Mean. The possible errors resulting from the assumption $\sigma_e = \frac{1}{6}(b - a)$ and the approximation $t_e = \frac{1}{6}(a + 4m + b)$ will be analyzed by comparing them with the *actual* values for the mean and standard deviation, assuming a beta distribution (on $[0, 1]$). The expressions for the actual mode, mean, and standard deviation are:

Mode: $m = \alpha/(\alpha + \beta)$,
Mean: $\mu = (\alpha + 1)/(\alpha + \beta + 2)$,
Standard Deviation: $\sigma = \sqrt{(\alpha + 1)(\beta + 1)/(\alpha + \beta + 2)^2(\alpha + \beta + 3)}$.

The mean and the standard deviation may be rewritten as functions of α and m, and these appear as the second terms in the error expressions below. The worst absolute error in the mean is

$$|\tfrac{1}{6}(4m + 1) - m(\alpha + 1)/(\alpha + 2m)|.$$

The worst absolute error in the standard deviation is

$$|\tfrac{1}{6} - \sqrt{m^2(\alpha + 1)(\alpha - \alpha m + m)/(\alpha + 2m)^2(\alpha + 3m)}|$$

The worst absolute error in the mean can be 33 per cent, and in the standard deviation 17 per cent. This occurs for extreme values of α and m. If we assume $1 \leq \alpha \leq 6$ and $|\tfrac{1}{2} - m| \leq \frac{1}{6}$, then the errors in the mean and standard deviation reduce to 4 per cent and 7 per cent, respectively.

Possible Error in the Three Time Estimates. Even if the random variable representing the duration of an activity is assumed to be beta-distributed, it is highly unlikely that any procedure could be devised to determine the exact parameters of the distribution, since, ultimately, any such procedure must rely on human estimates. Thus it is desirable to determine the contribution to the error in the PERT-calculated mean and the PERT-assumed standard deviation resulting from the PERT-estimating procedure itself.

In order to determine the magnitude and direction of possible errors in the estimates, it is assumed that the values a, m, and b are the actual values of the lower bound, mode, and upper bound, respectively, of a beta distribution. The estimates of these values are t_a, t_m, and t_b and it will be assumed that they could be incorrect to the following extent: $0.8a \leq t_a \leq 1.1a$; $0.9m \leq t_m \leq 1.1m$; $0.9b \leq t_b \leq 1.2b$. This is depicted in Figure 3.

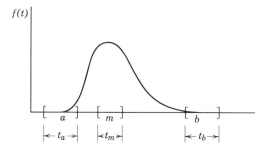

FIGURE 3.   Beta distribution with assumed errors in a, m, and b.

The sensitivity of the PERT expressions $t_e = \frac{1}{6}(a + 4m + b)$ and $\sigma_e = \frac{1}{6}(b - a)$ to incorrect estimates of a, m, and b can be seen below. We again assume that $a \leqq m \leqq \frac{1}{2}(a + b)$. The worst absolute error in the mean is

$$\frac{1}{b-a}\max\left[\left|\frac{(0.8a + 3.6m + 0.9b) - (a + 4m + b)}{6}\right|,\right.$$

$$\left.\left|\frac{(1.1a + 4.4m + 1.2b) - (a + 4m + b)}{6}\right|\right] = \frac{1}{60}\left[\frac{a + 4m + 2b}{b - a}\right]$$

The worst absolute error in the standard deviation is

$$\frac{1}{b-a}\max\left[\left|\frac{(0.9b - 1.1a) - (b - a)}{6}\right|, \left|\frac{(1.2b - 0.8a) - (b - a)}{6}\right|\right] = \frac{1}{30}\left[\frac{b + a}{b - a}\right]$$

Note that the value of the mode again affects the error in the mean, but not the standard deviation.

Summary of the Activity Section

As has been shown, the three factors discussed previously can each cause absolute errors in the PERT-calculated mean and the PERT-assumed standard deviation on the order of 30 per cent and 15 per cent of the range, respectively. The possible error caused by one of these factors—the human estimates of a, m, and b—was based on the assumption that these estimates would be incorrect to only a certain extent, i.e., ± 10 or 20 per cent of the range. These figures are thought to be conservative (unless the individuals are working to schedule), and the degree to which the estimates of a, m, and b are imprecise will vary with each activity and the particular individuals involved. Thus, the errors resulting from imprecise time estimates will likely be larger than the results obtained in this paper.

On the other hand, the errors in the mean and standard deviation can be either positive or negative, so some degree of cancellation can be expected to occur when all the activities are combined in a network.[5] Furthermore, since many of the cases considered—although theoretically possible—are rather extreme, the errors may be reduced from the 30 and 15 per cent stated above, to perhaps 5 or 10 per cent.

Addendum. It is interesting to note that the error analysis would have yielded approximately the same results if the PERT model had employed a triangular distribution instead of a beta. In addition, with a triangular distribution there would be no error in the expressions of the mean and standard deviation, since the mean and the standard deviation of a triangular distribution are given *exactly* by $t_e = \frac{1}{3}(a + m + b)$ and $\sigma_e = \sqrt{\frac{1}{18}[(b - a)^2 + (m - a)(m - b)]}$. These values would be used rather than the approximations used now.

Another possible factor in favor of the triangular distribution is that the possible range of its standard deviation, $(b - a)/\sqrt{18}$ to $(b - a)/\sqrt{24}$, is more centralized than the PERT-assumed standard deviation, $\frac{1}{6}(b - a)$, is in the total range of possible standard deviation, 0 to $(b - a)/\sqrt{12}$ (assuming, again, the three properties discussed previously).

When the mode and the range of a triangular distribution are specified (for example, by three time estimates—a, m, and b), the entire distribution is then determined. This, as we have noted previously, is not the case for the class of beta distributions. The class of beta distributions, then, is basically more flexible in being able to handle more activity data. However, the PERT procedure does not take advantage of this extra flexibility, since it is assumed that $\sigma_e = \frac{1}{6}(b - a)$.

Therefore, since there is no a priori justification for either function as an activity distribution, and since the actual standard deviations are unknown, the fact that the mean and the standard deviation can be given exactly for a triangular distribution make it an equally meaningful, and more manageable, distribution. It would be equally meaningful if its mean and standard deviation were used in a similar way to the approximate expressions used now; it would be more manageable if it was necessary to use the whole distribution, say in an analysis or a Monte Carlo study.

II. THE NETWORK

Network Considerations

Up to this point, this study has been on the level of the individual activities within a PERT network. Now attention will be directed to the network as a whole. After the mean and standard deviation of each activity have been computed, they can be used to determine some measure of the criticalness of all activities taken together and to aid in the estimation of the completion time distribution of the whole project.

As it has been shown, the possible errors in the individual activities could, by themselves, cause errors in the calculation of a project mean and standard deviation, although the extent and direction of these errors might be difficult to determine. However, even if the PERT data (i.e., the mean, standard deviation, and distribution) obtained for each activity are correct, significant errors can still be introduced into

---

[5] A Monte Carlo analysis of the effects of the three types of activity errors on actual networks might give some idea of the extent of possible cancellation, since the individual activity errors themselves could be calculated by the methods presented here.

the calculation of a network mean and standard deviation. As a result, probability statements concerning the various completion times of a project can also be incorrect.

The Project Distribution—PERT and Actual

The PERT procedure for obtaining the project completion time distribution may be stated as follows. Assume that in a network there are n different paths $P_1, P_2, \cdots$, P_n, which connect the origin node and the terminal node. Let $p_1, p_2, \cdots, p_n$ denote the n random variables that represent, respectively, the durations of the n paths P_1, $P_2, \cdots, P_n$. One of these random variables, say p_1, will have an expected value that is not less than the expected value of the other $n - 1$ variables, $p_2, p_3, \cdots, p_n$. Let the expected value and the standard deviation of p_i be $E(p_i)$ and σ_i, respectively. Thus, $E(p_1)$ is the expected or mean duration of path P_1, and σ_1 is its standard deviation.

PERT now uses $E(p_1)$ and σ_1 as the project mean and standard deviation, and assumes that the project duration is normally distributed and given by $f_p(t) = K\exp(-[t - E(p_1)]^2/2\sigma_1^2)$. Path P_1 is called the critical path.[6] In actuality, the project distribution is given by $F(t) = \Pr(\max_i p_i \leq t)$. Clearly, the expected value of the random variable $u = \max_i p_i$ is not less than the expected value of any one of the p_i. Hence, the PERT-calculated mean is generally less than, and never greater than, the true project mean. In general, the PERT-calculated standard deviation is greater than the actual standard deviation. If the distributions are symmetric, and with compact support, the standard deviation of the random variable u will be less than any of the σ_i. However, if the distributions are considerably skewed to the right (such as e^{-t}) the reverse may be true.

Computational Difficulties. In order to determine the error in the mean and standard deviation made by PERT in a particular network, it is necessary to calculate the actual project means and standard deviations from the data of each of the activities. The procedure used in this study to obtain the project mean and standard deviation relies exclusively on the calculation of an exact project distribution from the individial activity distribution by analytic methods.[7] Such a calculation is extremely difficult in all but a few simple networks, regardless of the distributions on the activities themselves. These difficulties are discussed in Appendix G of reference 4.

To get a feeling for the errors PERT makes by assuming that the project mean and standard deviation are given by $E(p_1)$ and σ_1, respectively, various simple networks are analyzed both analytically and according to the PERT procedure. In these calculations, it is assumed that the individual activity distributions are known exactly. This allows the determination of the errors made on the network level alone, without confounding them with possible errors made in the activities. Since calculations with beta and other continuous distributions are rather lengthy, the distributions used in the network analyses are, in general, discrete. Some results, however, have been obtained for the beta, uniform, and normal distributions.

---

[6] If there is more than one path with the largest expected value, PERT labels them all as critical paths, and uses the one with the largest standard deviation as P_1.

[7] An alternative method, which could provide a close approximation, would be to use Monte Carlo techniques.

Criticalness

To obtain a measure of the criticalness of each activity, PERT uses the critical path concept discussed above. Criticalness of an activity is a measure of the relative importance of the activity to the one-time completion of the over-all project. Some activities can obviously be delayed without delaying the project, while others cannot.

In PERT, only the means of the activity durations are used in determining the critical path. The stochastic element—the variance of the activity duration—is not incorporated. Thus, the model is reduced to a deterministic form. In a deterministic model (where no uncertainty in the activity durations is recognized) the longest path can be calculated by simple addition.

In a stochastic model, each path has a specific probability (in general, nonzero) of being the longest path at any particular time. However, if the network is large, the probability that any given path is the critical one may be very small. (An analogous situation would be one where a coin was tossed 1000 times. The most probable number of heads is 500, but the probability of getting exactly 500 heads is very small.) Thus, the most probable critical path may occur only rarely, and an activity that has a high probability of being on a longest path may not be on this most probable critical path. The following example may clarify these points.

Example of the Critical Path and the Critical Activity Concept. Consider the network depicted in Figure 4, with the customary time estimates of a, m, and b shown beside the corresponding activity. Using the PERT-calculated mean times (given in the circles below the activities), the PERT procedure would choose $ABDF$ as the critical path because it has the maximum sum of means (13). If the activity time estimates are assigned equal weight, for computational simplicity, calculations will show that path ABF has the probability 0.30 of being the longest path, and this is a larger probability than any of the other three paths. The probability of each activity being on the longest path is: AB, 0.58; AC, 0.42; BF, 0.30; $BD(F)$, 0.27; CF, 0.24; $CE(F)$, 0.19. Note that although path ABF is the most probable longest path, it does not contain activity AC which is more critical than activity BF, which is on this most prob-

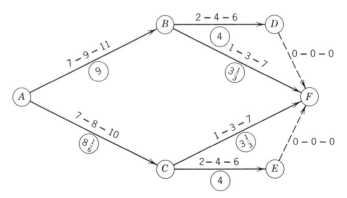

FIGURE 4.   PERT network showing activities with associated times.

able longest path. Similar results could be obtained by using distributions other than a uniform, such as a beta.

This example suggests that a critical activity concept may be more valid in a stochastic model than a critical path concept, especially since the PERT-calculated critical path is not even necessarily the most probable longest path. The computation of some sort of index of criticalness, such as that indicated above, would supplement (or possibly replace) the slack determination, as done in the present PERT procedure.

Possible Network Configuration-Based Errors—Some Examples

The possible errors in PERT networks depend on the particular network configurations; so generalizations can be made only to a very limited extent. The examples studied in this paper are of a very simple form because of the computational problems discussed earlier, and because larger networks are really no more 'typical' than small networks. They were chosen because they possessed some of the properties that may cause significant errors in regular PERT networks. The emphasis is on the factors causing the errors, the direction of the errors, and the magnitude of them. The subsection on network decomposition discusses the possible application of the results to much larger networks, such as those found in practice.

Examples—Simple Series and Parallel Networks. The first network configuration to be considered—a simple path—is one in which PERT makes no errors in calculating the project mean and standard deviation. This case, depicted in Figure 5, will actually occur when there is one path through a network that is so much longer than any of the other paths that all other paths have no effect whatsoever on the determination of the project completion time distribution. The Central-Limit Theorem is applicable here, and the correct way to obtain the project mean and standard deviation is by adding the activity means and variances (squared standard deviations), along this (critical) path, the same procedure that PERT uses. Thus, whenever the network reduces to one very much longer[8] path, the only PERT errors that can occur will be on the level of the individual activities.

Consider next the case where there are two paths of approximately the same length through the network. This results in the parallel configuration shown in Figure 6. One may wish to view a large number of intermediate nodes on each path (however,

FIGURE 5.   Series path.

---

[8]The meaning of 'very much longer' will be discussed in the subsections on the effect of slack and on network decomposition.

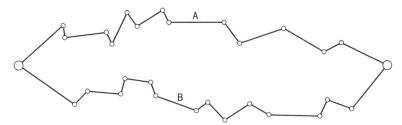

FIGURE 6. Two parallel paths.

there cannot be any connection between any of the nodes of the two paths). The PERT procedure will take as the mean and standard deviation of the project, the sum of the means and the square root of the sum of the variances along that path with the largest mean. However, if the other path has a mean very close to the first (for a limiting case assume they are equal), the activities on this second path (the one PERT ignores) will also be a major determinant of the project completion time distribution.

As an example, if paths A and B are beta-distributed on the interval $[0, 1]$ with parameters $\alpha = \beta = 1$, then each distribution has a mean at ½. However, the mean of the maximum time distribution of the two distributions is not at 0.50, but rather 0.63. Similarly, if the distributions are assumed to be normally and identically distributed, with mean μ and standard deviation σ, the mean of the maximum time distribution is $\mu + \sigma/\sqrt{\pi}$.

If a third path is present that is approximately as long as the other two, and has no cross connections with them, then there are three independent paths in parallel, and the error in the PERT-calculated mean increases. For example, in the beta distribution example above, the presence of a third path, with the same distribution as the other two, would raise the mean of the maximum time distribution to 0.69.

Examples—Parallel and Cross-Connected Networks. As indicated above, the more parallelism in a network, the larger will be the error in the PERT-calculated mean, other things being equal. However, there is a counter-balancing factor—correlation —which tends to offset the error resulting from parallelism. When activities are common to two or more paths, the paths are correlated. Thus, when one path has a very long duration, other paths that have activities in common with this first path are likely to have a long duration also.

The extent to which these two factors tend to compensate depends on the network configuration. Since parallelism tends to cause the actual mean to be larger than the PERT-calculated mean, the more parallelism will result in a larger discrepency. On the other hand, the more common are the activities in the network, the greater will be the tendency for the PERT-calculated mean and the actual mean to be closer together. A comparison between a parallel configuration and a common activity configuration is given in the following example.

Consider the four-event example in Figure 7. There are four activities, and the particular discrete distribution used on each activity can be identified by the corresponding mean on the network diagram. There are two paths, ABD and ACD, both

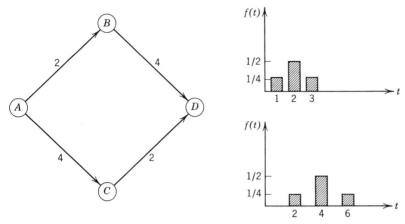

FIGURE 7. Four-event parallel network.

having a mean length of 6. The mean of the maximum time distribution is 6.89. Thus the error in the PERT-calculated mean is 12.9 per cent of the actual mean.

There are two possible ways a third path, with a mean length of 6, may be created by adding one more activity. In one case the path may be completely independent of the other two paths, thus resulting in a third parallel element, AD, as depicted in Figure 8(a). Alternatively, an activity BC can be added with a mean time of 2, thus creating path $ABCD$, shown in Figure 8(b). In both cases there are three paths, all of mean length 6, and the network has four events and five activities.

The addition of the third path in parallel [Figure 8(a)] leads to an increase in the deviation of the PERT-calculated mean (still 6) from the actual mean. The actual mean of the network in Figure 8(a) is 7.336; thus the error has increased to 18.2 per cent.

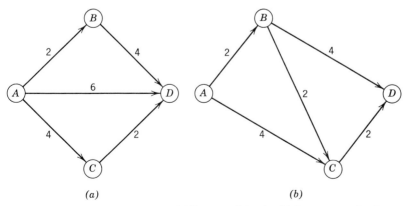

(a)   (b)

FIGURE 8. Three-paths networks. (a) Three parallel paths (b) Cross-connected paths

Figure 8(b), on the other hand, is a network configuration where there is a cross connection between two parallel paths. Since there are three paths, one would expect a larger error than in a similar network with only two paths (such as Figure 7), although not as large an error as in Figure 8(a), where the three paths are in parallel. The correlation (resulting from the common activities) in the network of Figure 8(b) does indeed have the effect discussed, and the mean of the maximum time distribution lies between these two bounds, being 7.074. The error as a per cent of the actual mean is 15.2 per cent.

Examples—Effect of Slack in Networks. The examples given in the two previous sections are extreme cases since all the paths have the same expected duration— hence, they are all critical paths. If the durations of some paths are shorter than the duration of the longest path, their effect on the project mean and standard deviation would not be as great. However, if they have a mean duration very close to the mean duration of the critical path, they would not be critical but they would have an effect almost as significant as the examples of the previous sections. The following examples shown in Figure 9 indicate the effect of slack in a path length.

The simple network has only two paths, ABC and AC. All activities are assumed to be normally distributed with standard deviation equal to 1, and the appropriate mean given on the diagram. It may be noted from the diagrams that various lengths were assumed for paths ABC and AC, ranging from both of them being of equal length, to path AC being only ¼ the length of path ABC. Table I summarizes the results.

This example indicates that the deviation of the PERT-calculated mean and standard deviation from the actual mean and standard deviation may be quite large when the paths are about equal in length, but the difference decreases substantially as the path lengths become farther apart.

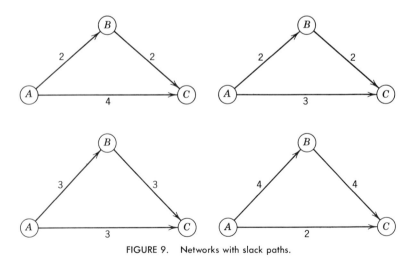

FIGURE 9.   Networks with slack paths.

TABLE 1
Summary of Results from Figure 9

Ratio of lengths: $\dfrac{\text{path AC}}{\text{path ABC}}$	$1/1$	$3/4$	$1/2$	$1/4$
PERT-calculated mean	4	4	6	8
Analytically-calculated mean	4.69	4.30	6.03	8.00
Per cent error (PERT from actual mean)	-17%	-8%	-0.5%	-0.00%
PERT-calculated standard deviation	1 or 1.414	1.414	1.414	1.414
Analytically-calculated standard deviation	1.015	1.149	1.364	1.414
Per cent error (PERT from actual std. dev.)	-1% or $+39\%$	$+23\%$	$+4\%$	$+0.00\%$

Examples—Errors in Small Networks. For more details on errors in networks see reference 4 where more examples are presented. In particular, a four-event-five-activity network for four different distribution forms (beta, normal, uniform, and discrete) is analyzed. This network, with discrete distributions, is then expanded to a five-event-eight-activity network. The PERT-calculated mean was, of course, always smaller than the actual mean, and the error in the PERT-calculated mean increased monotonically, from 7 to 14 per cent, as the network was expanded. The PERT-calculated standard deviation was first larger than the actual standard deviation, then became smaller. The error in the PERT-calculated standard deviation thus changed sign (from $+7$ to -9 per cent).

Examples—Combinations of Simple Series and Parallels. An analysis of combinations of the most elementary series and parallel elements discussed earlier allow the results to be generalized to a certain extent.

Consider the network in Figure 10, which is a simple connection of a series and parallel element.[9] As noted previously, no error will be made in combining activities along the series element. However, the two paths comprising the parallel element lead to error in the PERT calculation. The error in the whole network is at some intermediate value between these two extremes. The location of this intermediate value in the possible interval of error depends on which element is the dominant one. If the series element is dominant, the parallel error would not have much effect, and

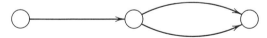

FIGURE 10.   Simple series-parallel network.

---

[9]Although this representation is not in strict accordance with PERT networking techniques, throughout this paper an arc may be considered as the resultant of numerous activities or sub-networks. The purpose of this type of representation is to highlight the symmetry of the particular network configuration.

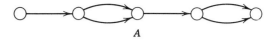

A

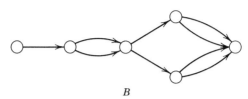

B

FIGURE 11. Series-parrallel arrangements. (A) = Arrangement of two identical series-parallel combinations. (B) = Arrangement of three identical series-parallel combinations.

the net error in the PERT-calculated mean for the whole network will be small. However, if the parallel combination is the dominant one, the error in the whole network may be nearly as large as it is in the parallel configuration alone.

Now suppose that two identical series-parallel combinations are joined together in series. The per cent error in the total will be the same as the per cent error in either individual one. This configuration is shown in Figure 11(a). However, if in another arrangement, three identical combinations are joined in a series-parallel arrangement, as shown in Figure 11(b), the error in the total will be greater than the error in any of the three individual series-parallel elements.

By this sort of procedure one might be able to determine the approximate error in simple series and parallel combinations, if the error in the individual elements were known along with the various relative values of the corresponding means.

Probability Statements

At this stage, it should be obvious that the PERT probability statements concerning the various possible project completion times may be considerably in error. Since these statements are based on a normal distribution having the PERT-calculated mean and standard deviation as parameters, and since it has been shown that these PERT calculations can be seriously in error, considerable doubt is cast on the validity of these statements. In addition, the normal approximation to the project distribution may be a poor one, since the parallelism in a network will tend to skew the distribution to the left.

Decomposition of Networks

Although the networks analyzed in this paper are very small, the results obtained are applicable, to some extent, to much larger networks. Most of the examples here have dealt with networks whose activities were all critical (i.e., they all made some contribution to the project distribution). However, in large networks many activities are not of a critical nature. Dropping all of the noncritical activities from consideration may reduce the network substantially. Another procedure would be to try to identify simple series and parallel elements in a network and to collapse parts of the network on the basis of the network configuration alone.

Examining this latter procedure first, one may find that some networks, or at least parts of them, are composed of simple series and parallel elements with few cross connections. Two activities in series can be treated as one larger activity by adding the two durations. Two activities in parallel can be treated as one activity by taking the maximum of the two durations. By such reduction, a large network can possibly be broken down into a number of small networks for which approximate results are known. Then, since the effects of combining errors in simple series and parallel arrangements are roughly known, an estimate of the error in the whole network may be obtained.

Unfortunately, this technique does not generally reduce the PERT network very much because of the numerous cross connections. However, if use is made of the time estimates given for each activity, and not just the network configuration alone, many noncritical activities can be eliminated from consideration. One method would be to compare the sum of the minimum times (i.e., optimistic estimates) of all activities along every path to a given node with the sum of the maximum times (i.e., pessimistic estimates) of all activities along every path to the same node. If the sum of the minimum times along one path is greater than the sum of the maximum times along another path, then the latter path cannot (by the definition of these times) be a determinant of the time distribution at that node. The latter path can thus be disregarded in the computation of the distribution at that node. The activities that are unique to this latter path can be removed from the analysis of distributions at the given node and all nodes further along.

Summary of the Network Section

The examples of this section demonstrate the possible sources of error in the PERT calculation of the project mean and standard deviation. They should also provide an indication of the magnitude and direction of the possible error in some very basic network configurations. The errors in the PERT-calculated mean and standard deviation for the examples studied were around 10 to 30 per cent.

The PERT-calculated mean will always be biased optimistically, but the PERT-calculated standard deviation may be biased in either direction. Precise statements about the magnitude of the errors, however, cannot be made since errors in the project mean and standard deviation vary with different network configurations. If there is one path through a network that is significantly longer than any other path, then the PERT procedure for calculating the project mean and standard deviation will give approximately correct results. However, if there are a large number of paths having approximately the same length, and having few activities in common, errors will be introduced in the PERT-calculated project mean and standard deviation. The more parallel paths there are through the network, the larger will be the errors. If, however, the paths share a large number of common activities, the errors will tend to be lower. The extent to which these two factors compensate depends on the particular network configuration.

The errors in the PERT-calculated project mean and standard deviation will tend to be large if many noncritical paths each have a duration approximately equal to the duration of the critical path. However, the more slack there is in each of the non-critical paths, the smaller will be the error.

Because of the possible errors in the PERT-calculated project mean and standard deviation, there may be correspondingly large errors in the probability statements that are based on these parameters.

It is suggested that for a stochastic model (such as PERT) a critical activity concept is more valid than, and probably as useful as, a critical path concept. This is based on the fact that the PERT-calculated critical path does not necessarily contain the most critical activities.

Networks very often contain many activities that are not of a critical nature. Eliminating these activities from consideration may reduce the network considerably without affecting to any large extent the final results. In general, if the sum of the minimum times along one path is greater than the sum of the maximum times along a parallel path, then the latter path will not influence the calculation of the time distribution at the common end node.

References

1. C. E. Clark, "The Greatest of a Finite Set of Random Variables," *Opns. Res.* **9**, 145–162 (1961).

2. D. R. Fulkerson, "Expected Critical Path Lengths in PERT Networks," Research Memorandum RM-3075-PR, The Rand Corporation, Santa Monica, Calif., March, 1962.

3. F. E. Grubbs, "Attempts to Validate Certain PERT Statistics or 'Picking on PERT'," *Opns. Res.* **10**, 912–915 (1962).

4. K. R. MacCrimmon and C. A. Ryavec, "An Analytical Study of the PERT Assumptions," Research Memorandum RM-3408-PR, The Rand Corporation, Santa Monica, Calif., December 1962 (DDC Number: AD 293 423).

5. J. E. Murray, "Consideration of PERT Assumptions," Conductron Corporation, Ann Arbor, Mich., April 1962.

6. "PERT, Program Evaluation Research Task," Phase I Summary Report, Special Projects Office, Bureau of Ordnance, Department of the Navy, Washington, D. C., July 1958.

APPENDIX D

Suggested course outline

This outline is derived from a number of courses presented at the University of California, Los Angeles, for undergraduates, graduates and practicing managers and from industrial seminars and workshops presented across the United States. The outline is basically intended for a one-semester course of 18 meetings, but it can be modified to suit other situations. (Chapter numbers refer to this text.)

Session	Topic	Assignments
1	Fundamentals	Chapter 1
2	Systematic planning for network systems	Chapter 2
3	Preparing the network plan	Chapter 3; Appendix C
4	Time estimates, network analysis, and scheduling	Chapters 4 and 5
5	Time/cost/manpower system	Chapters 6 and 7; Appendix F
6	Organizing for the integrated system	Chapter 8
7	Role of the computer	Chapter 9
8	Cost versus value of the network-based system	Chapters 10 and 11
9	Mid-term exam	
10	Multi-project integration	Chapter 12
11, 12	Class or individual projects; create	Chapters 13, 14, 15, and 16
13, 14	Work Breakdown Structure, develop networks, process with computer, update and analyze	
15	Problems and pitfalls	Chapter 17
16	Simulation and gaming	Chapter 18
17	Network-based systems and the management function; outlook for the system	Chapters 19 and 20
18	Final exam	

Additional selected references reflecting the most recent developments should be assigned as well. Detailed descriptions of particular systems or programs (such as the USAF PERT System Description and the IBM PERT/Cost System manual) are very useful. Practice exercises should be used wherever possible.

APPENDIX E

Selected bibliography

Hundreds of articles and papers and several books have appeared relating to network planning and critical path analysis. This bibliography lists those items which contain the original early material, and later references selected for their range of interest or significance. Only items which are in print and readily available are included. No doubt some references have been omitted which should appear; the authors regret that, due to their wish to be brief and perhaps also due to their lack of awareness of certain items, such omissions may have occurred.

Anon., "Administration,—Network Analysis System," Office of the Chief of Engineers, U.S. Army Headquarters, Regulation No. 1–1–11, Washington 25, D.C., March 15, 1963.

Anon., *Bibliography—PERT and Other Management Systems and Techniques*, PERT Orientation and Training Center, Bolling A.F. Base, Washington 25, D.C., June 1963. 71 pages.

Anon., "Control Center—Critical Path—Long-Range Modernization" (at Olin Metals Division's Brass Operations), Factory, vol. 120 (October 1962), pp. 110–113.

Anon., "CPM and Survival," editorial, *Engineering News-Record*, vol. 169, no. 3 (July 19, 1962), p. 116.

Anon., *CPM In Construction—A Manual for General Contractors*, The Associated General Contractors of America, 1957 E Street, N.W., Washington, D.C. 20006.

Anon., "Communications Shorthand for Management; the Critical Path Techniques," *Steel*, vol. 151 (November 9, 1962), pp. 74–78.

Anon., "The Critical Path Method—How To Use It," *Construction Methods & Equipment*, vol. 44 (May 1962), p. 130.

Anon., *NASA PERT and Companion Cost System Handbook*, Director of Management Reports, Office of Programs, NASA, Washington 25, D.C., October 30, 1962.

Anon., "NASA-PERT 'C' Computer Systems Manual," NASA, NPC 101-2A, U.S. Government Printing Office, Washington, D.C., 20402, September 1964. 60 cents.

Anon., "NASA PERT In Facilities Project Management," NASA, NPC 101-3, U.S. Government Printing Office, Washington, D.C., 20402, March 1965. 45 cents.

Anon., *"PERT Guide for Management Use,"* PERT Orientation and Training Center, Bolling A.F. Base, Washington 25, D.C., June 1963.

Anon., *PERT-MILESTONE SYSTEM*, Bureau of Naval Weapons, U.S. Navy, BUWEPS Instruction 5200. 13, Washington 25, D.C., December 19, 1961.

Anon., "PERT Summary Report Phase I," Navy Special Projects Office, U.S. Navy, Catalog No. D217.2: P 94/958, U.S. Government Printing Office, Washington, D.C., 20402, 1958. 25 cents.

Anon., "PERT Summary Report Phase II," Navy Special Projects Office, U.S. Navy, Catalog No. D217.2: P 94/958-2, U.S. Government Printing Office, Washington, D.C. 20402, 1958.

Anon., *PERT/TIME & PERT/COST Management Systems for Planning & Control*, Bureau of Ships, U.S. Navy, MIL-P-23189 (SHIPS), Military Specifications, March 2, 1962.

Anon., *Planning and Control Technique (PCT)*, Vol. I, II and III, Army Material Command, Management Engineering Training Agency, U.S. Army, Rock Island Arsenal, Rock Island, Ill., August 1963.

Anon., "Summary Minutes for Meeting of Contractor PERT Reporting Personnel," Navy Special Projects Office, U.S. Navy, GPO Catalog no. D217.2: P 94/5/9602, U.S. Government Printing Office, Washington, D.C., 20402, November 15–16, 1960. 15 cents.

Anon., *Supplement No. 1 to DOD and NASA Guide PERT COST Output Reports*, PERT Coordinating Group, Government Printing Office, Washington, D.C., 20402, March 1963. 40 cents.

Anon., "Line of Balance Technology," Office of Naval Materiel, U.S. Navy, (NAVEXOSP1851 Rev. 4–62), Washington, D.C., April 1962.

Anon., *Master Plans and Reports (PM2P)*, Vol. III, Army Materiel Command Headquarters, U.S. Army, AMC Reg. 11-16, November 1963.

Anon., *DOD & NASA Guide PERT COST Systems Design*, Joint publication of Office of Secretary of Defense and NASA, #D1.6/2, U.S. Government Printing Office, Washington, D.C., 20402, June 1, 1962. 94 pp., 75 cents.

Anon., *Glossary of Management Systems Terms (with Acronyms)*, PERT Orientation and Training Center, Department of Defense, NASA, Bureau of the Budget, Atomic Energy Commission, Federal Aeronautics Administration, et. al., Bolling A.F. Base, Washington 25, D.C., June 1963.

Anon., *USAF PERT SYSTEM:* Vol. I, "USAF PERT-Time System Description Manual"; Vol. II, "USAF PERT Time System Computer Handbook"; Vol. III, "USAF Pert Cost System Description Manual"; Vol. IV, "USAF PERT Cost System Computer Program Handbook"; Vol. V, "USAF PERT Implementation Manual," Hq. AFSC, U.S. Air Force, September 1963.

Antill, James M., and Ronald Woodhead, *Critical Path Methods in Construction Practice*, John Wiley, New York, 1965. 276 pp., $9.75.

Archibald, Russell D., "PERT/CPM Management System for the Small Subcontractor," *Technical Aids for Small Manufacturers*, No. 86 (March–April 1964). Small Business Administration, Washington 25, D.C., 4 pp.

Archibald, Russell D., "How PERT/CPM Saves Time and Cost," *Petroleum Management*, Vol. 37, no. 11 (November 1965), pp. 103–107.

Archibald, R. D. & A. T. Materna, "Dynamic Planning of Industrial Growth in Emerging Countries," *Management Sciences in The Emerging Countries*, Pergamon Press, London, 1965, pp. 79–103.

Archibald, R. D. and R. L. Villoria, "Product Planning and Evaluation," *Mechanical Engineering*, vol. 86, no. 6 (June 1964), pp. 48–51.

Ashachan, A., "Better Plans Come from Study of Anatomy of an Engineering Job," *Business Week*, no. 1513 (March 21, 1959), pp. 60–66.

Baumgartner, John Stanley, *Project Management*, Richard D. Irwin, Homewood, Illinois, 1963. 185 pages, $6.50.

Beckwith, R. E., "A Cost Control Extension of the PERT System, *IRE Transactions on Engineering Management*, vol. EM-9, no. 4 (December 1962), p. 147.

Berman, Herbert, "Try Critical Path Method to Cut Turnaround Time 20 Percent," *Hydrocarbon Processing and Petroleum Refiner,* vol. 4, no. 1 (January 1962), pp. 135–138.

Boehm, George A. W., "Helping the Executive Make Up His Mind," *Fortune,* vol. 65, no. 4 (April 1962), p. 128.

Case, James G., "PERT—A Dynamic Approach to Systems Analysis," *NAA Bulletin,* vol. 44, no. 7 (March 1963), p. 27.

Christensen, Borge N., "Network Models for Project Scheduling: Planning Phase," vol. 34, no. 11 (May 10, 1962), pp. 114–118; "Preliminary Scheduling Phase," vol. 34, no. 12 (May 24, 1962), pp. 173–177; "Advanced Scheduling Phase," vol. 34, no. 13 (June 7, 1962), pp. 132–138; "Preparing a Network Model," vol. 34, no. 15 (June 21, 1962), pp. 155–160; "Preparing Computer Data," vol. 34, no. 16 (July 5, 1962), pp. 105–111; "Choosing a Plan," vol. 34, no. 17 (July 19, 1962), pp. 136–140. *Machine Design.*

Clark, Charles E., "The PERT Model for the Distribution of an Activity Time," *Operations Research,* vol. 10 (May 1962), pp. 405–406.

Clark, C., "Optimum Allocation of Resources Among the Activities of a Network," *J. Industrial Engineering* 12 (January–February), pp. 11–17.

Cosinuke, W., "Critical Path Technique for Planning and Scheduling," *Chemical Engineering,* vol. 69 (June 25, 1962), pp. 113–118.

Dooley, et al. *Casebooks in Production Management,* vol. 3, Wiley, New York, 1964.

Driessnack, Capt. Hans H., "PERT on the C-141," *Aerospace Management,* vol. 5, no. 8 (August 1962), pp. 32–35.

Dusenbury, Warren, "Applying Advanced Science to Marketing and Ad Plans," *Printers' Ink,* vol. 292 (September 24, 1965), p. 15.

Falconer, David and Gale Nevill, "Critical Path Diagramming," *International Science and Technology,* no. 10 (October 1962), pp. 43–49.

Flagle, C. D., "Probability Based Tolerances in Forecasting and Planning," *Journal of Industrial Engineering,* vol. 12, no. 2, part 1 (May 1961), p. 97.

Freeman, Raoul J., "A Generalized Network Approach to Project Activity Sequencing," *IRE Transactions on Engineering Management,* vol. EM-7, no. 3 (September 1960), p. 103.

Fry, B. L., "Network-Type Management Control Systems Bibliography," Prepared for U.S. Air Force Project RAND, Memorandum RM-3074-PR, RAND Corporation, 1700 Main St., Santa Monica, California, February 1963. 199 pages.

Fulkerson, Dr. R., "A Network Flow Computation for Project Cost Curves," *Management Science,* vol. 7 (January 1961), pp. 167–178.

Fulkerson, D. R., "Increasing the Capacity of a Network: The Parametric Budget Problem," *Management Science,* vol. 5 (1959), pp. 472–483.

Geddes, Philip, "How Air Force Manages Weapons Acquisition," *Aerospace Management,* vol. 5, no. 7 (July 1962), pp. 18–22.

Glaser, L. B. and R. M. Young, "Critical Path Planning & Scheduling: Application to Engineering & Construction," *Chemical Engineering Progress,* vol. 57, no. 11 (November 1961), pp. 60–65.

Glassford, W., "Critical Path Scheduling," *Plant Administration and Engineering,* vol. 21 (October 1961), pp. 59–62.

Gritzner, C. L., J. P. Jones and J. M. Ellis, *Critical Path Scheduling in Maintenance,* Report No. K-1472, Union Carbide Nuclear Co., Oak Ridge Gaseous Diffusion Plant, Oak Ridge, Tennessee, April 10, 1961, available from the Office of Technical Services, U.S. Dept. of Commerce, Washington 25, D.C., 23 pages, 25 cents.

Harris, Remus, "New Product Marketing Plan Instituted by Agency VP; Predicts its Advanced Development by 1975," *Advertising Age,* vol. 36, no. 52 (December 27, 1965), p. 39.

Kast, Fremont E. and James E. Rosenzweig, *Science Technology, and Management; Proceed-*

ings of National Advanced-Technology Management Conference, Seattle, September 4–7, 1962. McGraw-Hill, New York, 1963. 368 pages, $7.95.

Kast, W. G., "Critical Path Method Ideal Tool for Plant Construction," Hydrocarbon Processing & Petroleum Refiner, vol. 41 (February 1962), pp. 123–130.

Kelley, J. E., Jr. and Morgan R. Walker, "Critical Path Planning & Scheduling," Proceedings of Eastern Joint Computer Conference, Boston, December 1–3, 1959, AIEE, New York, 1960, pp. 160–173.

Kelly, J., "Critical Path Planning & Scheduling: Case Histories," Operations Research, vol. 8 (1960), B109.

Kelly, J. E., Jr. and M. R. Walker, "Critical Path Planning and Scheduling, Factory, vol. 118 (July 1960), pp. 74–77.

Kelly, J. E., Jr., "Critical Path Planning and Scheduling: Mathematical Basis," Operations Research, vol. 9 (May 1961), 296–320.

Klass, Philip J., "PERT/PEP Management Tool Use Grows," Aviation Week, vol. 73, no. 22 (November 28, 1960), pp. 85–91.

Klein, H. E., "Psychoanalysis on the Production Line," Dun's Review and Modern Industry, February 1962, pp. 54–58.

Lasser, Daniel J., "Topological Ordering of a List of Randomly-Numbered Elements of a Network," Communications of ACM, vol. 4 (April 1961), pp. 167–168.

Loeber, Norman, "PERT for Small Projects," Machine Design, vol. 34, no. 25 (October 25, 1962), p. 134.

Lynch, Charles J., "Critical Path Scheduling," Production Engineering, vol. 32, no. 37 (September 1961), pp. 92–96.

Malcolm, D. G., J. H. Roseboom, C. E. Clark and W. Fazar, "Application of a Technique for Research & Development Program Evaluation," Operations Research, vol. 7 (September–October 1959), pp. 646–669.

Mauchly, J. W., "Critical Path Scheduling," Chemical Engineering, vol. 69, no. 8 (April 16, 1962), pp. 139–154.

Martino, R. L., "Finding the Critical Path," vol. I; "Applied Operational Planning," vol. II; "Allocating and Scheduling Resources," vol. III; American Management Association, 135 W. 50 St., New York, N.Y. 10020, 1964. $15 each.

McGee, A. A. and M. D. Markarian, "Optimum Allocation of Research/Engineering Manpower Within a Multi-Project Organization Structure," IRE Transactions on Engineering Management, vol. EM-9, no. 3 (September 1962), p. 104.

Miller, Norman C., Jr., "Maps for Managers Show Problem Areas of Big Defense Jobs," The Wall Stret Journal, vol. 158, no. 32 (August 16, 1961), p. 1.

Miller, Robert W., "How to Plan and Control with PERT," Harvard Business Review, vol. 40, no. 2 (March–April 1962), pp. 93–104.

Miller, Robert W., Schedule, Cost and Profit Control with PERT, McGraw-Hill, New York, 1963, 227 pp., $8.50.

Nakayama, Y., "PERT Milestone System Readied for All Navy Weapons," Aerospace Management, vol. 5, no. 4 (April 1962), pp. 46–50.

Norden, P., "On the Anatomy of Development Projects," IRE Trans-Engineering Management, vol. 7 (1960), pp. 34–42.

Pearlman, Jerome, "Engineering Program Planning & Control Through the Use of PERT," IRE Transactions on Engineering Management, vol. EM-7, no. 4 (December 1960), pp. 125.–134.

Peterson, R. J., "Critical Path Scheduling for Construction Jobs," Civil Engineering, vol. 38, no. 8 (August 1962), pp. 44–47.

Phillips, C. R. Jr., A Catalog of Computer Programs for PERT & Similar Management Systems, Operations Research, 8605 Cameron St., Silver Spring, Maryland, May 1, 1962. $1.00.

Raborn, W. F., "Lesson in Management-Program Evaluation & Review Technique," *Aviation Week and Space Technology*, vol. 76 (January 29, 1962), p. 21.

Sadow, R. M. and T. L. Senecal, "PERT in the Dyna-Soar," *Aerospace Management*, vol. 5, no. 12 (December 1962), pp. 18–23.

Sayer, J. S., J. E. Kelley, Jr., and M. R. Walker, "Critical Path Scheduling," *Factory*, vol. 118, no. 7 (July 1960), pp. 74–77.

Shultis, R. L., "Applying PERT to Standard Cost Revisions," *NAA Bulletin*, vol. 43, no. 13, sec. 1 (September 1962), pp. 35–43.

Wattel, Harold L. Ph.D., "Network Scheduling and Control Systems (CPM/PERT), the Dissemination of New Business Techniques," Hofstra University Yearbook of Business, vol. I, vol. II, January 1964, 329 pp.

Wynne, B. E., Jr., "Critical Path Method; An Effective Management Tool," *Controller*, vol. 30 (June 1962), pp. 258–264.

APPENDIX F

U.S. Government PERT/COST report formats and explanations

Because the U.S. Government has provided important leadership in the development and use of network-based management systems, the report formats of the government system are presented in this appendix, with the official definitions and explanations. Study of these reports will convey a good understanding of the government approach, but a potential user will also need to study the detailed implementing procedures provided by the various branches of the government in order to be able to comply fully with specific contractual requirements.

This Appendix is a condensation of *Supplement No. 1 to DOD and NASA Guide, PERT/COST Output Reports,* Pert Coordinating Group, Washington, D. C., March, 1963.

FOREWORD

These uniform PERT Cost Output Reports have been developed by the Technical Subcommittee of the Inter-Agency PERT Coordinating Group and are based on those contained in the DOD and NASA Guide, PERT Cost Systems Design, June 1962. The document represents experience from the Mauler, TFX, and Navy Implementation Teams as well as the Air Force Systems Command, Army Management Engineering Training Agency, Bureau of Ships, National Aeronautics and Space Administration, Atomic Energy Commission, Federal Aviation Agency, and similar organizations and agencies.

The forms have been approved by the PERT Coordinating Group and have been authorized for use on current contracts of the Department of Defense including:

TFX, Mauler, Subroc, Titan III, MMRBM, Lance, Polaris, certain contracts of the Bureau of Ships covered by a military specification and previously authorized by the PERT Coordinating Group.

Bureau of the Budget Number 22R 226 is authorized for this purpose.

It is the intent of the PERT Coordinating Group that this document can serve for

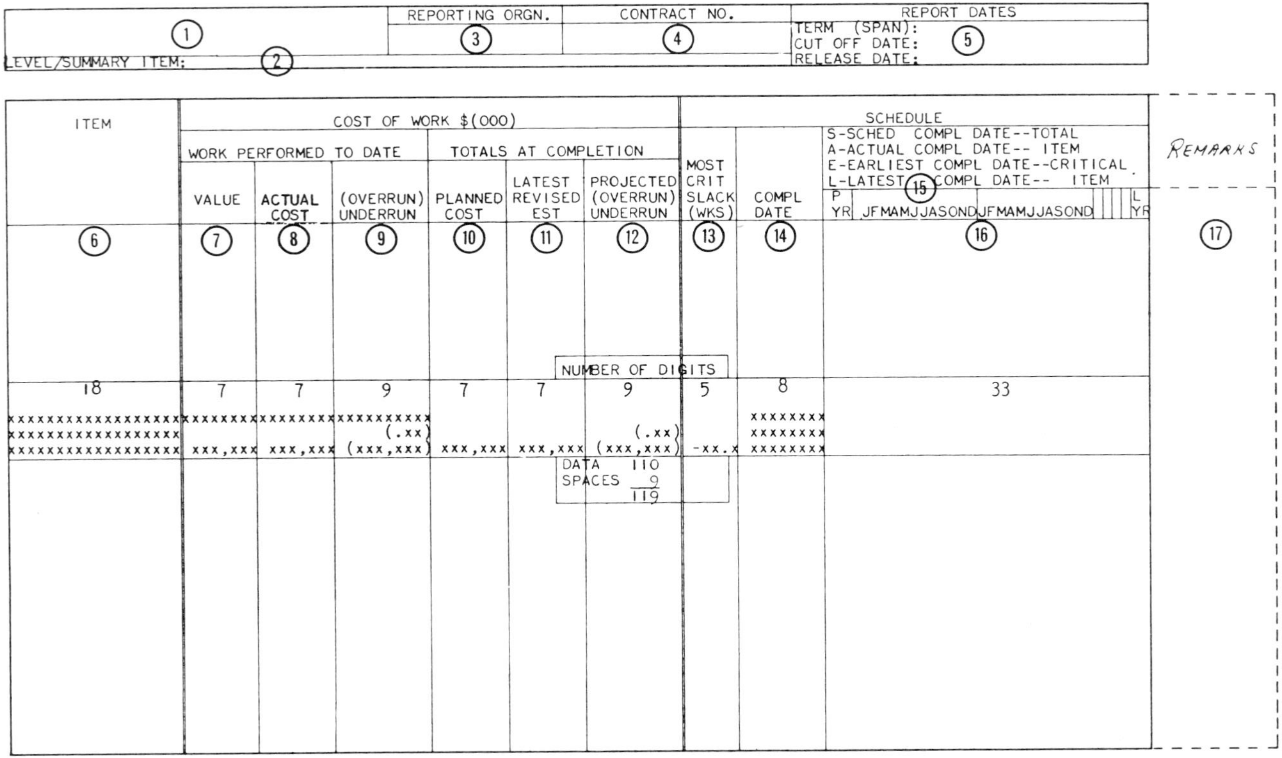

FIGURE 1. PERT COST management summary report.

all agency members as an exhibit in a contract, a reference or the basis of a specification. The Navy Specification No. Mil P-23189A(Navy) 25 Oct 62 is being revised to incorporate these forms. . . .

PERT COST MANAGEMENT SUMMARY REPORT

The PERT COST Management Summary Report shows current and projected schedule and cost status of the total program and of each of the major component items or elements within the program. The report is prepared at several levels of the work breakdown structure and for all contracts or a specified combination of contracts, depending upon the needs of management. The report may be machine produced, but when it is manually prepared, the necessary information is derived from the Program/Project Status Report.

The first line of each report shows total costs and significant schedule information for the summary item shown in title block ②. Subsequent lines show each subdivision of that summary item at the next lower level of the work breakdown structure; thus, each page of the report shows the time and cost status and all the next level backup information for a single summary item. Since each page of the report is a concise summary of one element of the program or project, the report is usually divided for distribution to appropriate government and contractor managers.

DEFINITIONS PERT COST MANAGEMENT SUMMARY REPORT

① The designation of the total (or a part of the total) system program or project that is identified with the reporting organization. For example, if reporting organization XYZ has the Missile and GHE part of weapon system ABC, the program or project definition would read:

ABC—Missile and GHE

② *LEVEL/SUMMARY ITEM:* The level number, noun description, and summary number of the summary item for which the report is being prepared.

③ *REPORTING ORGANIZATION:* The name or identification of the organization responsible for the work identified in the Contract Number ④ and Program/Project ① blocks.

④ *CONTRACT NUMBER:* The numeric designation of the contract(s) or agreement(s) included in each report (e.g., 33(600)28369A). When a report is prepared for a large program or project, several contracts may be included. Therefore, each contract number (or its representative code) would be indicated in this space. It may be noted that by sorting on contract number, a report can be prepared for each individual contract.

⑤ *REPORT DATES:*

Term (span): The beginning and ending date for the total increment being covered in the report. For example:

1 Jan 62 to 31 Dec 62
Total Program (Project)
Contract

REPORTING ORGN.	CONTRACT NO.	REPORT DATES
XYZ – A&S DIVN 22300	33(600)28369A	TERM (SPAN): TOTAL PROGRAM CUT OFF DATE: 30MAR63 RELEASE DATE: 10APR63

ABC – MISSILE AND GHE
LEVEL/SUMMARY ITEM: 3/BALLISTIC SHELL 22300

ITEM	COST OF WORK $(000) WORK PERFORMED TO DATE VALUE	ACTUAL COST	(OVERRUN) UNDERRUN	TOTALS AT COMPLETION PLANNED COST	LATEST REVISED EST	PROJECTED (OVERRUN) UNDERRUN	MOST CRIT SLACK (WKS)	COMPL DATE	SCHEDULE	REMARKS
BALLISTIC SHELL LEV 3 22300	19,600	20,500	(.05)(900)	35,200	39,650	(.13)(4,450)	0.0	10DEC64 31DEC63 31DEC63	E L / S	See Problem Analysis Rpt Items 1-3
NOSE FAIRING LEV 4 22310	27	25	.07 2	175	175		8.6	10DEC64 10JUN63 10AUG63	E L / S	Item 6
FIRST STAGE LEV 4 22320	6,700	6,400	.04 300	9,200	9,700	(.05)(500)	0.0	30APR64 31DEC63 31DEC63	E L / S	Items 9-12
SECOND STAGE LEV 4 22330	1,645	1,650	(5)	3,500	3,570	(.02)(70)	0.0	15JUN64 31DEC63 31DEC63	E L / S	Item 15

SCHEDULE
S-SCHED COMPL DATE--TOTAL
A-ACTUAL COMPL DATE-- ITEM
E-EARLIEST COMPL DATE--CRITICAL
L-LATEST COMPL DATE-- ITEM
P 1963 1964
Y J F M A M J J A S O N D J F M A M J J A S O N D

TIME NOW

FIGURE 2. PERT COST management summary report.

Cut off date: The accounting cut off date for the period of actual costs being reported.

Release date: The date that the report is to be released to management. In the event of subsequent rerun and redistribution of reports, it is permissible to suffix the report release date with a revision number.

⑥ *ITEM:* The level number, noun description, and summary number of each summary item on the work breakdown structure for which time information and cost information are presented in the report. The first item shown is the highest item for which the particular report is prepared and should be identical with the item named in the Level/Summary Item block ②. Three lines are available for each item description, and, if necessary, the top line may be extended into the Cost of Work columns.

⑦ *VALUE* (Work Performed to Date): The total planned cost for work completed within the summary item. This value is determined by summing the Planned Cost ⑩ for each completed work package. If a work package is in process, the part of its total planned cost which applies to work completed is approximated by applying the ratio of Actual Cost ⑧ to Latest Revised Estimate ⑪ for that work package.

⑧ *ACTUAL COST* (Work Performed to Date): The actual expenditures incurred plus any prespecified types of unliquidated commitments (unliquidated obligations or accrued liabilities) charged or assigned to the work packages within the summary item.

⑨ *(OVERRUN) UNDERRUN* (Work Performed to Date): The Value ⑦ for the work performed to date minus the Actual Cost ⑧ for that same work. When value exceeds actual cost, an underrun condition exists. When actual cost exceeds value, an overrun condition exists. The (overrun) underrun is also expressed as a percentage of the value of work performed to date immediately above the dollar amount. Parentheses are used as a notational device to indicate overruns. (Over)underruns in excess of one billion dollars print as 999,999.

⑩ *PLANNED COST* (Totals at Completion): The approved planned cost for the total summary item. This is the total of the planned costs for all work packages within the summary item.

⑪ *LATEST REVISED ESTIMATE* (Totals at Completion): The latest estimate of cost for the total summary item. This estimate is the sum of the actual costs plus estimates-to-complete for all the work packages in the summary item. This estimate is also known as anticipated final cost. For a completed item, the latest revised estimate equals the Actual Cost ⑧.

⑫ *PROJECTED (OVERRUN) UNDERRUN* (Totals at Completion): The Planned Cost ⑩ minus the Latest Revised Estimate ⑪ for the total summary item. When planned cost exceeds latest revised estimate, a projected underrun condition exists. When latest revised estimate exceeds planned cost, a projected overrun condition exists. The projected (overrun) underrun is also expressed as a percentage of the planned cost immediately above the dollar amount. Parentheses are used as a notational device to indicate (over)underruns. (Over)underruns in excess of one billion dollars print as 999,999.

⑬ *MOST CRITICAL SLACK (WEEKS):* The slack, in weeks, associated with the "E" and "L" notations shown in the Schedule Completions section ⑯. This represents the worst slack (least algebraic) with respect to designated program or project end points for any of the activities within the summary item.

⑭ *COMPLETION DATE:* The day, month, and year of the "S," "A," "E," and "L" positions shown in the Schedule Completions section ⑯.

⑮ *SCHEDULE CALENDAR:* A calendar time reference for display of schedule completions. The calendar contains one division for all prior years, two years divided by months, four years by years, and one division for all later years. When the calendar is printed by a computer, one space is left between the months before and after the Cut Off Date ⑤. A "Time Now" line is printed in this space. If the cut off date falls between the 10th and the 30th of a month, that month is considered to be the "past month" and it appears to the left of the Time Now line. If the cut off date falls between the 1st and 10th of a month, that month is considered to be the "next future month" and it appears to the right of the Time Now line. Each year the calendar is adjusted so that two years, by months, appear ahead of the Time Now line.

⑯ *SCHEDULE COMPLETIONS:* Two types of schedule completions are displayed in this section:

a. The scheduled (S) or actual (A) completion of all work contained within the summary item shown in the item column.

b. The earliest (E) and latest (L) completion for the *most critical* schedule element or effort with respect to designated program or project end points within that summary item.

The symbol "S" is used to show the scheduled completion date of all work within the item. The "S" is located under the calendar position of the directed date (T_D) or the scheduled completion date (T_S) if no T_D is established for the last activity within the summary item. If T_S has not been established for the end of the total item, "S" is placed at the calendar position which represents the earliest completion date (S_E) for the last activity in the item. When the total item has been completed, the symbol "A" is placed under the calendar position of the actual completion date for the item.

The "E" and "L" symbols represent the earliest completion date (S_E) and latest completion date (S_L) for the most critical schedule element or effort within the item with respect to designated program or project end points. The most critical element within an item may or may not be the same as the last scheduled item. This will depend on whether there are critical interfaces within the item which pose more serious constraints from a program or project point of view than the completion of a total item itself. The most critical element is the one with the worst slack (least algebraic) within the item. The "E" and "L" positions, therefore, portray the earliest completion date and the latest completion date for that activity within the summary item with the worst slack status. When several activities have the same worst slack condition (for instance, when they are all on the same path), the "E" and "L" positions reflect the last activity on that path.

⑰ *REMARKS:* Notations made by an analyst to indicate critical cost and schedule conditions within summary items. Reference may be made, by paragraph number, to the Problem Analysis Report for a detailed analysis of the critical conditions. The heading for this area of the report is not computer printed.

PERT COST PROBLEM ANALYSIS REPORT

The Problem Analysis Report is a narrative report prepared to supplement the Management Summary Report as well as other reports which identify significant problems.

The report contains three basic sections:

a summary analysis of the total contractor's portion of the program covered by the Management Summary Report;

an analysis of tasks where current or potential problems exist. Problems may be schedules, costs, technical performance, or combinations of these;

a narrative description of:

the nature of the problem;
the reasons for cost and/or schedule variance;
the impact on the immediate task;
the impact on the total program; and
the corrective action: what action, by whom, when, and expected effect.

Additional instructions for preparation of this report will be established by the Government and the contractors for each program or project.

PERT COST PROGRAM/PROJECT STATUS REPORT

The Program/Project Status Report is a comprehensive computer-produced output report. It is organized to reflect the end item work breakdown structure and provides time and cost information from the work package level up to the top of the program or project.

For each work package and summary item shown on the report there is a line of item description followed by a line of significant time and cost information. The first line presents data for the summary item shown in the title block ②. Subsequent lines show all subdivisions of that item down to the work package levels. (Work packages may appear at different levels of the work breakdown structure.)

The primary purpose of the Program/Project Status Report is to back up the Management Summary Report. The two reports contain similar information, but whereas the Management Summary Report highlights information for a manager, this report retains detail for an analyst. The Management Summary Report is divided for distribution and the Program/Project Status Report remains intact as reference material for the entire portion of the program or project for which reports are prepared.

The standard sorting procedure for this report arranges summary items and work packages in the order determined by the work breakdown structure. However, other sorting sequences may be used; e.g., a sequential listing of work packages by charge number; a listing of only completed work, in-process work, or future work, etc.

LEVEL/SUMMARY ITEM: ①   ②

	REPORTING ORGN. ③	CONTRACT NO. ④	REPORT DATES ⑤
			TERM (SPAN): CUT OFF DATE: RELEASE DATE:

IDENTIFICATION				TIME STATUS			COST OF WORK $(000)					
							WORK PERFORMED TO DATE			TOTALS AT COMPLETION		
CHARGE OR SUMMARY NUMBER ⑥	L E V E L ⑦	FIRST EVENT NO. ⑧	LAST EVENT NO. ⑨	SCHED OR ACT (A) COMPL DATE ⑩	EARLIEST & LATEST COMPL DATE ⑪	MOST CRIT SLACK (WKS) ⑫	VALUE ⑬	ACTUAL COST ⑭	(OVERRUN) UNDERRUN ⑮	PLANNED COST ⑯	LATEST REVISED ESTIMATE ⑰	PROJECTED (OVERRUN) UNDERRUN ⑱
18	2	9	9	8	8	9	7	7	9	7	7	9
					NUMBER OF DIGITS	-xx.x			(.xx)			(.xx)
xxxxxxxxxxxxxxxxxx	xx	xxxxxxxxx	xxxxxxxxx	xxxxxxxx	xxxxxxxx	xxxxxxxxx	xxx,xxx	xxx,xxx	(xxx,xxx)	xxx,xxx	xxx,xxx	(xxx,xxx)
xxxxxxxxxxxxxxxxxx		xxxxxxxxx	xxxxxxxxx		xxxxxxxx							

DATA SPACES: 109 / 10 / 119

FIGURE 3. PERT COST program/project status report.

478

		IDENTIFICATION			TIME STATUS			COST OF WORK $(000)					
								WORK PERFORMED TO DATE			TOTALS AT COMPLETION		
CHARGE OR SUMMARY NUMBER	LEVEL	FIRST EVENT NO.	LAST EVENT NO.	SCHED OR ACT (A) COMPL DATE	EARLIEST & LATEST COMPL DATE	MOST CRIT SLACK (WKS)		VALUE	ACTUAL COST	(OVERRUN) UNDERRUN	PLANNED COST	LATEST REVISED ESTIMATE	PROJECTED (OVERRUN) UNDERRUN
FIRST STAGE 22320	4	12000999	12000199	30APR64	31DEC63 / 31DEC63	0.0	12000612	6,700	6,400	.04 / 300	9,200	9,700	(.05) {500}
INSTRUMENTATION 22322	5	12000700	12000400	10JAN64	31DEC63 / 31DEC63	0.0	12000612	165	172	(.04) / (7)	415	430	(.04) (15)
POWER CABLE ASSY. 22323	5	12000899	12000800	15FEB64	15JUN63 / 15JUN63	0.0	12000783	270	200	.26 / 70	1,250	1,180	.06 / 70
ELECTRICAL DESIGN 32164	6	12000700	12000420	25JUL63	10JUN63 / 25JUN63	2.1	12000682	110	112	(.02) / (2)	205	209	(.02) (4)
ELECTRICAL DESIGN 32165	6	12000869	12000860	12JAN64	15JUN63 / 15JUN63	0.0	12000783	22	20	.10 / 2	175	175	
MANUFACTURING 52073	6	12000690	12000410	22AUG63	10JUN63 / 25JUN63	2.1	12000682	55	60	(.11) / (5)	125	137	(.10) (12)
TESTING 78340	6	12000622	12000400	10JAN64	31DEC63 / 31DEC63	0.0	12000612				85	84	.01

FIGURE 4. PERT COST program/project status report.

DEFINITIONS PERT COST PROGRAM/PROJECT STATUS REPORT

[① through ⑤ previously defined.—Authors.] . . .

⑥ *CHARGE OR SUMMARY NUMBER:* The noun description and charge or summary number of each work package or summary item for which time information and cost information are presented in the report. For a work package, the charge number is the contractor or government charge number (shop order number, work order number) used to identify the work package for purposes of estimating and accumulating costs. The title or short description of the charge number is printed immediately above the number itself. For the summary item, the summary number is the identification of an end item on the work breakdown structure above the work package level. The title or description of the summary item is also printed directly above the summary number.

⑦ *LEVEL:* The number of the level on the work breakdown structure at which the charge or summary number appears.

⑧ *FIRST EVENT NUMBER:* The number of the first event in time (based on S_E) for the work package or summary item. This event number defines the beginning of the work package or summary item in relation to the network.

⑨ *LAST EVENT NUMBER:* The number of the last event in time (based on S_E) for the work package or summary item. This event number defines the end of the work package or summary item in relation to the network.

⑩ *SCHEDULED OR ACTUAL COMPLETION DATE:* The calendar date on which all the work contained in the work package or summary item is scheduled for completion or was actually completed. The scheduled completion date (T_S) is established by management as an internal control on the completion of the work. If no scheduled completion date has been established for the work package or summary item, the column is blank. The actual completion date (T_A) is the date on which all work in the work package or summary item has been completed. When the date in this column is an actual completion date, an "A" is printed in front of the date.

⑪ *EARLIEST COMPLETION DATE (S_E) AND LATEST COMPLETION DATE (S_L):* The earliest calendar date on which the work package or summary item can be completed and the latest completion date on which the work package or summary item can be scheduled for completion without delaying the completion of the program or project. When the work package or summary item has been completed, this column is blank.

The earliest completion date (S_E), printed on the upper line, is calculated by:

summing the scheduled elapsed time (t_s) values for activities on the longest path from the beginning of the program or project to the end of the work effort; and then adding this sum to the calendar start date of the program or project.

The latest completion date (S_L), printed on the lower line, is calculated by:

summing the scheduled elapsed time (t_s) values for activities on the longest path from the end of the work effort to the end of the program or project; and then subtracting this sum from the calendar end date of the program or project.

If the longest path contains activities which are not scheduled, expected elapsed time (t_e) values for the unscheduled activities will be processed as scheduled elapsed time (t_s) values in the calculation of S_E and S_L.

(12) *MOST CRITICAL SLACK (WEEKS):* The worst (least algebraic) slack with respect to the designated program or project end points, in weeks, for any of the activities within the work package or summary item. This slack is based on a comparison of S_L minus S_E for each activity. The slack indicated will not necessarily be the difference between the S_L and S_E for the *end* of a work package or summary item since the worst slack situation may be associated with an activity *within* the work package or summary item. The number of the network event at the end of the worst slack path within the work package is printed below the slack value. If the work package or summary item has been completed, this column is blank. . . .

(14) *ACTUAL COST* (Work Performed to Date): The actual expenditures incurred plus any prespecified types of unliquidated commitments (unliquidated obligations or accrued liabilities) charged or assigned to a work package. For summary items, the appropriate work package data is summed. [(13) and (15) through (18) previously defined.—Authors.] . . .

PERT COST ORGANIZATION STATUS REPORT

The Organization Status Reports provide operating level contractor managers with detailed information breakdown from the available store of data in the PERT COST computer program.

Several types of reports may be produced within this format by changing the sorting sequence of Charge Number (6), Responsible Organization (7), Performing Organization (8), and Resource Code (9).

Following are several examples of possible reports:

Responsible Organization 1, Charge Number 2, Performing Organization 3, Resource Code 4. This report shows, for each responsible organization, all work packages which are within its responsibility and a breakout of organizations and skills which will actually perform the work (Figure 6).

Performing Organization 1, Charge Number 2, Responsible Organization 3, Resource Code 4. This report shows, for each performing organization, that portion of each work package assigned to it for accomplishment, with a further identification of the organization responsible for each work package and the resources required.

Performing Organization 1, Charge Number 2. This report is another version of the above. It shows less detail and is more suitable for higher levels of management (Figure 7).

Charge Number 1, Performing Organization 2. This report is a work package listing (shop order ledger) commonly used as an accounting aid.

Totals are shown on the reports for the first and second sort categories only.

(1)	REPORTING ORGN.	CONTRACT NO.	REPORT DATES
	(3)	(4)	TERM (SPAN): CUT OFF DATE: RELEASE DATE: (5)
LEVEL/SUMMARY ITEM: (2)			

IDENTIFICATION				MANHOURS				DIRECT COSTS $(000)				TIME	
				WORK TO DATE	TOTALS AT COMPLETION			WORK TO DATE	TOTALS AT COMPLETION				
CHARGE NUMBER	RESP ORGN	PERF ORGN	RES CODE	ACTUAL	PLANNED	LATEST REVISED ESTIMATE	PROJECTED (OVERRUN) UNDERRUN	ACTUAL	PLANNED	LATEST REVISED ESTIMATE	PROJECTED (OVERRUN) UNDERRUN	MOST CRIT SLACK (WKS)	SCHED OR ACT (A) COMPL DATE
(6)	(7)	(8)	(9)	(10)	(11)	(12)	(13)	(14)	(15)	(16)	(17)	(18)	(19)
						NUMBER OF DIGITS							
18	6	6	4	7	7	7	9	7	7	7	9	5	8
xxxxxxxxxxxxxxxxxx xxxxxxxxxxxxxxxxxx	xxxxxx xxxxxx	xxxxxx xxxxxx	xxxx xxxx	xxx,xxx	xxx,xxx	xxx,xxx	(.xx) (xxx,xxx)	xxx,xxx	xxx,xxx	xxx,xxx	(.xx) (xxx,xxx)	-xx.x	xxxxxxxx
						DATA SPACES 107 12 119							

FIGURE 5. PERT COST organization status report.

ABC – MISSILE AND GHE
LEVEL/SUMMARY ITEM: 3/BALLISTIC SHELL 22300

REPORTING ORGN.	CONTRACT NO.	REPORT DATES
XYZ – A&S DIVN	33(600)28369A	TERM (SPAN): TOTAL PROGRAM CUT OFF DATE: 30MAR63 RELEASE DATE: 10APR63

IDENTIFICATION				MANHOURS				DIRECT COSTS $(000)				TIME	
				WORK TO DATE	TOTALS AT COMPLETION			WORK TO DATE	TOTALS AT COMPLETION				
CHARGE NUMBER	RESP ORGN	PERF ORGN	RES CODE	ACTUAL	PLANNED	LATEST REVISED ESTIMATE	PROJECTED (OVERRUN) UNDERRUN	ACTUAL	PLANNED	LATEST REVISED ESTIMATE	PROJECTED (OVERRUN) UNDERRUN	MOST CRIT SLACK (WKS)	SCHED OR ACT(A) COMPL DATE
ELECTRICAL DESIGN, INSTRUMENTATION 32164	2217	2217	E1	16,900	30,000	31,100	(1,100)	41	90	94	(4)	2.1	25JUL63
		4422	E2	16,800	20,000	20,100	(100)	40	60	60			
			A10	3,500	7,000	7,000		12	25	25			
			M60		5,000	5,000		5	15	15			
		5514	D1	1,200	3,300	3,300		5	10	10			
			D	2,800				9					
TOTAL								112	205	209	(.02) (4)		
ELECTRICAL DESIGN, PWR CABLE ASSY 32165	5514	5514	D1	2,200	4,200	4,200		6	12	12		4.2	15JUL63
TOTAL								1,300	2,600	2,500	.04 100		

FIGURE 6. PERT COST organization status report by responsible organization, charge number, performing organization, responsibility code.

REPORTING ORGN.	CONTRACT NO.	REPORT DATES
XYZ - A&S DIVN	33(600)28369A	TERM (SPAN): TOTAL PROGRAM CUT OFF DATE: 30MAR63 RELEASE DATE: 10APR63

ABC - MISSILE AND GHE
LEVEL/SUMMARY ITEM: 3/BALLISTIC SHELL 22300

IDENTIFICATION				MANHOURS				DIRECT COSTS $(000)				TIME	
				WORK TO DATE	TOTALS AT COMPLETION			WORK TO DATE	TOTALS AT COMPLETION				
CHARGE NUMBER	RESP ORGN	PERF ORGN	RES CODE	ACTUAL	PLANNED	LATEST REVISED ESTIMATE	PROJECTED (OVERRUN) UNDERRUN	ACTUAL	PLANNED	LATEST REVISED ESTIMATE	PROJECTED (OVERRUN) UNDERRUN	MOST CRIT SLACK (WKS)	SCHED OR ACT(A) COMPL DATE
32163		2217		9,200	25,000	25,000		- 22	75	75		3.6	15JUL63
32164				33,700	50,000	51,200	(1,200)	81	150	154	(4)	2.1	25JUL63
TOTAL								382	825	831	(.01)(6)		
32163		4422		500	1,200	1,200		2	7	7		3.6	15JUL63
32164				3,500	7,000	7,000		12	25	25		2.1	25JUL63
TOTAL								297	622	622			

FIGURE 7. PERT COST organization status report by performing organization, charge number.

DEFINITIONS PERT COST ORGANIZATION STATUS REPORT

[① through ⑤ previously defined.—Authors.] . . .

⑥-⑨ The sorting sequence for these identification columns is indicated in the report title. Information will appear only in those columns listed in the title.

⑥ *CHARGE NUMBER:* [Previously defined.—Authors.] . . .

⑦ *RESPONSIBLE ORGANIZATION:* The contractor's organization responsible for management of the work package ⑥.

⑧ *PERFORMING ORGANIZATION:* The contractor's department or organization which will perform work on the work package.

⑨ *RESOURCE CODE:* The contractor's code for a particular manpower skill or material type.

⑩-⑬ *MANHOURS:* Cost information shown in this area of the report may be used for services and facilities, such as computer usage, as well as for direct labor. No totals are shown in these columns.

⑩ *ACTUAL* (Work to Date): The actual manhour expenditures assigned to a work package or work package subdivision.

⑪ *PLANNED* (Totals at Completion): The approved planned manhours for the work package or work package subdivision. [⑫ & ⑬ Previously defined.—Authors.] . . .

⑭-⑰ *DIRECT COSTS $(000):* Cost information in this area of the report represents materials and other direct costs as well as the direct labor dollar value of costs shown in ⑩-⑬. Total dollar costs (including overhead) may be used when they are more appropriate to a contractor's normal operation than direct costs. [⑭ through ⑲ previously defined.—Authors.] . . .

PERT COST FINANCIAL PLAN AND STATUS REPORT

The Financial Plan and Status Report provides data for a monthly comparison (at any given level) of actual costs and/or latest revised estimates against planned costs, and thus serves as a tool for monitoring the financial plans.

Historical (prior month) cumulative costs are shown for each charge number. Both incremental and cumulative costs by charge number are shown for each future month within the time period identified in the Report Dates block ⑤.

The report is prepared for higher levels of management by printing only totals for each month (Figure 10).

The Cost of Work Report (Display) may be prepared from data available in the Financial Plan and Status Report.

DEFINITIONS PERT COST FINANCIAL PLAN AND STATUS REPORT

[① through ⑤ previously defined.—Authors.] . . .

⑥ *MONTH:* The accounting time period for which (or through which) estimates and actuals are shown.

⑦ *CHARGE NUMBER:* [Previously defined.—Authors.] . . .

⑧ *ACTUAL* (Incremental Cost): The actual expenditures incurred plus any pre-specified types of unliquidated commitments (unliquidated obligations or accrued

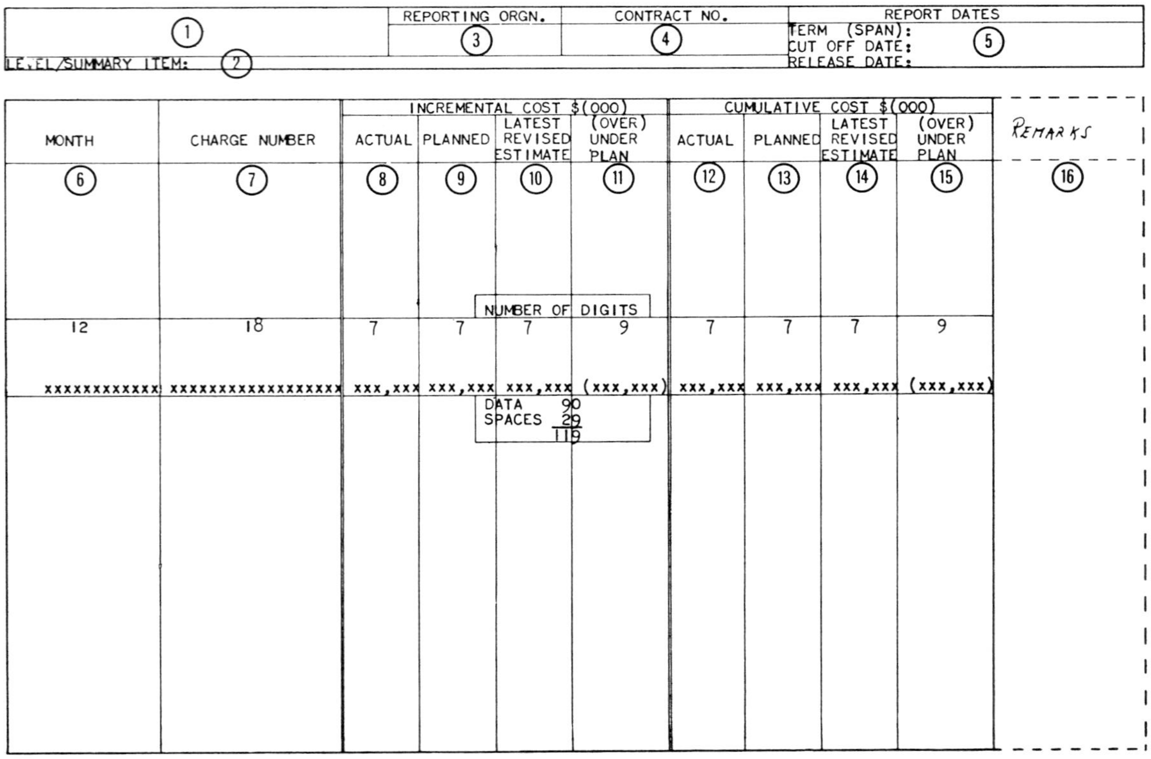

FIGURE 8. PERT COST financial plan and status report.

ABC — MISSILE AND GHE
LEVEL/SUMMARY ITEM: 4/FIRST STAGE, BALLISTIC SHELL 22300

	REPORTING ORGN.	CONTRACT NO.	REPORT DATES
	XYZ – A&S DIVN	33(600)28369A	TERM (SPAN): TOTAL PROGRAM CUT OFF DATE: 30MAR63 RELEASE DATE: 10APR63

MONTH	CHARGE NUMBER	INCREMENTAL COST $(000)				CUMULATIVE COST $(000)				Remarks
		ACTUAL	PLANNED	LATEST REVISED ESTIMATE	(OVER) UNDER PLAN	ACTUAL	PLANNED	LATEST REVISED ESTIMATE	(OVER) UNDER PLAN	
PRIOR	32163					24	24	24		Value of Work Performed to date
	32164					92	93	92		1) CUM to CUT OFF $6,700,000
	52072					12	12	12		2) Latest Month $275,000
	52073					2	2	2		
	78339									
	TOTAL					6,150	6,200	6,150	50	(Over)/under-run for Work Performed to date $300,000
MAR63	32163	–	1	–		25	25	25		
	32164	20	19	20	(1)	112	112	112	(1)	
	52072	3	2	3	(1)	15	14	15	2	
	78339	2	2	2		2	4	2		
	TOTAL	250	300	250	50	6,400	6,500	6,400	100	
APR63	32163		1	–			26	26		
	32164		2	6	(4)		16	21	(5)	
	TOTAL		98	140	(42)		6,598	6,540	58	
TOTAL PERIOD					(42)	6,400	9,200	9,700	(500)	

FIGURE 9. PERT COST financial plan and status report by month, charge number.

487

REPORTING ORGN,	CONTRACT NO.		REPORT DATES
XYZ – A&S DIVN	33(600)28369A	TERM (SPAN): TOTAL PROGRAM	
		CUT OFF DATE: 30MAR63	
		RELEASE DATE: 10APR63	

ABC – MISSILE AND GHE
LEVEL/SUMMARY ITEM: 4/FIRST STAGE, BALLISTIC SHELL 22320

MONTH	CHARGE NUMBER	INCREMENTAL COST $(000)				CUMULATIVE COST $(000)				Remarks
		ACTUAL	PLANNED	LATEST REVISED ESTIMATE	(OVER) UNDER PLAN	ACTUAL	PLANNED	LATEST REVISED ESTIMATE	(OVER) UNDER PLAN	
PRIOR						6,150	6,200	6,150	50	Value of work performed to date
MAR63		250	300	250	50	6,400	6,500	6,400	100	1) Cum to cut off $6,700,000
APR63			98	140	(42)		448	378	70	2) Latest Month $275,000 (Over) Adjustment for work behind sched. $300,000
TOTAL PERIOD						6,400	9,200	9,700	(500)	

FIGURE 10. PERT COST financial plan and status report by month.

liabilities) charged or assigned during the indicated month ⑥. This value is shown for individual Charge Numbers ⑦ when they are included in the report. This column is used only for the month preceding "cut off date."

⑨ PLANNED: (Incremental Cost): The approved planned cost for the indicated time period ⑥. This value is shown for individual Charge Numbers ⑦ when they are included in the report. No information appears in this column for prior months.

⑩ LATEST REVISED ESTIMATE (Incremental Cost): The latest estimate of cost for the indicated time period ⑥. This value is shown for individual Charge Numbers ⑦ when they are included in the report.

⑪ (OVER) UNDER PLAN (Incremental Cost): The Planned Cost ⑨ minus the Latest Revised Estimate ⑩. When planned cost exceeds latest revised estimate, a projected underplan condition exists. When latest revised estimate exceeds planned cost, a projected overplan condition exists. Parentheses are used as a notational device to indicate an overplan condition. No information appears in this column for prior months.

⑫ ACTUAL (Cumulative Cost): The actual expenditures incurred plus any pre-specified types of unliquidated commitments (unliquidated obligations or accrued liabilities) charged or assigned during the period from the beginning of the program or project to the end of the indicated Month ⑥. This value is shown for individual Charge Numbers ⑦ when they are included in the report.

⑬ PLANNED (Cumulative Cost): The approved planned cost during the period from the beginning of the program or project to the end of the indicated Month ⑥. This value is shown for individual Charge Numbers ⑦ when they are included in the report.

⑭ LATEST REVISED ESTIMATE (Cumulative Cost): The latest estimate of cost during the period from the beginning of a program or project to the end of the indi-cated Month ⑥. This value is shown for individual Charge Numbers ⑦ when they are included in the report. This estimate is the sum of actual costs plus estimates through the end of the indicated month. For the period prior to the cut off date, the latest revised estimate equals the Actual ⑫.

⑮ (OVER) UNDER PLAN (Cumulative Cost): The Planned Cost ⑬ minus the Latest Revised Estimate ⑭. When planned cost exceeds latest revised estimate, a projected underplan condition exists. When latest revised estimate exceeds planned cost, a projected overplan condition exists. Parentheses are used as a notational device to indicate overplans.

⑯ REMARKS: This column contains the remaining data needed to make the Finan-cial Plan and Status Report the sole source of information for plotting the Cost of Work Display. This data, (which may be transferred from the Program/Project Status Report), is:

Value of Work Performed to Date 1) Cumulative—(from column ⑬ Program/Project Status Report) 2) Latest Month—(from column ⑬ Program/Project Status Report this month minus column ⑬ Program Project Status Report last month)

(Over) Underrun to Date (from column ⑮ of the Program/Project Status Report).

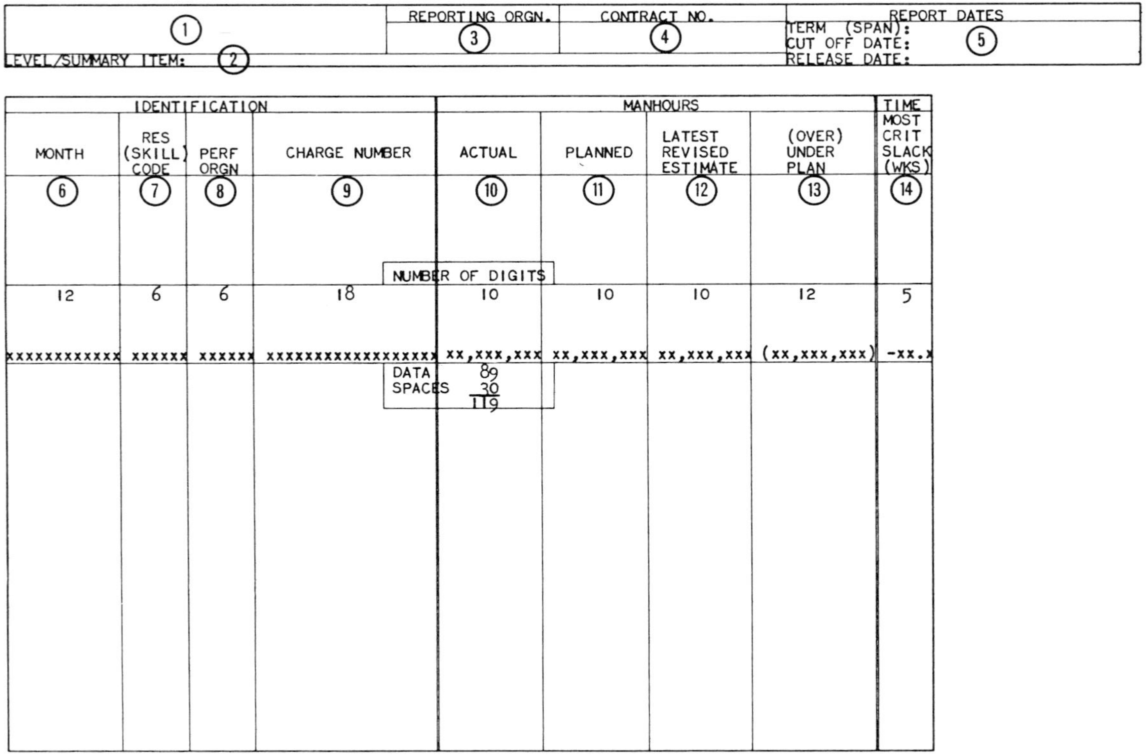

	IDENTIFICATION			MANHOURS				TIME
MONTH (6)	RES (SKILL) CODE (7)	PERF ORGN (8)	CHARGE NUMBER (9)	ACTUAL (10)	PLANNED (11)	LATEST REVISED ESTIMATE (12)	(OVER) UNDER PLAN (13)	MOST CRIT SLACK (WKS) (14)
NUMBER OF DIGITS								
12	6	6	18	10	10	10	12	5
DATA SPACES								
xxxxxxxxxxxx	xxxxxx	xxxxxx	xxxxxxxxxxxxxxxxxx	xx,xxx,xxx	xx,xxx,xxx	xx,xxx,xxx	(xx,xxx,xxx)	-xx.x

89
30
119

LEVEL/SUMMARY ITEM: (1) (2) REPORTING ORGN. (3) CONTRACT NO. (4) REPORT DATES (5)
TERM (SPAN):
CUT OFF DATE:
RELEASE DATE:

FIGURE 11. PERT COST manpower loading report.

PERT COST MANPOWER LOADING REPORT AND DISPLAY

The Manpower Loading Report and the Manpower Loading Display are intended for use by contractors to report manpower loading for various levels of summary within the program. The Manpower Loading Report lists actual, planned, and latest estimated monthly manhours for the desired level of summary by the type of manpower.

The Manpower Loading Display is a graphical presentation of the data contained in the Manpower Loading Report and is manually prepared.

The "type of manpower" is one of (or a combination of) the contractor's resource codes. These codes often identify types of materials, services, and facilities for which cost estimates have been made in hours, but which may not be significant in an analysis of manpower application. Therefore, the Manpower Loading Report is frequently prepared only for certain specified resource codes (skill categories).

The report is prepared for higher levels of management by printing only totals for each month (Figure 13). When the Government requires reporting in categories other than those identified by contractors' resource codes, the report is prepared by grouping resource codes within the specified categories by use of a translation table.

The sequence of sort and the categories included in the report are indicated in the report title. In addition to the examples shown, the report may be prepared by Performing Organization ⑧, Month ⑥, and Resource Code ⑦, to show organizational loading.

DEFINITIONS PERT COST MANPOWER LOADING REPORT AND DISPLAY

[① through ⑤ previously defined.—Authors.] . . .

⑥–⑨ The sorting sequence for these identification columns is indicated in the report title. Information will appear in only those columns listed in the title.

⑥ *MONTH:* The accounting time period for which estimates and actuals are shown.

⑦ *RESOURCE (SKILL) CODE:* The contractor or government organization code for a particular manpower skill.

⑧ *PERFORMING ORGANIZATION:* The contractor or government organization which will perform work on the work package.

⑨ *CHARGE NUMBER:* [Previously defined.—Authors.] . . .

⑩ *ACTUAL* (Manhours): The actual manhour expenditures incurred or assigned to a work package or work package subdivision. This information may appear only as a total figure when charge numbers are not shown in the report.

⑪ *PLANNED* (Manhours): The manhours planned for a work package or work package subdivision during the indicated month. This information may appear only as a total figure when charge numbers are not shown in the report.

⑫ *LATEST REVISED ESTIMATE* (Manhours): The latest estimate of manhours for a work package or work package subdivision during the indicated month. This information may appear only as a total figure when charge numbers are not shown in the report.

⑬ *(OVER) UNDERPLAN* (Manhours): The Planned Manhours ⑪ minus the Latest Revised Estimate ⑫. When planned manhours exceed latest revised estimate, a pro-

	REPORTING ORGN.	CONTRACT NO.	REPORT DATES
			TERM (SPAN): TOTAL PROGRAM
ABC - MISSILE AND GHE	XYZ - A&S DIVN	33(600)28369A	CUT OFF DATE: 30MAR63 / RELEASE DATE: 10APR63
LEVEL/SUMMARY ITEM: 3/BALLISTIC SHELL 22300			

IDENTIFICATION				MANHOURS				TIME
MONTH	RES (SKILL) CODE	PERF ORGN	CHARGE NUMBER	ACTUAL	PLANNED	LATEST REVISED ESTIMATE	(OVER) UNDER PLAN	MOST CRIT SLACK (WKS)
PRIOR	E1	2217	32163	800	2,100	800	1,300	0.0
			32166	13,000	14,000	13,000	1,000	2.1
		4422	32166	2,200	2,200	2,200		16.2
			32163	400	400	400		0.0
			32166	600	600	600		16.2
TOTAL				175,000	179,000	175,000	4,000	
MAR63		2217	32163	400	400	400		0.0
			32164	3,900	4,100	3,900	200	2.1
TOTAL				95,000	97,000	95,000	2,000	
APR63		2217	32163		400	400		0.0
			32164		4,500	4,500		2.1
TOTAL					86,000	98,000	(12,000)	
TOTAL				270,000	850,000	856,000	(6,000)	

FIGURE 12. PERT COST manpower loading report by resource, month, performing oranization, charge number.

	IDENTIFICATION			MANHOURS				TIME
MONTH	RES (SKILL) CODE	PERF ORGN	CHARGE NUMBER	ACTUAL	PLANNED	LATEST REVISED ESTIMATE	(OVER) UNDER PLAN	MOST CRIT SLACK (WKS)
PRIOR	EI			175,000	179,000	175,000	4,000	
MAR63				95,000	97,000	95,000	2,000	
APR63					86,000	98,000	(12,000)	
TOTAL				270,000	850,000	856,000	(6,000)	

REPORTING ORGN. XYZ - A&S DIVN

CONTRACT NO. 33(600)28369A

ABC - MISSILE AND GHE
LEVEL/SUMMARY ITEM: 3/BALLISTIC SHELL 22300

REPORT DATES
TERM (SPAN): TOTAL PROGRAM
CUT OFF DATE: 30MAR63
RELEASE DATE: 10APR63

FIGURE 13. PERT COST manpower loading report by resource, month.

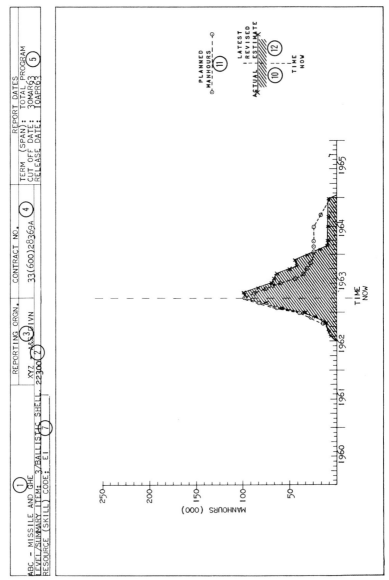

FIGURE 14. PERT COST manpower loading display.

jected underplan condition exists. When latest revised estimate exceeds planned manhours, a projected overplan condition exists. Parentheses are used as a notational device to indicate an overplan condition.

⑭ *MOST CRITICAL SLACK (WEEKS):* [Previously defined.—Authors.] . . .

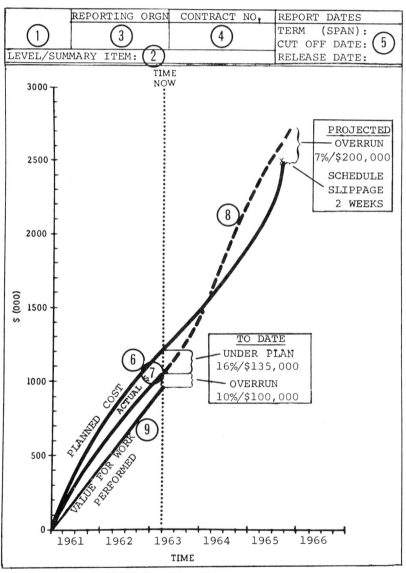

FIGURE 15.   PERT COST cost of work report.

PERT COST COST OF WORK REPORT

The Cost of Work Report is a graphical equivalent of the Financial Plan and Status Report with the additional feature of showing the distribution of actual costs and the value for work performed to "time now."

The Cost of Work Report is manually prepared each month from data contained in the Financial Plan and Status Report. The Cost of Work Report provides a comparison of:

Projected cost vs. planned cost at completion.
Value for work performed vs. actual cost to date.
Planned rate of expenditure vs. actual rate of expenditure to date.
Planned rate of expenditure vs. latest estimated rate of expenditure to completion.

DEFINITIONS PERT COST COST OF WORK REPORT

[① through ⑤ previously defined.—Authors.] . . .

⑥ *PLANNED COST:* The planned cost for the Summary Item ② plotted cumulatively by month. Values are plotted each month from the Financial Plan and Status Report, column ⑬.

⑦ *ACTUAL COST:* The actual cost for the Summary Item ② plotted cumulatively by month. This line is developed by plotting, each month, the new cumulative actual cost from the Financial Plan and Status Report, column ⑫.

⑧ *LATEST REVISED ESTIMATE:* The latest estimate of cost for the Summary Item ② plotted cumulatively by month from "time now" to program or project completion. This value is available from the Financial Plan and Status Report, column ⑭.

⑨ *VALUE FOR WORK PERFORMED TO DATE:* The planned cost for work completed within the Summary Item ② plotted cumulatively by month. This line is developed by plotting, each month, the new value of work performed to date from the Financial Plan and Status Report Remarks ⑯ (or from the Program/Project Status Report ⑬).

PERT COST COST OUTLOOK REPORT

The Cost Outlook Report shows (for any given level and summary item) the projected cost status at work completion. It also shows what the projected cost was at every cycle previous to the current one, thus providing for the recognition of trends.

Each month, new projections which provide new entries for the Cost Outlook Report are obtained from the Management Summary Report. The Cost Outlook Report is manually prepared by periodically plotting the projections obtained. These projections may be plotted by month for two years, after which the report is redrawn to show previous projections condensed by year.

Limit lines, established by the manager for each program or project, identify the values of (over) underrun which require a narrative analysis to be included in the Problem Analysis Report.

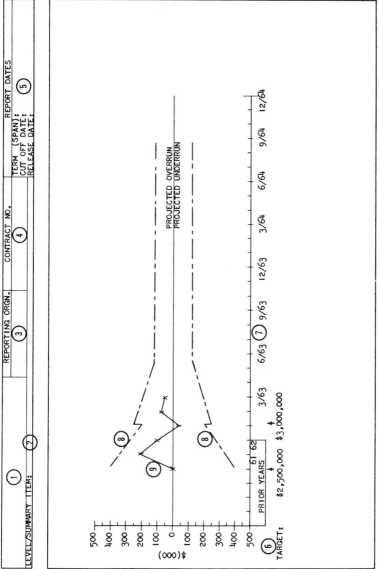

FIGURE 16. PERT COST cost outlook report.

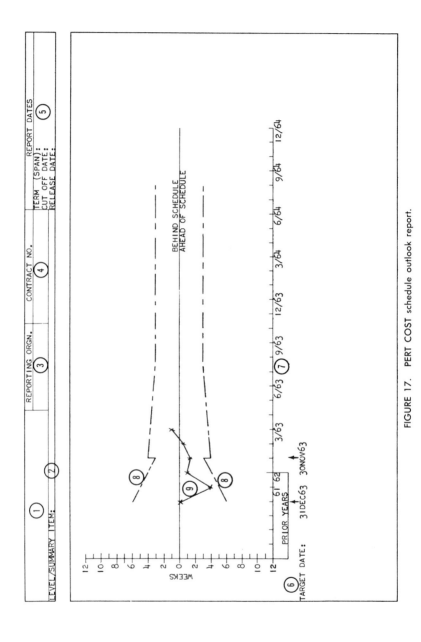

FIGURE 17. PERT COST schedule outlook report.

DEFINITIONS PERT COST COST OUTLOOK REPORT

[① through ⑤ previously defined.—Authors.] . . .

⑥ *TARGET:* The planned cost for the Summary Item ② identified in the title block. An arrow indicates on the Calendar ⑦ the date when the target value was established.

⑦ *CALENDAR:* The calendar shows two years of projected values by month and six years of condensed historical information. Managers may elect to use other time scales.

⑧ *LIMIT LINES:* Limit lines, established for each program or project, identify the values of (overrun) underrun which require that a narrative analysis be included in the Problem Analysis Report.

⑨ *PROJECTED (OVERRUN) UNDERRUN:* This value, from the Management Summary Report Projected (Overrun) Underrun ⑫, is plotted each month.

PERT COST SCHEDULE OUTLOOK REPORT

The Schedule Outlook Report shows (for any given level and summary item) the projected schedule status at work completion. It also shows what the projected schedule status was at every cycle previous to the current one, thus providing for the recognition of trends.

Each month, new projections are obtained from the Management Summary Report which provide new entries for the Schedule Outlook Report. This report is manually prepared by periodically plotting the projections obtained. These projections may be plotted by month for two years, after which the Schedule Outlook Report is redrawn to show previous projections condensed by year.

Limit lines, established by the manager for each program or project, identify the values of schedule status which require a narrative analysis to be included in the Problem Analysis Report.

DEFINITIONS PERT COST SCHEDULE OUTLOOK REPORT

[① through ⑤ previously defined.—Authors.] . . .

⑥ *TARGET DATE:* The planned scheduled completion date for the Summary Item ② identified in the title block. An arrow indicates on the Calendar ⑦ the date when the target value was established.

⑦ *CALENDAR:* The calendar shows two years of projected values to be plotted by month and six years of condensed historical information. Managers may elect to use other time scales.

⑧ *LIMIT LINES:* Limit lines, established for each program or project, identify the values of schedule slippage which require that a narrative analysis be included in the Problem Analysis Report.

⑨ *PROJECTED SCHEDULE STATUS:* This value, from the Most Critical Slack ⑬ of the Management Summary Report, is plotted each month.

(1)	REPORTING ORGN.	CONTRACT NO.	REPORT DATES
LEVEL SUMMARY ITEM: (2)	(3)	(4)	TERM (SPAN): CUT OFF DATE: (5) RELEASE DATE:

IDENTIFICATION	MANHOURS					TOTAL COST $(000)				
	TO DATE		TOTALS AT COMPLETION			WORK TO DATE		TOTALS AT COMPLETION		
COST CATEGORY	PLANNED	ACTUAL	PLANNED	LATEST REVISED ESTIMATE	PROJECTED (OVERRUN) UNDERRUN	PLANNED	ACTUAL	PLANNED	LATEST REVISED ESTIMATE	PROJECTED (OVERRUN) UNDERRUN
(6)	(7)	(8)	(9)	(10)	(11)	(12)	(13)	(14)	(15)	(16)
			NUMBER OF DIGITS							
18	11	11	11	11	9	7	7	7	7	9
xxxxxxxxxxxxxxxxxx	xxxxxxxxxxx xxx,xxx,xxx	xxxxxxxxxxx xxx,xxx,xxx	xxxxxxxxxxx xxx,xxx,xxx	xxxxxxxxxxx xxx,xxx,xxx	(.xx) (xxx,xxx)	xxx,xxx	xxx,xxx	xxx,xxx	xxx,xxx	(.xx) (xxx,xxx)
			DATA SPACES	108 11 119						

FIGURE 18. **PERT COST** cost category status report.

PERT COST COST CATEGORY STATUS REPORT

The Cost Category Status Report presents a grouping of functional, hardware, or other significant cost elements in specified categories for reporting purposes.

These cost categories are established by relating work packages or elements of cost within work packages to the specified categories. Thus, no distortion of the work breakdown structure is required to segregate these data.

Any cost categories which satisfy this relationship to the work breakdown structure may be established for a prognam or project, but once established, they must remain as originally defined for the life of the program or project.

The Cost Category Status Report provides for each cost category a manpower and total dollar comparison of:

planned vs. actual expenditure to date.
planned vs. latest revised estimate at completion.

DEFINITIONS PERT COST COST CATEGORY STATUS REPORT

[① through ⑯ previously defined.—Authors.] . . .

PERT MILESTONE REPORT

The PERT Milestone Reports present schedule information for selected network events which represent major milestones of accomplishment toward completion of the program or project.

The reports are tiered, like the Management Summary Report, for several levels of management. However, the Milestone Report represents key network events that are of major significance in achieving the program or project objectives, whereas the Management Summary Report flags critical areas and work completions.

Together, the Milestone Report and the Management Summary Report provide the most comprehensive status information available in the PERT COST System.

DEFINITIONS PERT MILESTONE REPORT

[① through ⑤ previously defined.—Authors.] . . .

⑥ *MILESTONE DESCRIPTION:* The network event number and nomenclature which are selected as milestones. Two lines are available for description.

⑦ *SLACK:* The slack, in weeks, associated with the network event (Milestone) ⑥. This is the time difference between the *"E"* and *"L"* dates shown in the Schedule ⑨.

⑧ *DATE:* The day, month, and year of the *"S," "A," "E," "L,"* or *"M"* positions shown in the Schedule ⑨.

⑨ *SCHEDULE CALENDAR:* A calendar time reference for display of schedule completions. The calendar contains one division for all prior years, two years divided by months, four years by years, and one division for all later years. A "Time Now" line appears between the next future month and the month of the cut off date.

REPORTING ORGN.	CONTRACT NO.	REPORT DATES
ABC – MISSILE AND GHE	XYZ – A&S	TERM (SPAN): TOTAL PROGRAM
LEVEL/SUMMARY ITEM: 3/BALLISTIC SHELL 22300	33(600)28369A	CUT OFF DATE: 30MAR63 RELEASE DATE: 10APR63

IDENTIFICATION	MANHOURS					TOTAL COST $(000)				
	TO DATE		TOTALS AT COMPLETION			WORK TO DATE		TOTALS AT COMPLETION		
COST CATEGORY	PLANNED	ACTUAL	PLANNED	LATEST REVISED ESTIMATE	PROJECTED (OVERRUN) UNDERRUN	PLANNED	ACTUAL	PLANNED	LATEST REVISED ESTIMATE	PROJECTED (OVERRUN) UNDERRUN
1010 ENG A&S	117,000	120,000	220,000	225,000	{ (.02) (5,000) }	2,500	2,700	3,500	3,900	{ (.01) (400) }
1012 DEV D&T	172,000	172,000	380,000	380,000		2,100	2,100	4,300	4,300	
2010 EQUIP FA&C	211,000	212,000	420,000	421,000	(1,000)	3,100	3,800	5,200	5,900	{ (.01) (700) }
3010 ENG FLD SPT	63,000	61,000	170,000	169,000	1,000	980	900	2,800	2,700	{ .04 100 }
TOTAL						18,620	20,500	35,200	39,650	{ (.13) (4,450) }

FIGURE 19. PERT COST cost category status report.

502

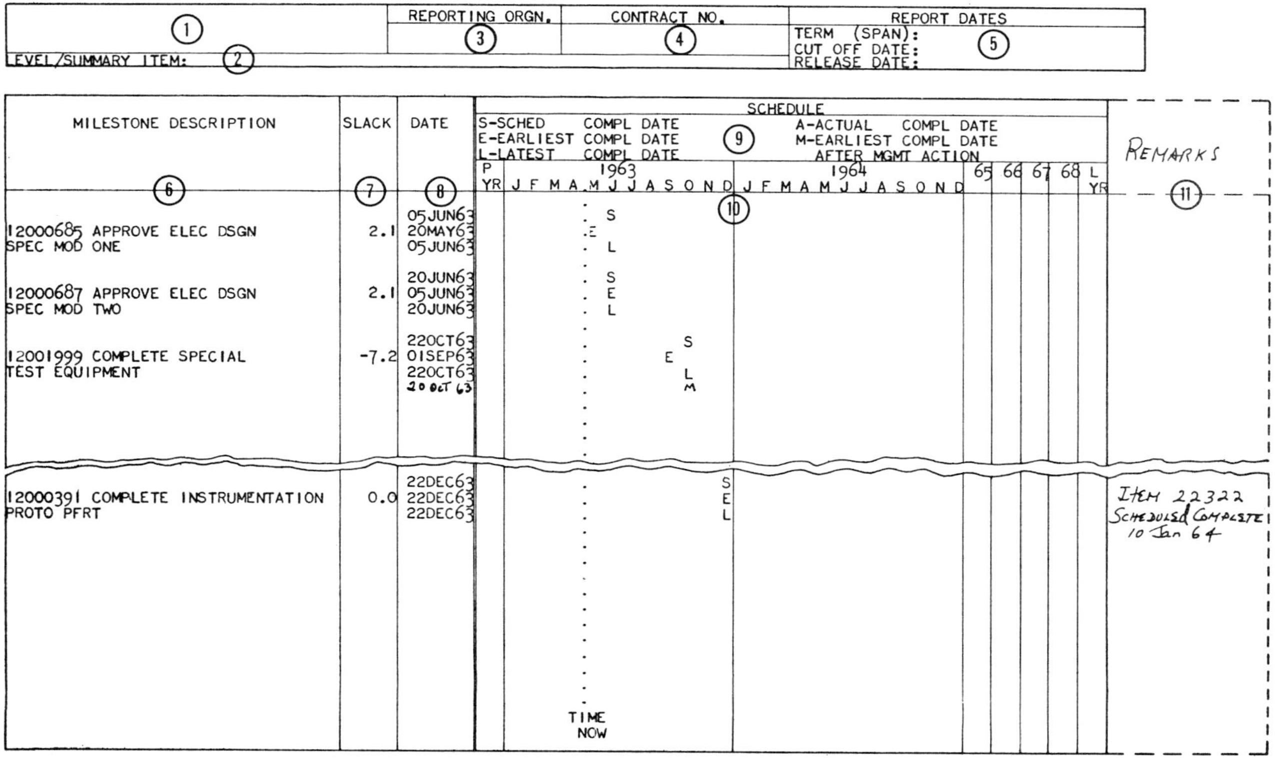

FIGURE 20. PERT milestone report.

503

⑩ *SCHEDULE COMPLETIONS:* The scheduled *"S,"* Actual *"A,"* Earliest *"E,"* and Latest *"L"* completion dates for the network event (Milestone) in colume ⑥ with respect to designated program or project end points. *"M"* may be entered by an analyst to indicate a revised completion date anticipated as a result of management action.

Index